Only in Africa The Ecology of Human Evolution

That humans originated in Africa is well-known. However, this is widely regarded as a chance outcome, dependent simply on where our common ancestor shared the land with where the great apes lived. This volume builds on from the 'Out of Africa' theory, and takes the view that it is only in Africa that the evolutionary transitions from a forest-inhabiting frugivore to savanna-dwelling meat-eater could have occurred. This book argues that the ecological circumstances that shaped these transitions are exclusive to Africa. It describes distinctive features of the ecology of Africa, with emphasis on savanna grasslands, and relates them to the evolutionary transitions linking early ape-men to modern humans. It shows how physical features of the continent, especially those derived from plate tectonics, set the foundations. This volume adequately conveys that we are here because of the distinctive features of the ecology of Africa.

Norman Owen-Smith headed the Centre for African Ecology at the University of the Witwatersrand before his retirement as Emeritus Professor there. He is an A-rated scientist and Fellow of the Royal Society of South Africa. He received Gold Medals from the Zoological Society of Southern Africa and the South African Association for the Advancement of Science, Wildlife Excellence Award from the Southern African Wildlife Management Association, Honorary Life Membership in the Ecological Society of America, and was awarded a Harry Oppenheimer Fellowship in 2005. He has written or edited six books, including *Megaherbivores: The Influence of Very Large Body Size on Ecology* (Cambridge University Press, 1988) and *Adaptive Herbivore Ecology: From Resources to Populations in Variable Environments* (Cambridge University Press, 2002).

Only in Africa The Ecology of Human Evolution

NORMAN OWEN-SMITH
University of the Witwatersrand

Shaftesbury Road, Cambridge CB2 8EA, United Kingdom

One Liberty Plaza, 20th Floor, New York, NY 10006, USA

477 Williamstown Road, Port Melbourne, VIC 3207, Australia

314–321, 3rd Floor, Plot 3, Splendor Forum, Jasola District Centre, New Delhi – 110025, India

103 Penang Road, #05–06/07, Visioncrest Commercial, Singapore 238467

Cambridge University Press is part of Cambridge University Press & Assessment, a department of the University of Cambridge.

We share the University's mission to contribute to society through the pursuit of education, learning and research at the highest international levels of excellence.

www.cambridge.org
Information on this title: www.cambridge.org/9781108832595

DOI: 10.1017/9781108961646

First published 2021

A catalogue record for this publication is available from the British Library

ISBN 978-1-108-83259-5 Hardback
ISBN 978-1-108-95966-7 Paperback

This book is dedicated to my extended family: my much-loved wife Margie, daughters Trishya and Lynne, and all of my former students.

Contents

Foreword

Picking through a pile of elephant dung with a twig, quizzically chewing a leaf from a newly identified plant, interrupting a conversation to clap binoculars to his eyes and scrutinise something moving in the distance ... that is the characteristic behaviour of Norman Owen-Smith in his preferred habitat, African savanna. His keen curiosity about nature is combined with an incisively analytical mind and seemingly inexhaustible capacity for integrating his observations and ideas into the published literature. Driven by his enthusiasm for science, it was inevitable that Norman would achieve the international recognition that he has today as a leader in the field of animal ecology. Now, this book exemplifies his meticulous style of investigation, analysis and inference. It is a personal synthesis of Norman's observations, notes and photographs collected during a rich career involving extensive travels across Africa. Generously integrated with information and ideas from the literature, the book not only reviews African savanna ecology and human evolution, but also articulates an important message for humanity. The message – that it is *only in Africa* that humans could have evolved – is as provocative as it is compelling from the line of reasoning laid out with characteristic rigour in this book.

It all stems from a sequence of unconnected events and circumstances, like an asteroid slamming into a planet. That happens all the time, somewhere in the universe, but when a particularly big one slammed into Earth about 66 million years ago the implications for the biosphere were profound. Non-flying dinosaurs – and many other groups of organisms – quickly died out, opening a game-changing evolutionary opportunity for our mongoose-sized mammalian ancestors. Over the next 60 million years, while the mammals radiated into a diversity of taxa over a widening range of body sizes and trophic niches, plate tectonics moved the continents around and broke them up. Different assemblages developed on different continents that had different shapes and positions on the Earth's crust. By the time the common ancestor of humans and chimpanzees evolved, it occupied a species range in the middle of the big and geologically complex continent we call Africa, positioned squarely over the equator. That was crucial, because climatic cycles caused by the eccentric Earth-around-Sun orbit drove advances and retreats of the equatorial forest, taking 100,000 years per cycle. When the forest advanced, it covered mountains, plains and valleys; when it retreated, it left refugia of montane forest in a matrix of savanna. That forest–savanna ecotone, advancing and retreating at a rate suiting the speciation of large mammals, was the 'species pump' that drove the dramatic radiation of wildlife (as we now view it) in

African savannas. And among those radiating species there just happened to be a bipedal savanna-adapted large primate making its evolutionary appearance in just the right place and at just the right time. It coexisted with perennial supplies of big carcasses to scavenge from, diverse plant communities to gather from, and dry grass and wood to fuel lightning fires that could be weaponised and harnessed, sparking a complex social system.

Such is the sequence of events and circumstances that allowed the highly branched tree of human evolution to take root, and *could only have,* in African savannas, as Norman Owen-Smith explains here. Paradoxically, the biodiversity of African savannas – particularly the megafauna that sustained our ancestral scavengers – is now being shredded by the very species it spawned. This important book, by outlining the unique significance of African savannas for humanity, should leave readers with added resolve to conserve our ancestral habitat.

Johan T. du Toit
Utah State University

Preface

That humans originated in Africa is generally accepted, but this is considered to be a chance outcome, dependent simply on where our common ancestor with the great apes lived. The story that I will develop in the pages of this book is that it is *only in Africa* that the evolutionary transitions from a forest-inhabiting frugivorous ape to savanna-dwelling meat-eater could have occurred. Distinctive features of the ecology of Africa shaped these transitions. They are only partly recognised in the literature on human origins and inadequately covered in textbooks of ecology.

The concept for this book emanated from my awareness that the evolutionary origin of the animals that I studied – Africa's large mammalian herbivores – was coupled in time with the divergence of our earliest human predecessors from the ancestor that they shared with the great apes. Prior to 12 million years ago, during the Miocene epoch, Africa was thronged by a variety of strange beasts, most of them very large and many elephant- or rhino-like, browsing in quite dense woodlands. Monkeys and apes were around in profusion, but adapted especially for a lifestyle in the tree-tops. Following a vague hiatus in the fossil record, the fauna that took shape some 5 million years ago at the start of the Pliocene was radically different. A diverse assemblage of medium-sized ungulates (hoofed mammals) had evolved, many of them adapted for grazing grass rather than browsing tree leaves. The animal that I studied for my doctoral degree, the white rhinoceros (*Ceratotherium simum*), had emerged and other very large herbivores also showed signs of switching towards grass-based diets. Moreover, some of the apes showed adaptations for upright locomotion to traverse widening spaces between the trees, reinforcing evidence of the spread of savanna grasslands. How had environmental conditions changed to foster the contemporaneous evolutions of grasslands, grazers and bipedal apes? I recall while out in the gloom of an African night, brightened by moonlight and a starry milky way, watching a white rhino bull munching grass in the shadows, and noting that one star was moving – a satellite recently placed in orbit. I tried to comprehend the enormous sweep in time covering the transition from the world that generated this primeval-looking animal to the object circuiting the Earth that had been created by the human descendants of certain apes.

The connections were not covered in standard ecology texts because grasses and grazers were not as prominent outside of Africa where the authors lived. Indeed, Africa's savannas were interpreted as formations disturbed by humans from their 'natural' wooded state, through clearing and the promotion of fires. But humans of various forms had inhabited Africa for millions of years and recurrent fires were a necessary outcome of the accumulation of seasonally dry grass, however ignited. Now, towards the end of my career, there is time to look back and identify the connections that made Africa's ecology different in several ways from that of other continents. Hence this book is basically a compendium of savanna ecology, within which the evolution of the lineage leading to modern humans – to us – is embedded.

Recognition of Africa's distinct features crystallised from the bold claim by geologists that plate tectonics – movements of the continents – were ultimately responsible for the origins of humankind.[1] The ecology of its plants and animals was connected somehow to the physical features of landforms, geology, climate and soils that took shape following the splitting of the supercontinent called Gondwana in the early Jurassic period as far back as 180 Ma. I needed to expand my comprehension by reading into the literature of these other fields of science. The linkages that I recognised are explained in the chapters of this book.

My experiences have made me unusually well-grounded in Africa for this synthesis. During my doctoral research, I walked almost daily following white rhinos while they went about their lives for 3.5 years. Their world was my world – the undulating landforms, soils and plants and how they changed with wet and dry seasons. I shared the dung-heap established by the neighbouring white rhino bull situated conveniently close to the caravan where I lived, parked under a spreading thorn tree. I became fascinated by the patterns formed by plants, particularly the shifts in grass species within a few metres. Having been grounded in this diversity, while furthering my studies in the USA I found it hard to accept that uniform vegetation formations could have existed over vast areas of North America as climax states of closed forest or treeless grassland. Names given to soil types did not capture the links I had discerned between soils, vegetation and where I was most likely to find white rhinos. During my subsequent academic career, my academic horizons became widened to encompass other parts of southern Africa, coupled with visits to parts of the continent further north. Nevertheless, this book inevitably expresses a southern perspective on Africa's ecology.

The book begins with a set of chapters identifying the physical features distinguishing Africa from other continents, forming Part I. They establish that Africa is unusually high-lying, relatively dry, subject to widespread volcanism, has widely varying river flows and lake levels and soils that are unusually fertile for the tropics. Part II covers the ecology of the vegetation, establishing the mechanisms contributing to the predominance of grasses. Part III covers the large mammalian herbivores that are central to the story and have been the focus of most of my academic research along with my students. They establish how the diversity and abundance of medium-large grazers in particular is founded on features of the grasses, leading into the wider roles of these herbivores in ecosystem dynamics. The chapters forming Part IV place the evolutionary origins and subsequent adaptive transitions of the hominin lineage within Africa's ecological contexts. Finally, I contemplate what the role of Africa's unique biological heritage might be in the future. The broad sweep covered by this book does not allow me scope for delving into detail on any one topic. Hence, at the end of each chapter, I suggest a few publications for further reading in greater depth. The text is supported by a profusion of colour illustrations, largely gleaned from my own travels. Ecology expressed in words can be boring. The images are intended to relate the words to the world beyond.

My focus is specifically on the ecology of Africa's savannas. The ecology of its forests is beautifully covered by Vande Weghe[2] with numerous colour illustrations. That of its southern deserts is treated by Lovegrove and Siegfried.[3] The comparisons I draw with other continents remain superficial, merely noting the features that seem distinctive of Africa. For an ecological perspective on all forms of open ecosystems worldwide, encompassing savannas, grasslands and shrublands, I refer you to Bond.[4]

Many people contributed to my intellectual advancement and hence to this book, in various ways. Rudi Bigalke opened the opportunity for me to study white rhinos under the auspices of the Natal Parks Board, before I became a biologist. John Emlen provided academic guidance into behavioural ecology at the University of Wisconsin, where I arrived to switch fields from chemistry to zoology. Salmon Joubert invited me to study kudus, believed to be centrally involved in spreading anthrax in Kruger Park. John Skinner supported my postdoctoral fellowship through the Mammal Research Institute of the University of Pretoria. Brian Huntley drew me into the South African Savanna Ecosystem programme undertaken in the Nylsvley Nature Reserve and the opportunity provided to move ecology into the computer era, traipsing behind tame kudus. Brian Walker hosted me in the Centre for Resource Ecology at the University of the Witwatersrand and challenged me about how to link behavioural ecology to ecosystem management. Tony Starfield led me into computer modelling, a perspective that has permeated my scientific approach. Wayne Getz expanded this orientation into the metaphysiology of biomass dynamics, linking behavioural responses to their outcomes for population and community processes. Joseph Ogutu inducted me into the statistics of handling big data through model selection while he was a postdoc in my group. Martin Haupt offered me two GPS collars for testing on disappearing sable antelope and opened the field of movement ecology, which stopped me from retiring when I should have. Barend Erasmus, Robyn Hetem, Francesca Parrini and Melinda Boyers joined me in one last team study tagging animals in the vast Kalahari region of Botswana, which has kept me working on abundant data long into retirement. To my PhD students, I acknowledge how you have widened my knowledge: Susan Cooper, Johan du Toit, Ignas Heitkonig, Peter O'Reagain, Laurence Watson, Mark Vandewalle, Jonas Chafota, Angela Gaylard, Adrian Shrader, Randal Arsenault, Michelle Henley, Kirsten Neke, Joanne Shaw, Valerio Macandza, Joe Chirima and Gabi Teren; plus also the postdocs, belatedly supported by my university: Andrew Kennedy, Joseph Ogutu, Steve Henley, Sander Oom, James Cain, Sophie Grange, Jason Marshal, Yoganand Kandasamy, Jodie Martin, Lochran Traill, Sze Wing Yiu and Melinda Boyers. Marco Anson contributed the original artwork illustrating Chapter 17. Very special thanks go to my life partner, Margie Loffell and our two girls, Trishya and Lynne, who put up with having all of our family holidays diverted to where my students were based in some game reserve.

I thank Manuel Dominguez-Rodrigo for reading especially the chapters on early hominin ecology to ensure that they did not depart too far from the facts. Earlier drafts of other chapters were read and improved by Bob Scholes, Sally Archibald, Tim O'Connor, Kathy Kuman, Marion Bamford, Tyler Faith, Anabelle Cardoso, David Morgan and Lochran Traill.

I acknowledge the support that I received from the team at Cambridge University Press: Dominic Lewis, Aleksandra Serocka, Jenny van der Meijden and Sara Brunton.

Africa is my home continent and I hope that I have adequately conveyed the special features of its ecology to you in the pages that follow. All photographs were taken by the author unless otherwise acknowledged.

References

1. Gani, MR; Gani, NDS. (2008) Tectonic hypotheses of human evolution. *Geotimes* 53:34–39.
2. Vande Weghe, JP. (2004) *Forests of Central Africa: Nature and Man*. Protea Book House, Pretoria.
3. Lovegrove, B; Siegfried, R. (1993) *The Living Deserts of Southern Africa*. Fernwood Press, Simons Town.
4. Bond, WJ. (2019) *Open Ecosystems: Ecology and Evolution Beyond the Forest Edge*. Oxford University Press, Oxford.

Abbreviations

BCE	Before the Current Era (replacing BC)
CAM	crassulacean acid metabolism
CE	Common Era (replacing AD)
CEC	cation exchange capacity
CV	coefficient of variation
ENSO	El Niño–Southern Oscillation
ESA	Earlier Stone Age
GR	Game Reserve
ITCZ	Intertropical Convergence Zone
ka	thousand years ago
kyr	thousand years
LGM	Last Glacial Maximum, ~20,000 years ago
LSA	Later Stone Age
Ma	million years ago
MAR	mean annual rainfall
MSA	Middle Stone Age
mtDNA	mitochondrial DNA
NP	National Park
NR	National Reserve
SNP	single nucleotide polymorphism
TEB	total exchangeable bases

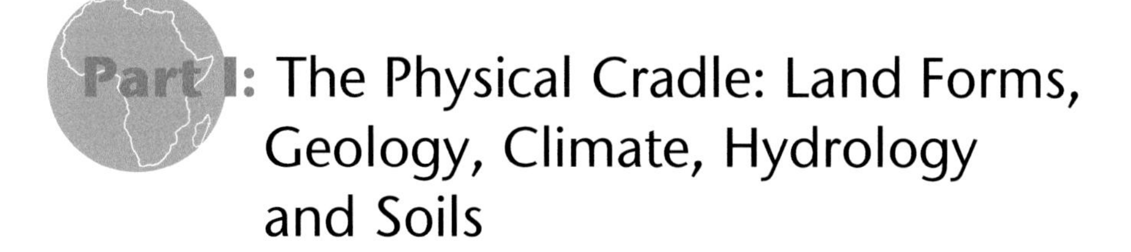

Part I: The Physical Cradle: Land Forms, Geology, Climate, Hydrology and Soils

Africa emerged from the middle of the supercontinent called Gondwana, splitting from the land masses that became South America, Australia, India and Antarctica (Figure I.1). The rupture was initiated by massive outpourings of flood basalts, which commenced 183 million years ago (Ma) during the early Jurassic period in what is now southern Mozambique. The volcanic overlay spread inland from there at least as far as south-western Zambia. By 160 Ma, a widening trough separated Africa from eastern Antarctica and Madagascar, filled by the proto-Indian Ocean. In the west, the separation of South America from Africa began with lava eruptions in what is now Namibia, initiated around 123 Ma, and the South Atlantic Ocean began opening. Unencumbered by adjoining land masses, Africa drifted slowly northward, and rotated a little anticlockwise. The location of the equator shifted from the southern Sahara region towards its current middle position, with similar portions of the continent to its north and south. Once the continents eventually halted their drift, South America lay almost 3000 km from the nearest point of Africa, while Australia ended up almost 10,000 km distant on the other side of the Indian Ocean. The Tethys Sea separated Africa from Europe.

Following its parting from the other southern continents, Africa's high-lying land surface became progressively worn down by erosion, lowering the hilltops and filling in the valleys. By 66 Ma, when the Cretaceous period ended with the demise of the dinosaurs, a gently undulating plain had been formed over most of the continent. This is known as the African erosion surface. The only mountain ranges lay in the far south and far north. The Cape Fold Mountains were formed during the Permian ~250 Ma, when Africa's land mass pressed against Antarctica, while the Atlas Range was formed much later where Africa's drift northward butted against Eurasia. Freed from the adjoining land masses, Africa's coastal margins tilted upward. With the passage of time, the coastal escarpments became eroded back by as much as 200 km in the east and 50 km in the west. Material removed from the high country accumulated in the Kalahari, Congo and Chad basins and extended shorelines especially in the east and south. Through the interior, low hills emerged where more resistant rocks intruded. Africa's surface probably resembled Australia, worn down and

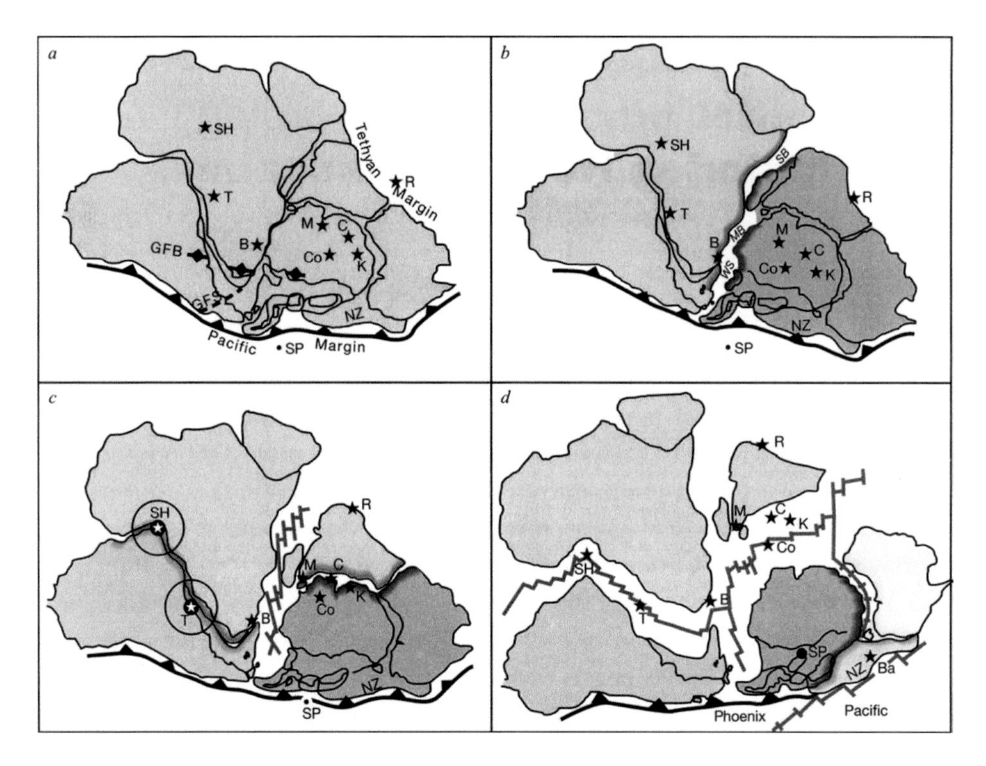

Figure I.1 Stages of the breakup of the supercontinent called Gondwana. (a) 200 Myr, (b) 160 Myr, (c) 130 Myr, (d) 100 Myr (from Storey (1995) *Nature* 377: 301).

depleted of nutrients. Fossil deposits preserve remains of the mammal-like reptiles and early dinosaurs that thronged Gondwana, but record little of the later dinosaurs present in Africa during the late Jurassic and Cretaceous periods. The fragments of their bones were mostly swept from the eroding land surface.

This static situation ended around 45 Ma, when a plume from deep within the Earth's mantle began pushing up in the north-east under present-day Ethiopia, initiating volcanic eruptions. The doming of the land surface caused faults to develop by 25 Ma, which opened the Red Sea and propagated southward through the continent. Massive outpourings of volcanic lava produced layers upon layers of basalt. The raised land surface deflected rain-bearing winds and altered river courses.

This tectonic activity not only reshaped the physical features of the continent; it nurtured the plants and animals that evolved on its surface, including the walking, meat-eating, cerebral and culturally sophisticated ape that became us. The chapters forming this first section of the book establish the interconnected consequences of the tectonic uplift for land forms, climates, rivers and soils. They present largely an east-side story, but expanded coast-to-coast in the south. West-central and western Africa remained mostly low-lying and rather similar to tropical regions of South America, Australia and Asia in their ecology, and thus feature less in the story.

Suggested Further Reading

McCarthy, T; Rubidge, B. (2005) *The Story of Earth & Life. A Southern African Perspective on a 4.6-Billion-Year Journey*. Struik Nature, Cape Town.

Partridge, TC. (2010) Tectonics and geomorphology of Africa during the Phanerozoic. In Werdelin, L; Sanders, WJ (eds) *Cenozoic Mammals of Africa*. University of California Press, Berkeley, pp. 3–17.

Chapter 1: High Africa: Eroding Surfaces

The formation of High Africa began in the Afar region of north-east Ethiopia around 30 Ma, brought about by the billowing mantle plume. By ~20 Ma, the continental uplift had propagated into South Africa, where the land surface was raised by about 250 m in the east and 150 m in the west. The elevated surface was subjected to a renewed phase of erosion, which produced the Post-African I (or Miocene) landscape. More substantial uplift took place starting around 10 Ma and accelerating after 5 Ma, especially along the eastern side of the continent. The land surface rose by as much as 1500 m in eastern Africa and by up to 900 m in eastern parts of South Africa.[1] This generated the Post-African II (or Pliocene) cycle of erosion, not readily distinguished from the preceding phase I. The product is the high interior plateau extending from Ethiopia through eastern Africa and broadening westward across southern Africa (Figure 1.1; Box 1.1 explains the continental divisions used throughout the book). The elevated continental interior profoundly influenced climates, exposed bedrock, altered river courses, shifted lakes and affected soil formation, as will be outlined in the chapters that follow.

Land Surfaces

Remnants of the African erosion surface persist in South Africa's Highveld region, Manica region of central Zimbabwe, above the Muchinga escarpment in Zambia and over parts of eastern Africa despite blanketing there under lava flows (Figure 1.2).[2] Distinctions between the African and Post-African surfaces are particularly striking in the southern Highveld and Karoo regions of South Africa, where flat-topped hills capped by resistant bedrock retain remnants of the African surface on their crests (Figure 1.2D). Further north in parts of Zimbabwe, Angola and northern Mozambique, erosion has exposed inselbergs composed of basement granite (Figure 1.2E). While erosion lowered the interior surface, it pushed back coastal escarpments. Sandy sediments accumulated in the Kalahari and Congo basins in the west and extended the eastern coastline outward (Figure 1.2F).

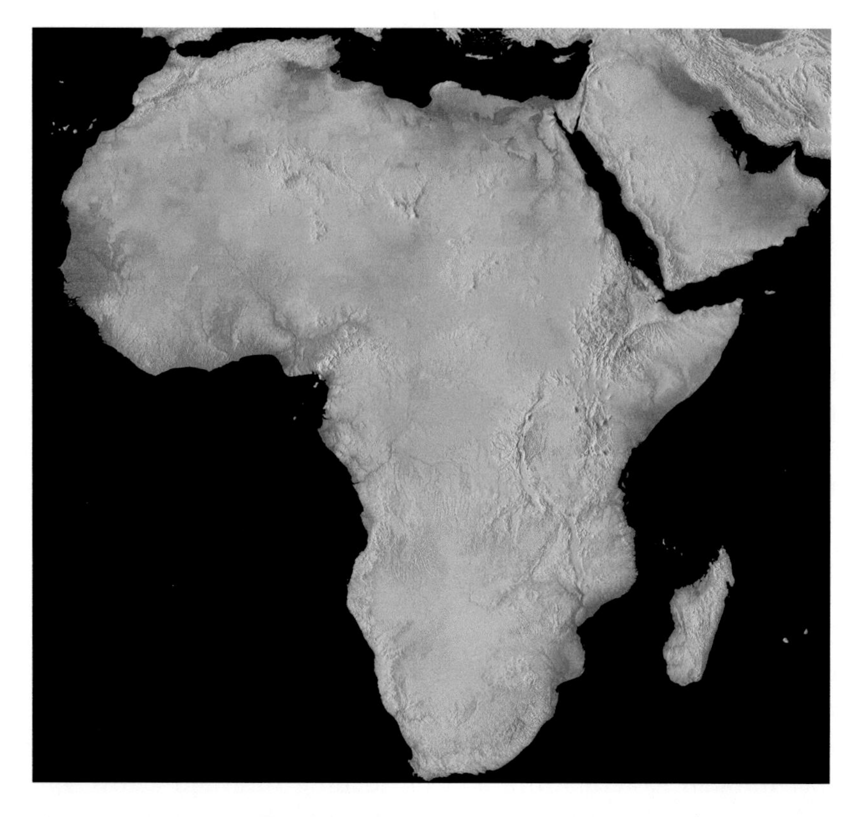

Figure 1.1 Topographic map showing the extent of High Africa stretching continuously from Ethiopia through southern Africa, shown in yellow.

Box 1.1 Geographical Subdivisions of Africa Distinguished in the Text

1. Eastern Africa – the equatorial region encompassing Kenya, Tanzania and Uganda
2. South-Central Africa – the tropical region extending through Zambia, Zimbabwe, Malawi, northern Mozambique and Angola
3. Southern Africa – the region extending into the subtropics encompassing South Africa, Botswana, Namibia and southern Mozambique
4. North-eastern Africa – southern parts of Sudan, Ethiopia and Somalia
5. Western Africa – the region extending from Cameroon to Senegal
6. Northern Africa – the Mediterranean region extending from Egypt to Morocco

Figure 1.2 Land surfaces. (A) The gently undulating Africa surface generated by the early Miocene, represented in Athi-Kaputiei Plains in Kenya; (B) eroding eastern edge of South Africa's Highveld plateau; (C) elevated highlands in eastern Zimbabwe, representing marginal uplift following the breakup of Gondwana; (D) flat-topped hill retaining the Africa surface above the Post-African or Pliocene erosion surface in South Africa's Karoo region; (E) granite inselbergs exposed by erosion into the basement shield in south-eastern Zimbabwe; (F) depositional surface of Kalahari sand in north-western South Africa.

South Africa's Highveld plateau attains its maximum elevation of 2332 m near Dullstroom ~200 km from Johannesburg above the eastern (or Transvaal) escarpment, while the Lesotho highlands above the Drakensberg escarpment rise to 3482 m and retain remnants of the Gondwana surface on hilltops. In the south-west, highlands near the Namibian capital Windhoek reach 2606 m, while Angola's western escarpment rises ~2500 m above the coastal plain, counteracting the general lowering westward. Serengeti National Park (NP) in Tanzania ranges in elevation from 1200 to 2000 m, while the Ngong Hills near Nairobi reach 2460 m. Plateau regions of Ethiopia exceed 3000 m in elevation.

In contrast, western Africa is mostly low-lying, with only localised high country. The Jos Plateau in central Nigeria forms a tableland at a mean elevation of 1280 m, with its highest point 1829 m above sea level. Further west, the Guinean highland reaches a maximum altitude of merely 1538 m. Other highland regions exist deep within the Sahara. The plateau region of Cameroon and adjoining Nigeria connects with the eastern African highlands via the Ubangi-Shari region of the Central African Republic. Every major city in Africa within the eastern and southern interior lies more than 1000 m above sea level, while no city in the west approaches this elevation (Table 1.1). Higher eminences are all volcanic cones associated with rift valley formation, except for the Ruwenzori range, which is an upthrust block within the Western Rift.

Rift Valley Formation

The rifting that began in Ethiopia eventually spread through northern Mozambique, spanning a distance of ~6000 km (Figure 1.3). The downward subsidence in the trough was counterbalanced by raised rift shoulders due to the local pressure release. In some regions there was only a single fault, generating a 'half-graben' rather than a two-sided full graben. The rift valleys accumulated sedimentary deposits, and thus played a crucial role in preserving fossils of past faunas. They trace much of our knowledge of the course of human evolution. Volcanic cones rose beside the rifts and fissure eruptions spread volcanic deposits more widely. Minerals contained in the lava deposits can be used to date the time line of evolution.

In Ethiopia, the rift depression dividing the Simien Mountains in the north-west from the Bale Mountains in the south-east forms a valley 50-km wide. On its margin, Ras Dashen in the Simien range reaches an altitude of 4624 m, while the floor of the Danakil depression in neighbouring Eritrea lies 125 m below sea level. The Eastern (or Gregory) Rift extended through northern Kenya after 12 Ma and reached northern Tanzania by 5 Ma, where it fades out. The Western or Albertine Rift branched off along the border of the Congo DRC with Uganda, Rwanda, Burundi and western Tanzania. Incipient signs of rifting appeared in the Semliki region of Uganda around 8 Ma, but its current configuration was attained only after 3 Ma. The Western Rift continued propagating through southern Tanzania and Malawi to reach the Mozambican coast

Table 1.1 Altitudes of major African cities situated away from coastal regions and their mean annual and dry season rainfall totals (source: Wikipedia and climate-data.org)

City	Country	Altitude (m)	Mean annual rainfall (mm)	Dry season rainfall (mm)
Johannesburg	South Africa	1753	790	67
Windhoek	Namibia	1728	359	12
Gaborone	Botswana	1014	457	28
Lilongwe	Malawi	1050	860	13
Harare	Zimbabwe	1490	831	23
Lusaka	Zambia	1277	831	3
Huambo	Angola	1721	1366	39
Dodoma	Tanzania	1120	564	2
Nairobi	Kenya	1795	869	139
Kampala	Uganda	1190	1293	457
Goma	Congo DRC	1460	1192	381
Kigali	Rwanda	1567	1000	226
Addis Ababa	Ethiopia	2200	1143	111
Juba	South Sudan	550	941	101
Abuja	Nigeria	840	1389	49
Yaoundé	Cameroon	726	1643	355
Bangui	Central African Republic	369	1535	290
Ndjamena	Chad	298	481	0
Niamey	Niger	218	505	2
Bamako	Mali	350	953	5

Note: Dry season rainfall is averaged over the five driest months.

near Beira. A minor south-western offshoot extended through Zambia's Luangwa Valley into northern Botswana, ending in the faults blocking drainage from the Okavango Delta. Most rift activity took place between 9 and 5 Ma in the north-east, then again 1–2 Ma extending through the south, associated with the two phases of tectonic uplift.

Lake Naivasha in the rift floor lies at an elevation of 1884 m, higher than the elevation of the city of Nairobi to the east (Figure 1.4A). The lowest region of the rift near Lake Turkana in northern Kenya is only 375 m above sea level. The rift margins in the Turkana Basin are separated by ~300 km, but become narrowed to under 60 km near Nairobi.[3] Approaching its terminus in northern

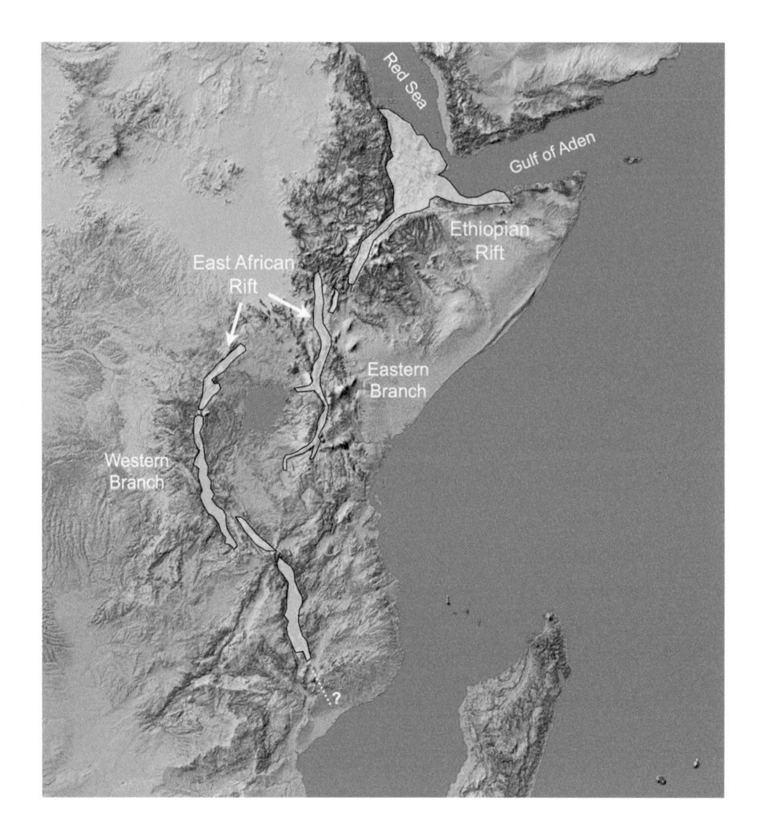

Figure 1.3 Map of the African Rift Valley System extending through eastern Africa from Ethiopia in the north to Mozambique in the south (from Wood & Guth, www.geology.com/articles/east-africa-rift).

Tanzania, the Eastern Rift splits into three arms. The western arm extends through Lake Eyasi, the central one reaches south of Lake Manyara, and the eastern one goes past the town of Moshi near Kilimanjaro. The road heading towards Serengeti ascends the margin of the Eastern Rift while passing Lake Manyara (Figure 1.4B).

The Western Rift is somewhat narrower than the Eastern Rift and includes several deep basins filled by large lakes. The Ruwenzori Mountains were formed as an upthrust block within the rift subsidence, with Mount Stanley reaching 5120 m above sea level. Lake Kivu's surface lies at 1460 m, while the surface elevation of Lake Tanganyika to the south is much lower at 773 m. Lake Tanganyika attains a maximum depth of 1470 m, meaning that its floor lies well below sea level. The Western Rift is still widening at a rate of a few millimetres per year, portending a split of the Somali plate including much of eastern Africa from the rest of the continent in some distant future. The Virunga volcanoes lie to the east of this rift, with Mount Karisimbi attaining

Figure 1.4 African Rift Valley views. (A) Eastern Rift Valley descending from the Ngong Hills in Kenya; (B) Eastern Rift shoulder above Lake Manyara in Tanzania; (C) rift wall rising beyond Lake Naivasha; (D) arid floor of the Eastern Rift near the equator in Kenya.

an elevation of 5109 m. Mounts Nyamulagira and Nyiragongo are still active. The low volcanic cones and associated crater lakes in Queen Elizabeth National Park in Uganda were generated quite recently.

Numerous volcanoes are allied with the Eastern Rift, most of them on adjoining platforms rather than within the rift subsidence. Mount Elgon (4321 m), situated on the border between Kenya and Uganda, was formed around 22 Ma before local rifting began. Mount Kilimanjaro (5895 m) rose to its full height between 2.5 and 1 Ma. Mount Kenya (5199 m) formed earlier around 2.6 Ma and was initially much higher than it is today. Ngorongoro Crater, with its rim rising to 2380 m, is the remnant of a volcano blasted open by a tremendous explosion around 2 Ma, producing the world's largest caldera. Oldoinyo Lengai, situated in the Ngorongoro highlands, still spews carbonatite tuffs over earlier lava flows.

Volcanic cones occur also in the Cameroon highlands, with Mount Cameroon rising to an altitude of 4040 m. Other volcanoes formed the islands of Sao Tome and Principe in the Gulf of Guinea. Southern Africa's land surface has remained much more stable. There have been no volcanic eruptions more recent than the early Jurassic period when Gondwana broke up.

Overview

Africa was predominantly high-lying following its separation from Gondwana, with its surface becoming eroded to a gently undulating plain rimmed by retreating scarps. Its interior elevation was raised further by tectonic uplift during the Miocene and Pliocene, especially through the east and south, while western Africa remained mostly low-lying. The eastern region became disrupted by ramifying rift valleys, with volcanic cones emerging on their margins. These troughs accumulated the sediments that preserve most of the fossil record of evolution since the Miocene. The Kalahari basin in the west accumulated predominantly loose sands that do not retain fossils, while in the east sediments carried by rivers extended coastlines.

In contrast, most of South America east of the Andes mountain ranges is low-lying. The plateau region of southern Brazil is divided from the highlands of Venezuela and Colombia by the vast Amazon basin. A transect strip at 30°S through southern Brazil and adjoining Argentina, excluding the Andes, averages under 200 m in elevation, compared with 1200 m for a corresponding strip across southern Africa. Within North America, a distance of 3000 km must be traversed from the east coast to the Rocky Mountains before elevations exceeding 1000 m are encountered. Most of Europe and northern Asia is low-lying, except where the Alps and other fold mountains intrude. Much of the Indian peninsula exists as an undulating plateau under 1000 m in elevation, although hills in the Western Ghats approach 3000 m above sea level. To the north of India, the Tibetan plateau reaches 4500 m, while several peaks in the Himalaya and adjoining mountain ranges rise above 8000 m. Australia remains mostly the low-lying plain developed by the end of the Cretaceous. Only a narrow region of the Great Dividing Range in the east exceeds 1000 m, while the highest peak, Mount Koskiusko in the Snowy Mountains, attains merely 2440 m. Western Africa resembles South America and other continents in its mostly low-lying terrain. The extent of the East African Rift System is unrivalled in any other continent.

Africa's disrupted topography affects local rainfall patterns, as will be described in the following chapter. Its eroding land surface exposes soils to bedrock influences below as well as above escarpment rims. In these ways 'High Africa' constitutes a foundation for Africa's distinctive ecology, as will be outlined in subsequent chapters.

Suggested Further Reading

Burke, K; Gunnell, Y. (2008) The African erosion surface: a continental-scale synthesis of geomorphology, tectonics, and environmental change over the past 180 million years. *Geological Society of America Memoirs* 201:1–66.

Macgregor, D. (2015) History of the development of the East African Rift System: a series of interpreted maps through time. *Journal of African Earth Sciences* 101:232–252.

References

1. Partridge, TC. (2010) Tectonics and geomorphology of Africa during the Phanerozoic. In Werdelin, L; Sanders, WJ (eds) *Cenozoic Mammals of Africa*. University of California Press, Berkeley, pp. 3–17.
2. Burke, K; Gunnell, Y. (2008) The African erosion surface: a continental-scale synthesis of geomorphology, tectonics, and environmental change over the past 180 million years. *Geological Society of America Memoirs* 201:1–66.
3. Mathu, EM; Davies, TC. (1996) Geology and the environment in Kenya. *Journal of African Earth Sciences* 23:511–539.

Chapter 2: Climate: Rainfall Seasonality

The climatic feature most relevant to Africa's ecology is precipitation or, more specifically, rainfall. Conditions cold enough to produce snow occur only in Africa's southern and northern extremes, and on a few of the highest peaks in between. Moreover, it is not merely the total annual amount of rainfall that is important, but particularly its seasonal distribution. Plant growth ceases for part of the year not because it is too cold, but because it is too dry. It does matter, though, whether the dry months occur during the cool (winter) period or warm (summer) period, recognising that concepts of summer and winter do not apply near the equator. Nevertheless, the annual seasonal cycle takes the form of an alternation between a wet season and a dry season even in the equatorial zone.

Seasonal variation in temperature and rainfall is governed by the tilt in the Earth's axis of rotation, which determines which hemisphere receives most solar radiation during different stages of the Earth's rotation around the sun. In mid-December, the sun is overhead at noon at latitude 23.5°S (the Tropic of Capricorn) in the southern hemisphere. In mid-June, it shines directly down at midday over 23.5°N (the Tropic of Cancer). Where the sun is overhead at noon draws together the easterly trade winds from the north and south, generating the Intertropical Convergence Zone, or ITCZ, in tropical latitudes. Where the ITCZ is located shifts north and south of the equator during the course of the year. Near the equator, this produces two rainfall peaks, timed shortly after the sun's transit overhead – one in November and one in April. Towards the subtropics, there is a single rainy season timed during the warmer summer months. During the winter months, high pressures developed by descending air produce clear sunny days. Ocean temperatures also influence rainfall, because where the surface water is cooler less water evaporates. This results in deserts on the west coasts of southern Africa, South America and Australia.

Local topography has a further modifying influence on rainfall received. Higher-lying areas deflect air upwards, causing moisture to condense locally while producing rain shadows leeward. Drier conditions develop westward because the easterly winds have deposited much of their moisture by then. Winds also affect ocean currents and consequent sea surface temperatures, with ramifying effects on precipitation around the globe. The El Niño–Southern Oscillation (ENSO) is controlled by the Humboldt current, which conveys cold Antarctic water northwards off the western coast of South America. This affects atmospheric pressure cells located over Africa and hence the development of rain-forming clouds during the southern wet season. The rain that southern Africa does not receive falls further north, generating wetter conditions in eastern Africa.

Current Climates

Africa is prevalently drier than other continents, apart from Australia. The regions of Africa receiving a mean annual rainfall (MAR) above 2000 mm, supporting rainforest, are located mostly in low-lying regions of west-central Africa near the coast. They include coastal Cameroon and adjoining parts of Nigeria, plus the region extending from Ivory Coast to Guinea further west (Figure 2.1). Some regions of the central Congo Basin receive similarly high rainfall. Elsewhere, rainfall amounts of this magnitude occur only locally on mountain slopes.

Over most of High Africa, MAR ranges between 500 and 1000 mm (see Table 1.1). Amounts below 600 mm occur close to the equator in the Eastern Rift valley and on the Serengeti Plains, situated in the shadow of the Ngorongoro highlands. Northern Kenya is dry due to the deflection of the north-easterly rain-bearing winds by the Ethiopian highlands. Rainfall diminishes towards desert conditions along the west coast of southern Africa, culminating in the Namib Desert. In the north-west, conditions grade from the Sahel into the Sahara Desert.

Within the African tropics, rainfall is fairly evenly distributed through the year only in parts of the central Congo basin, where moist air is drawn from both the Indian and Atlantic oceans (Figure 2.2). Nevertheless, quite substantial amounts of rain are received during the dry season months in much of western Africa (Table 1.1). In eastern Africa close to the equator, the dry season lasts from May through October and there is a variable lull in rainfall through January–February. Most rain falls in March–April during the northward movement of the ITCZ. The duration of the dry season gets shortened to 3 months in parts of Uganda near the Western Rift. Southward through Zambia, Zimbabwe and Malawi, as well as in parts of Angola in the west, the dry season is lengthened to 7 months and the pre-rain period from October into November is intensely hot as well as dry.

In the summer rainfall region of South Africa, 85 percent or more of the annual rainfall typically falls during a wet season lasting from October or November through March or April. The cold winter months from May through

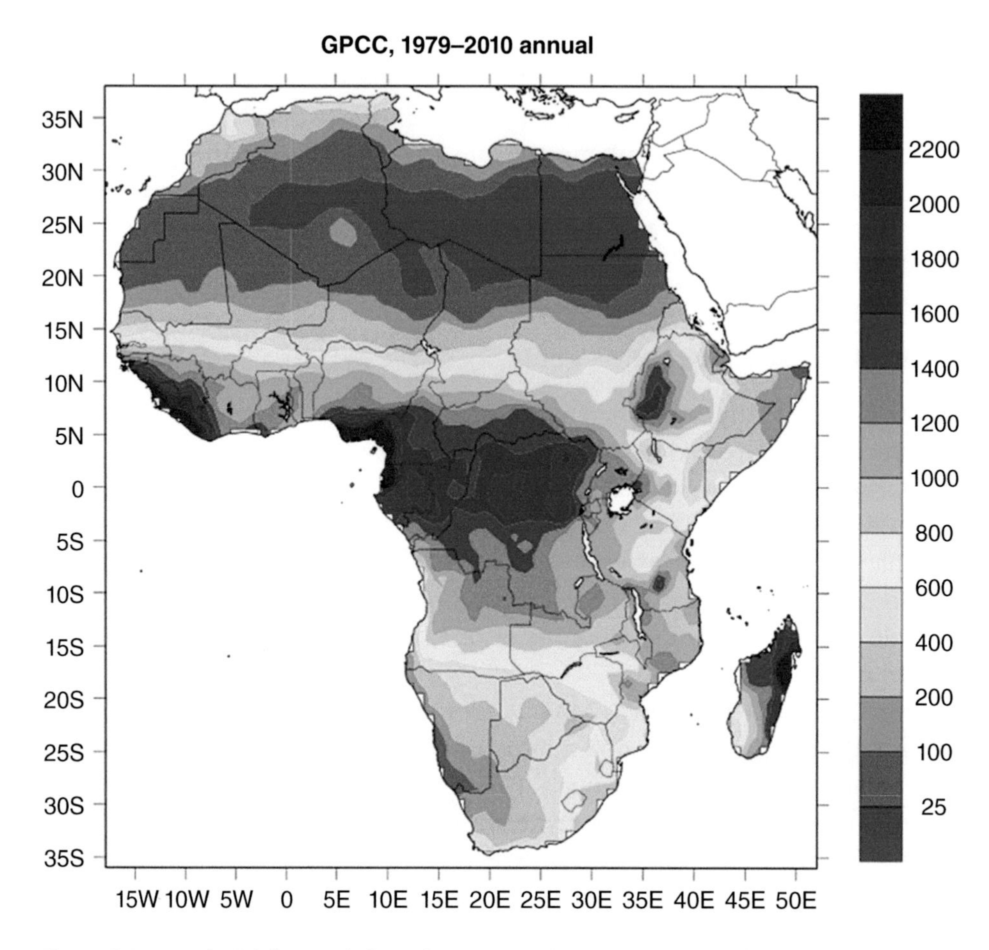

Figure 2.1 Annual rainfall map of Africa, showing prevalence of mean annual rainfall totals under 800 mm through eastern and southern Africa, broken by a moist corridor extending from coastal Mozambique through Malawi (map produced by the Global Precipitation Climatology Center, from A. Siebert in *Geography Compass*, June 2014).

July or August are dry and sunny because high-pressure conditions prevail and suppress cloud development. Hot 'berg' winds blow from the interior during August and September, promoting the spread of fires shortly before the rainy season commences. In western Africa, north-easterly trade winds blowing over the interior during the dry season generate dry and dusty conditions known as the Harmattan. Images captured from space by satellite depict the north to south alternation of green vegetation across the continent (Figure 2.3).

The MAR totals hide the substantial variation in the annual amounts of rainfall actually received. Over much of eastern and southern Africa, the typical range of variation is from less than half to nearly twice the MAR, generating a coefficient of variation, or CV (standard deviation divided by

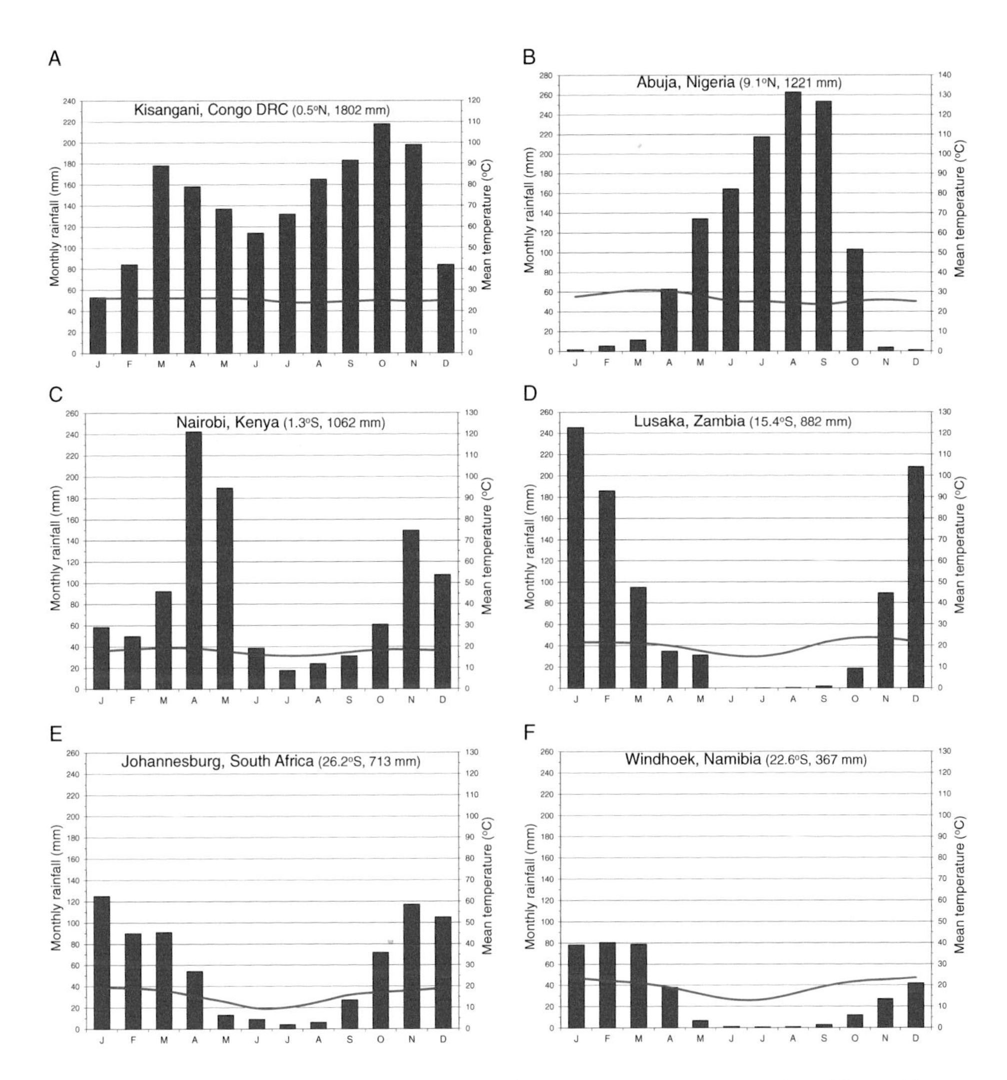

Figure 2.2 Seasonal rainfall patterns and associated temperature regimes for various African cities showing variation in the duration and intensity of the dry season (latitude and mean annual rainfall in brackets). Moisture deficits arise when the monthly rainfall in millimetres dips below the mean monthly temperature in degrees Celsius (panels drawn using data from climate-data.org).

the mean) of around 25 percent. The CV widens as the mean annual total diminishes, and vice versa. Arid savanna regions may go for a year or longer without rainfall.

The timing of the first rains of the wet season and when the dry season sets in also varies from year to year, with important ecological consequences. Hence the blocks of months representing wet and dry season conditions can differ from year to year. Rain received during the normally dry season months alleviates the

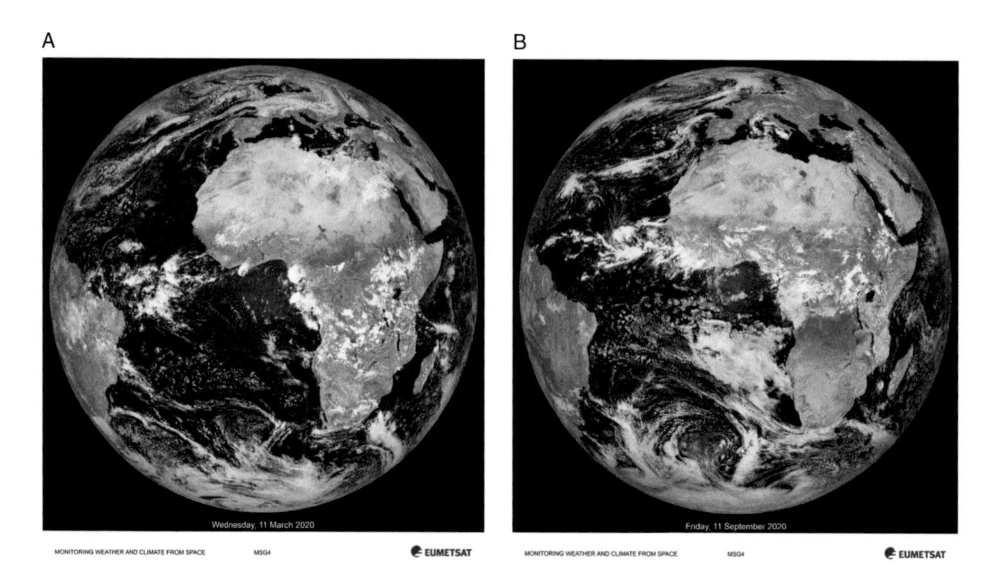

Figure 2.3 Contrasting shifts in cloud cover and greenness in southern versus northern sections of Africa between (A) March 2020 and (B) September 2020. Note the broad dry region in the south during September at the end of the dry season and the dry region extending through western Africa in March (© EUMETSAT 2020).

dryness of the grass cover, but has little or no effect on trees. Dry spells may also interrupt the normally rainy season months. During my field study in the Hluhluwe-iMfolozi Park, the state of the grassland in January 1970, normally the peak wet month, was as dry as normally experienced during July in the mid dry season. Herbivores must cope somehow with the resultant variability in food availability. Storm fronts also vary in their spatial tracks. Even during severe droughts, some regions may receive adequate rainfall and alleviate the shortfall in grass growth for animals able to move towards the greener regions.[1]

The effectiveness of rainfall for plant growth depends further on how temperature and wind conditions control evaporative losses. Within High Africa, temperatures do not rise very high even near the equator, being tempered by the elevation coupled with persistent cloud cover. The maximum midday temperature generally ranges between 25 and 30°C year-round, and seldom much exceeds 30°C. Temperature conditions are less pleasant in coastal regions, where they are coupled with high humidity. In more southern and northern latitudes, midday temperatures frequently rise above 40°C in the period preceding the rains. The highest temperature recorded in the north of Kruger NP, just within the tropics is 48°C, while in subtropical Hluhluwe-iMfolozi Park the highest temperature I experienced was 44°C. The range in temperature between day and night can also be quite wide in the interior, typically spanning ~20°C during the dry season months.

Along the eastern coast of South Africa, the south-flowing Agulhas current brings tropical conditions as far as 30°S. In the west, the cold Benguela current produces coastal fog and promotes cloudy conditions in the dry season, reaching as far north as Gabon in west-central Africa. Overnight frosts, as well as occasional snow, occur during winter in the southern interior of South Africa, and at any time of the year in the highest mountains of eastern Africa and Ethiopia. Sub-zero temperatures may also develop overnight in interior Botswana and adjoining parts of Zimbabwe, producing 'black' frosts because of the low humidity at that time of the year.

Over most of Africa, more water potentially evaporates than falls as rain. Climatograms relating monthly rainfall to prevailing temperatures indicate periods with negative water balance, if scaled appropriately (Figure 2.2). Potential evapotranspiration (including water lost through plants) reaches around 1600 mm annually at a latitude of 25°S, diminishing to 1300 mm nearer the equator because of the greater cloud cover.[2] Positive water balances enabling plant growth are restricted to the wet season months, and perhaps to only a portion of this period. Wind accelerates water losses through evaporation and further restricts plant growth. Windy conditions feature especially in the plateau and escarpment regions of southern Africa. During intense storms, the deluge of water can exceed the infiltration capacity of soils, so that much water runs off into gullies, streams and rivers, accentuating erosion.

Topographic variation in elevation and surface terrain generates quite wide variation in annual rainfall amounts over short distances.[3] For instance, while MAR on the rim of the Ngorongoro Crater in Tanzania reaches up to 1700 mm, it falls to ~500 mm on the plains to the west where Olduvai Gorge is situated. The MAR within the Serengeti–Mara ecosystem spans an overall range from ~500 mm in the south-east, in the shadow of the Crater Highlands, to 1100 mm in the far north over a distance of 250 km. Tsavo East NP, situated a few degrees south of the equator and 200–300 km inland from the Kenyan coast, receives a MAR of only 530 mm, while Mombasa on the coast has a MAR of 1200 mm. The Hluhluwe-iMfolozi Park, situated in the escarpment footslopes of eastern South Africa, encloses a rainfall gradient from under 600 mm in the low-lying south-west to nearly 1000 mm in the hilly north-east over a distance of 30 km.

Past Climates

Indicators of past climates are obtained from hydrogen and oxygen isotope ratios preserved in air pockets retained in glacial ice, sediments accumulated in the adjoining oceans, local soil features and records of lake levels. Global climates since the end of the Cretaceous period 66 Ma have shifted extremely from conditions so warm that both the South and North Poles were free of ice, and times so cold that massive ice sheets covered much of the northern continents.[4] Cyclic alternations in prevailing temperatures became manifested particularly over the last two million years. These cycles are generated by changing features of the Earth's orbit, tilt and spin relative to the sun, named the Milankovitch cycles (Box 2.1). The climatic oscillations and trends had important implications for the evolution of humans and their hominin ancestors.

Carbon dioxide levels in the atmosphere affect how much of the heat radiation received from the sun is retained (the greenhouse effect) and hence contributes to warmth. During the Eocene epoch ~53 Ma, atmospheric CO_2 concentrations were projected to lie between 1000 and 2000 ppm, dropping to around 400 ppm by the commencement of the Pliocene 5 Ma. Around the Last Glacial Maximum (LGM) 20 ka, CO_2 levels dropped as low as 180 ppm. Following industrialisation and the burning of fossil fuels, atmospheric CO_2 levels have recently risen above 400 ppm again. Local temperature regimes depend additionally on how ocean currents redistribute heat. When temperatures are colder globally, conditions are generally drier, because less moisture evaporates from the oceans.

Following the breakup of Gondwana ~180 Ma, most of the world remained particularly hot and wet from the mid-Cretaceous period into the Eocene.[5] Temperature conditions at the start of the following Oligocene epoch remained almost 10°C warmer than those prevailing today (Figure 2.4).[6] In Africa the climatic equator, i.e. where movements of the ITCZ are centred, lay further north than at present because of the latitudinal location of Africa at that time. Consequentially, wet conditions prevailed through much of what is now the Sahara Desert. A cooling trend occurred during the Oligocene, by up to 8°C.[7] Radical changes in the barriers represented by the continents caused the drift of major sea currents to shift from north-to-south to east-to-west. The Southern Ocean circulation passing between Australia and Antarctica became operational around 35 Ma, when South America finally split off from Antarctica, opening the Drake Passage. As a consequence, an ice sheet developed in the eastern Antarctic.[6] The Miocene until 15 Ma remained 5°C warmer than present-day conditions, but cooled by 2–3°C between 15 and 12.5 Ma, shortly preceding the spread of savanna grasslands. The uplift of the Tibetan plateau, which commenced rising ~50 Ma following the collision of drifting India with the rest of Asia, contributed further to drying in eastern Africa by deflecting moisture brought to eastern Africa by the Indian Ocean monsoon.[8] Freshly exposed rock surfaces in the Himalayas sequestrated carbon in limestone deposits on ocean floors, lowering atmospheric carbon dioxide levels and contributing further to global cooling.

Box 2.1 Milankovitch Cycles

A Serbian astronomer named Milankovitch recognised how orbital variations in the Earth's movements generate cyclic oscillations in climatic conditions (Figure B2.1). Three contributions interact. (1) The eccentricity in the Earth's orbit, affecting its nearness to the sun, oscillates with a period of around 100 kyr affecting the intensity of the solar radiation impinging on the Earth. (2) The tilt, or obliquity, of the Earth's axis of rotation, which varies between 21.8° and 24.4° with a period of 41 kyr, affects which hemisphere receives most sunshine in particular seasons. This modifies how extreme the seasonal contrasts are, amplified when the tilt is greatest. (3) The precession or wobble

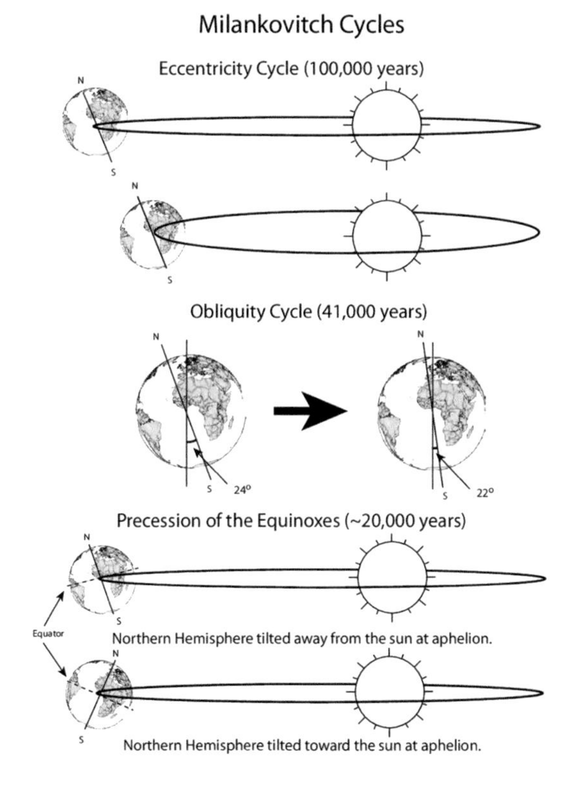

Figure 2B.1 Aspects of orbital geometry generating the Milankovitch cycles in global climate. *Eccentricity* changes the shape of the orbit on a 100,000-year cycle from a circular to a more elliptical shape. *Obliquity* is the change of the angle of the Earth's axis of rotation, which ranges from 22° to 24° from normal, and occurs on a 41,000-year cycle. *Precession* or 'wobble' in the axis affects the Earth's aspect relative to the sun in particular seasons, with a period varying between 19,000 and 23,000 years (from Ross et al. (2010) *Climate Change Past, Present & Future: A Very Short Guide*. Paleontological Research Institution Special Publication No. 38; download).

continues

Box 2.1 (cont)

in the Earth's axis of rotation, in conjunction with the orbital eccentricity, determines when during the Earth's orbit the sun is vertically overhead at noon. This increases the seasonal contrast in one hemisphere and decreases it in the other, affecting regional rainfall particularly in the tropics and subtropics with a 19–23 kyr period.

The Tethys Sea in the north lost its connection with the Indian Ocean between 11 and 7 Ma, accentuating aridity in the Sahara region. In the south-west, the cold Benguela current developed, suppressing moisture drawn from the Atlantic Ocean and producing the Namib Desert by ~7 Ma. Around 5.6 Ma, the Mediterranean Sea dried up completely, following closure of its connection with the North Atlantic Ocean, promoting extreme aridity in the north of Africa. Temperature conditions then stabilised during the Pliocene from 5 Ma to 3.5 Ma, after the Mediterranean Sea had filled again. Between 2.8 and 2.6 Ma, the gap between South and North America in the region of Panama, which had connected the Atlantic and Pacific Oceans, closed. This initiated the formation of glacial ice around the North Pole and produced a sharp drop in global temperatures, initiating the transition from the Pliocene into the Pleistocene epoch.

The climatic oscillations between glacial advances and interglacial interludes, characterising the Pleistocene from a northern hemisphere perspective, then set in. Initially this coupled cycle periods of 41 kyr, governed by changes in the obliquity of the Earth's axis of rotation, and ~21 kyr, related to the precession wobble in this axis (see Box 2.1). The progressive cooling and drying trend continued (Figure 2.4).[9] Tectonic movements associated with rift valley formation further influenced local precipitation in eastern Africa. After 0.8 Ma beginning in the mid-Pleistocene, the period between glacial peaks lengthened to 100 kyr, and global temperature fluctuations became widened to as much as 5°C between full glacial and interglacial conditions (Figure 2.4). Cold glacial conditions lasted around 100 kyr, while warmer interglacial interludes persisted only ~10–15 kyr. Rapid deglaciation took place within merely a few thousand years approaching the warm peaks. Through the course of the Pleistocene, the cold extremes, and hence inferred dryness in Africa, reached unprecedented levels, while little change took place in interglacial temperature conditions.[10] The most extreme lows in glacial cold occurred around 650 ka, 450 ka, 150 ka and during the LGM

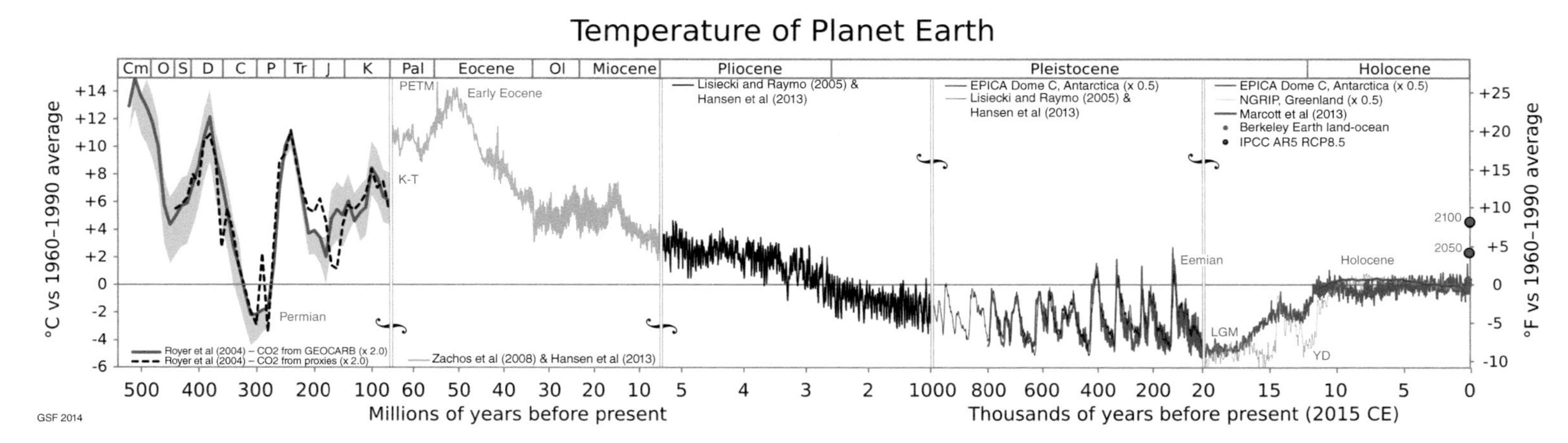

Figure 2.4 Global temperature variation through time inferred from various sources, with the scale increasingly finely partitioned within five periods: (i) through the Palaeozoic and Mesozoic eras from 520 to 65 Ma (red); (ii) through the Cenozoic era from 65 to 5.3 Ma (green); (iii) through the Pliocene and early Pleistocene from 5.3 to 1.0 Ma (black), (iv) through the late Pleistocene from 1.0 to 0.2 Ma (blue), and (v) through the period following the Last Glacial Maximum into the Holocene. Note how global mean temperature has declined progressively while the amplitude of temperature fluctuations has widened (from Wikipedia, assembled using various data sources).

20 ka. Temperature conditions during the peak glacial advances around 550 ka, 350 ka and 250 ka were not as extreme. Global temperature regimes additionally affected the Hadley cells governing where air rising over the climatic equator settled in the subtropics. This circulation contracts towards the equator during cold extremes and expands towards higher latitudes during interglacial interludes, affecting local rainfall governed by the ITCZ.

Finer-scale reconstructions of past temperature and moisture conditions are available for Africa after 130 ka, during the late Pleistocene.[11] The region of south-central Africa from Lake Malawi to Lake Tanganyika experienced a 'mega-drought' from 135 to 75 ka, as conditions cooled after the preceding interglacial, perhaps exacerbated by a northward shift in the ITCZ.[12,13,14,15] MAR could have been reduced to as little as 60 percent of current levels through this region. Lake temperatures in eastern Africa spanned a 4°C range between glacial and interglacial extremes during the late Pleistocene, compared with 6–7°C for terrestrial temperatures. At Olduvai Gorge, local rainfall possibly varied as widely as 200–700 mm.[16] These extremely variable conditions shortly preceded the major exodus of humans from Africa.

The wobble in the Earth's rotation varying in period between 19 and 23 kyr generates oscillations between wet and dry extremes taking place within as little as 2000 years, separated by more stable periods lasting 8000 years.[17] Projected annual rainfall in the vicinity of the Tswaing Crater, situated near the northern margin of the South African Highveld, varied between 500 and 870 mm with a 23 kyr period over the past 250 kyr.[18] During the LGM, small local glaciers developed in the Lesotho highlands because winter rainfall penetrated further inland than at present, even as far as the Free State and Gauteng provinces.[19]

In the Kalahari region of Botswana, rainfall dropped to as little as 40 percent of the current mean during the LGM.[20] Rainfall rose but fluctuated widely during the cool conditions that ensued there between 16.6 and 12.5 ka after the LGM.[21] A brief cool interval between 12.9 ka and 11.7 ka, known in Europe as the Younger Dryas, was associated with intensified aridity over most of southern Africa. Between 7000 and 4500 years ago, conditions became ~2°C warmer than at present during a period known as the Holocene altithermal. It was associated with lowered rainfall through South Africa and southern Zimbabwe, while the Kalahari region and further north remained wetter than at present.[18] During the 'Little Ice Age' in Europe, which extended from 1300 to 1810 CE, temperatures in South Africa fell by about 1–2°C.

Although most of Africa was drier when northern continents were colder during glacial advances, exceptions occurred in parts of southern Africa. Marine sediments show that the coastal hinterland in the south-east was subject to greater erosion, indicating wetter conditions, during glacial periods, possibly due to strengthening of the Agulhas current.[22] The flow of the Zambezi River surged between 16 and 12 ka, indicating higher rainfall in its catchment, although the northern hemisphere still remained cool at that time.[23]

In the north, the Sahara Desert became extremely dry during the late Miocene 7–11 Ma, following closure of the Tethys Sea.[24] The drying trend there intensified further after 2.8 Ma.[5] North Africa was relatively humid for a period between 133 and 117 ka during interglacial conditions, then again from 100 to 75 ka, when south-central Africa was especially dry.[15] The Sahara supported a lush savanna thronged with animals (even crocodiles) from 9000 until 4500 years ago, after which desert conditions took hold again.

Overview

Africa is relatively dry compared with other continents spanning tropical latitudes. Most of Africa's eastern region receives under 1000 mm of rainfall annually, and parts even less than 650 mm, the functional threshold between moist and dry savannas (see Chapter 7). Dry seasons span several months with little or no rainfall and occur during the cooler period of the year away from the equator. Mean annual rainfall in South America tends to be twice that recorded at similar latitudes within Africa.[2] Easterly trade winds convey heavy rain deep into the low-lying interior of tropical South America, while Africa's regional rainfall totals get attenuated to the west of its uplifted rim, especially in rift valley depressions and in the southern subtropics. In Australia, rainfall remains low through most of the interior away from the eastern seaboard and fluctuates even more erratically between years than in tropical and subtropical Africa. In temperate latitudes of Eurasia and North America, precipitation in the form of winter snow moistens soils in spring. Even where there is no snow, rainfall is more effective for promoting plant growth because of the prevalently cooler conditions.

Past climatic conditions, particularly in rainfall, would have been more widely variable than during the interglacial interlude that we are currently experiencing, despite the developing disruption by global warming. Global temperature regimes became progressively colder and hence drier from the late Miocene through the Pliocene and Pleistocene, influenced by declining atmospheric CO_2 levels and oceanic circulation patterns. Major climatic transitions occurred around 2.63 Ma when cyclic variation with a 41 kyr period took hold, associated with the onset of Arctic glaciation, and around 0.8 Ma, when the period of the glacial–interglacial alternations lengthened to ~100 kyr. Within these regimes, shorter-term oscillations with a ~21 kyr period further affected rainfall variability in Africa. As a consequence of the reduced rainfall, dry seasons became more intensely dry and wet season totals fluctuated more extremely from one year to the next than at present. Major droughts would formerly have been both more extreme and more frequent. Somehow our hominin ancestors coped through the dry seasons when plant growth ceased.

Rainfall amounts and seasonal variation therein had ramifying effects on river flows and lake levels. Rainfall also strongly influences soil properties, in interaction with the geological substrate. These interrelationships will be described in the following chapters.

Suggested Further Reading

Burke, K; Gunnell, Y. (2008) The African erosion surface: a continental-scale synthesis of geomorphology, tectonics, and environmental change over the past 180 million years. *Geological Society of America Memoirs* 201:1–66.

Feakins, SJ; deMenocal, PB. (2010) Global and African regional climate during the Cenozoic. In Werdelin, L; Sanders, WJ (eds) *Cenozoic Mammals of Africa*. University of California Press, Berkeley, pp. 45–55.

Levin, NE. (2015) Environment and climate of early human evolution. *Annual Review of Earth and Planetary Sciences* 43:405–429.

Sepulchre, P, et al. (2006) Tectonic uplift and Eastern Africa aridification. *Science* 313:1419–1423.

References

1. Malherbe, J, et al. (2020) Recent droughts in the Kruger National Park as reflected in the extreme climate index. *African Journal of Range & Forage Science* 37:1–17.
2. Walling, DE. (1996) Hydrology and rivers. In Adams, WM, et al. (eds) *The Physical Geography of Africa*. Oxford University Press, Oxford, pp. 103–121.
3. Partridge, TC. (1997) Late Neogene uplift in eastern and southern Africa and its paleoclimatic implications. In Ruddiman, WF (ed.) *Tectonic Uplift and Climate Change*. Springer, New York, pp. 63–86.
4. Zachos, J, et al. (2001) Trends, rhythms, and aberrations in global climate 65 Ma to present. *Science* 292:686–693.

5. Burke, K; Gunnell, Y. (2008) The African erosion surface: a continental-scale synthesis of geomorphology, tectonics, and environmental change over the past 180 million years. *Geological Society of America Memoirs* 201:1–66.
6. Denton, GH. (1999) Cenozoic climate change. In Bromage, TG; Schrenk, F (eds) *African Biogeography, Climate Change, and Human Evolution.* Oxford University Press, Oxford, pp. 94–114.
7. Feakins, SJ; deMenocal, PB. (2010) Global and African regional climate during the Cenozoic. In Werdelin, L; Sanders, WJ (eds) *Cenozoic Mammals of Africa.* University of California Press, Berkeley, pp. 45–55.
8. Sepulchre, P, et al. (2006) Tectonic uplift and Eastern Africa aridification. *Science* 313:1419–1423.
9. Maslin, MA; Christensen, B. (2007) Tectonics, orbital forcing, global climate change, and human evolution in Africa: introduction to the African paleoclimate special volume. *Journal of Human Evolution* 53:443–464.
10. Maslin, M. (2017) *The Cradle of Humanity.* Oxford University Press, Oxford.
11. Singarayer, JS; Burrough, SL. (2015) Interhemispheric dynamics of the African rainbelt during the late Quaternary. *Quaternary Science Reviews* 124:48–67.
12. Scholz, CA, et al. (2007) East African megadroughts between 135 and 75 thousand years ago and bearing on early-modern human origins. *Proceedings of the National Academy of Sciences of the United States of America* 104:16416–16421.
13. Tierney, JE, et al. (2008) Northern hemisphere controls on tropical southeast African climate during the past 60,000 years. *Science* 322:252–255.
14. Burnett, AP, et al. (2011) Tropical East African climate change and its relation to global climate: a record from Lake Tanganyika, Tropical East Africa, over the past 90+ kyr. *Palaeogeography Palaeoclimatology Palaeoecology* 303:155–167.
15. Blome, MW, et al. (2012) The environmental context for the origins of modern human diversity: a synthesis of regional variability in African climate 150,000–30,000 years ago. *Journal of Human Evolution* 62:563–592.
16. Magill, CR, et al. (2013) Ecosystem variability and early human habitats in eastern Africa. *Proceedings of the National Academy of Sciences of the United States of America* 110:1167–1174.
17. Trauth, MH, et al. (2007) High- and low-latitude forcing of Plio–Pleistocene East African climate and human evolution. *Journal of Human Evolution* 53:475–486.
18. Tyson, PD, et al. (2001) Late Quaternary environmental change in southern Africa. *South African Journal of Science* 97:139–150.
19. Engelbrecht, FA, et al. (2019) Downscaling last glacial maximum climate over southern Africa. *Quaternary Science Reviews* 226:105879.
20. Shaw, PA; Thomas, DSG. (1996) The quaternary palaeoenvironmental history of the Kalahari, Southern Africa. *Journal of Arid Environments* 32:9–22.
21. Cordova, CE, et al. (2017) Late Pleistocene–Holocene vegetation and climate change in the middle Kalahari, Lake Ngami, Botswana. *Quaternary Science Reviews* 171:199–215.

22. Simon, MH, et al. (2015) Eastern South African hydroclimate over the past 270,000 years. *Scientific Reports* 5.
23. Schefuss, E, et al. (2011) Forcing of wet phases in southeast Africa over the past 17,000 years. *Nature* 480:509–512.
24. Zhang, ZS, et al. (2014) Aridification of the Sahara desert caused by Tethys Sea shrinkage during the Late Miocene. *Nature* 513:401–404.

Chapter 3: Water in Rivers, Lakes and Wetlands

Despite its prevalently low rainfall, Africa is not lacking in lakes and wetlands, and contains the longest river in the world. However, water flow in rivers can vary hugely in response to seasonal and annual variation in rainfall. Lakes can form and dry, reacting not only to rainfall variation but also to rift valley subsidence and other tectonic movements affecting water flow over quite short periods. Rivers have also changed their courses in response to ructions of the land surface.

Rivers bring surface water to regions that may otherwise be waterless. Groundwater seeping from channels supports flanking woodlands, which may retain green foliage when the rest of the landscape is brown and dry. Nevertheless, water-dependent herbivores prefer to seek their surface water needs away from major rivers, in sandy channels retaining water below the surface and from temporary pools in depressions known as pans. These are less likely to house lurking crocodiles and have less surrounding vegetation cover where lions can hide in ambush.

The flow of rivers has further biological consequences for terrestrial animals. No river crossing Africa's interior plateau is deep and wide enough to block the movements of hominins or other animals between the north-eastern and southern extremes of the continent; at least, not during low flow in the dry season. Even the biggest east-flowing river, the Zambezi (Figure 3.1B), is easily forded upstream of Victoria Falls at low water. The White Nile, flowing north-west from its source in Lake Victoria (Figure 3.1A), continues northward after joining the Blue Nile emanating from the mountainous highlands of Ethiopia. The mighty Congo and other rivers having substantial flow year-round all trend westward.

Rivers

Africa's five major rivers, in terms of mean volume of water discharged at their mouths, are the Congo, Niger, Ogooue, Zambezi and Nile, in that order (Table 3.1). The Congo is by far the largest and second globally in water volume amalgamated from its tributaries draining the tropical forests of the

Figure 3.1 African rivers. (A) Nile River below Murchison Falls in Uganda, deep and fast-flowing; (B) Zambezi River below Kariba Dam, broad and deep; (C) Limpopo River spread over a vast sandy bed; (D) Mara River in Kenya with deeply cut banks; (E) Letaba River in Kruger NP, flowing gently over a sandy bed; (F) Kidepo River in northern Uganda, seasonally without surface water.

Congo Basin. Nevertheless, the amount of water discharged from its mouth averages only 20 percent of that carried by the Amazon River in South America. Second in water flow in Africa is not the Zambezi, but rather the Niger, arising in the west in the Guinean highlands and curving eastward then southward to enter the Atlantic Ocean in the 'armpit' of Africa in Nigeria. The Ogooue River draining the forests of Gabon is also bigger than the Zambezi in terms of mean water flow. The Nile River comes fifth. It is the longest river in

Table 3.1 African rivers listed in order of mean flow volume at their mouths (source: Wikipedia)

Name	Length (km)	Mean discharge (m^3/s)	Range (m^3/s)	Mouth
Congo	4371	41,200	23,000–75,000	Atlantic
Niger	4180	5589	500–27,600	Atlantic
Ogooue	1200	4706	–	Atlantic
Zambezi	2574	3400	920–18,600	Indian
Nile	6853	2830	–	Aswan
Sanaga	976	1985	234–6950	Atlantic
Volta	1500	1210		Atlantic
Senegal	1086	650	$<$50–2250	Atlantic
Rovuma	800	475	–	Indian
Orange	2200	365	–	Atlantic
Limpopo	1750	170	–	Indian

the world, but is way down the ranking in water flow. Its discharge listed in Table 3.1 is the amount entering the Aswan Dam, because only half of this actually reaches its mouth into the Mediterranean Sea.

The next biggest African river flowing into the Indian Ocean, after the Zambezi, is the Rovuma, forming the border between Tanzania and Malawi, but it carries only 15 percent of the mean discharge from the Zambezi. Third largest in the east is the Limpopo, with considerably lower flow levels. All of the major rivers draining savanna regions show huge seasonal fluctuations in flow, with the lowest level in the dry season amounting to 5 percent or less of the maximum flow during the wet season (Table 3.1). Several large rivers have catchments in the Angolan highlands where the MAR is as high as ~1400 mm. The Kunene turns westward towards the Atlantic Ocean along the northern border of Namibia, while the Okavango flows south-east into a spreading alluvial fan in Botswana from which only a trickle flow emerges. The Kwanga flows north-west to enter the Atlantic Ocean near Luanda. South Africa's two largest rivers, the Gariep (or Orange) flowing westward toward the Atlantic Ocean and the Tugela flowing eastward into the Indian Ocean, both have their catchments in the Lesotho highlands. Many other rivers originating in Africa's interior plateau flow only seasonally, and merely retain pools in their sandy beds during the dry season (Figure 3.1F).

No African river is navigable by boat very far upstream from its mouth, due to sedimentation in lower reaches and rapids or waterfalls in upper courses. Before reaching the sea, the Congo River cascades over the Inga Falls and associated rapids, with the drop of 96 m over 15 km blocking navigation

A

B

Figure 3.2 (A) Victoria Falls in July with highest flow following the end of the wet season; (B) Victoria Falls in November with lowest flow at the end of the dry season.

through this barrier. Boat travel resumes upstream from Kinshasa and Brazzaville as far as Kisangani well into the interior. River travel up the Zambezi had been blocked by the Cahora Bassa Gorge and associated rapids in Mozambique, before this section was submerged beneath the waters of a huge dam. The Zambezi plunges 108 m over Victoria Falls on the border between Zimbabwe and Zambia, with its width becoming constricted from 1708 m to merely 110 m at this point. The breadth of its curtain of falling water at high flow, masked by shifting clouds of spray, is greater than that of any other waterfall worldwide. By the late dry season, it gets attenuated to remnant trickles over sections of its width (Figure 3.2). Below the falls, the Zambezi continues through zigzag gorges representing the locations of previous waterfalls cut through the local basalt. The White Nile traversing Uganda becomes similarly restricted through the Murchison Falls gorge where it enters the Western Rift depression. The Tugela Falls, dropping in total 983 m over the Drakensberg escarpment in South Africa, is rivalled in height only by the Angel Falls in Venezuela.

The courses followed by many African rivers have changed over geological time, in response to tectonic movements and headward erosion. Following the breakup of Gondwana, the Karoo and Kalahari rivers flowing westward from the southern African interior became joined to form the lower section of the Gariep River.[1] At that time the Limpopo accumulated the flow of several rivers emanating in the Angolan highlands, including the upper Zambezi. It was vastly bigger than its present meagre flow, as shown by its wide valley (Figure 3.1C).[2] The Okavango River flowed into a vast Lake Mababe in Botswana until as recently as 8.5 ka, before tectonic blockages generated the Okavango Delta.[3] The Savuti channel, branching from a bend where the Kwando River emanating from Angola hits the fault line, has fluctuated in its onward flow towards the Savuti Marsh, formerly part of Lake Mababe, over the past few centuries. Fossil drainage lines in the Kalahari, currently infilled with sediment and bone dry, testify to the former presence of quite large rivers in this currently arid region.

In eastern Africa, the Pliocene uplift diverted rivers that had flowed westward from the region of the Kenya–Uganda border to drain instead into the depression filled by Lake Victoria, forging a connection with the Nile. In western Africa, the Niger River was formed by the joining of two previously separate rivers. One had ended in a lake north-east of Timbuktu, while the other had its original source south of Timbuktu. The Congo River earlier had its mouth located about 200 km north of its current position.

Although many of the lesser African rivers cease surface flow during the dry season, water generally remains in their sandy beds. Many more would have flowed only seasonally during glacial advances when rainfall was generally lower than at present. This reinforced the lack of barriers to animal dispersals between eastern and southern Africa. Only the lower Zambezi River currently remains formidably deep during the dry season. South Africa's biggest river, the Gariep (or Orange), was readily forded by ox-waggons at designated 'drifts'. Thus, many animal species, or their geographic replacements, have distributions extending across Africa from the Cape to the Maghreb region bordering the Mediterranean coast.

Lakes and Wetlands

Lake levels in Africa have fluctuated hugely, contributing to local swings between arid and moist conditions. Lake Victoria, currently Africa's largest lake in area (Figure 3.3A), is quite shallow and has dried up completely several times, most recently between 18 and 14 ka after the LGM.[4] At other times it covered an even larger area than at present, extending into the western corridor of Serengeti NP. Lake Chad, in north-central Africa, was larger in area than current Lake Victoria as recently as 5 ka and had an outlet to the Atlantic Ocean then. Its progressive contraction in recent years indicates the increasing aridity in its catchment. Lakes Afar, Omo, Turkana and Baringo formed after 5 Ma within the northern section of the rift valley extending from Ethiopia into Kenya.[5] Lake Tanganyika, located in the Western Rift, is currently the second deepest lake in the world and third largest in water volume, with parts of its floor below sea level. Lake Malawi is nearly as deep, but nevertheless became reduced to shallow puddles around 0.75 Ma.[6] Other large lakes in the Western Rift include Kivu, Edward (Figure 3.3B), George and Albert. Lakes situated in the Eastern Rift, including Bogoria, Nakuru, Elmenteita, Magadi, Natron, Manyara and Eyasi, are shallow and highly saline. Lake Naivasha is exceptional in containing fresh water, presumably due to a hidden outlet.

Figure 3.3 African lakes. (A) Lake Victoria in Tanzania, vast but shallow; (B) Lake Edward in the Western Rift in Uganda, huge and deep; (C) Lake Nakuru in Eastern Rift in Kenya, shallow and saline; (D) Lake Malawi in the southern end of the Eastern Rift in Malawi, deep.

The Okavango Delta (strictly, an alluvial fan), located in northern Botswana, constitutes one of the six major wetlands of the world (Figure 3.4A,B).[7] It was formed after a seismic fault blocked onward flow of water from the Okavango River, perhaps as recently as 120 ka. The river entering its 'pan-handle' branches into numerous distributaries. Only a variable trickle flows onward via the Botete River into the Makgadikgadi depression, now mostly a dry salt pan. The remainder of the water entering the Delta evaporates. Water flowing through the Delta is exceptionally clear, because it comes from the Angolan highlands where the Kalahari sand cover lacks clay. Due to the delay in water passage through the various channels, the main inflow of water from the catchment during January–February generates peak flooding within the Delta only in July, during the middle of the dry season. Sediment accumulation in the form of sand eventually causes former channels bounded by papyrus to become islands. In earlier times, the paleo-lakes Makgadikgadi, Mababe and Ngami combined may have exceeded Lake Victoria in their total area, perhaps as recently as 17 ka.[3,7,8] A wetland exists

Figure 3.4 African wetlands. (A) Okavango Delta interfaced with savanna, Botswana; (B) Okavango Delta islands formed around termite mounds, Botswana; (C) Sudd Swamp bordering Nile River, Sudan (image from UNEP); (D) Busanga Plains in Kafue NP, Zambia (photo: Dean Polley, image provided by Wilderness Safaris); (E) aerial view of pan with elephants approaching, Botswana; (F) large pan kept full by pumping, with elephants, in Hwange NP, Zimbabwe.

in Namibia across its border with Botswana where the Kwando River hits a fault line and changes its flow direction to become the Linyanti and eventually join the Zambezi.

The Nile River flows through a vast swamp called the Sudd upstream of Khartoum in Sudan (Figure 3.4C). Within Mali, the Niger River forms a region of braided streams, marshes and lakes before reaching Timbuktu. Wetlands elsewhere in Africa include the Kafue, Barotse and Busanga flats (Figure 3.4D) as well as Lake Bangweulu in Zambia, the Gorongoza floodplain in Mozambique, and Rukwa-Katavi floodplain in Tanzania.

Seasonally waterlogged valleys called dambos (or vleis in South Africa) are a feature especially of south-central Africa where miombo woodlands prevail on basement granitic soils. They can develop into quite extensive wetlands where their drainage is blocked, for example at Nylsvley in the northern Limpopo Province of South Africa.

Pan depressions varying in size are a common feature of drier regions of Africa, accumulating water run-off during the wet season and becoming dry sometimes quite late into the dry season (Figure 3.4E,F).[9] In years with high rainfall, larger pans may retain water year-round, effectively sealed by impervious clay or calcrete crusts beneath. Numerous pans are a feature especially of the south-western Kalahari and may be self-perpetuating through the trampling activities of large herbivores.[10] There were numerous small pans retaining pools for varying periods in the Hluhluwe-iMfolozi Park where I did my white rhino study. Following any well-trampled animal trail usually led me to one of these water sources. Which of the pans retained water longest varied from year to year. Pans can keep herbivore concentrations away from perennial water sources in rivers and lakes for much of the year.

The fluctuations in water levels in lakes suggest changing rainfall conditions in their local catchments, but respond also to earth movements. Lakes Turkana and Baringo filled and emptied repeatedly between 2.69 and 2.58 Ma and again around 1.7 Ma, during the early Pleistocene.[5,11] Rapid swings between low- and high-water levels over relatively short timescales occurred, responding to the 21 kyr cycle driven by precession in the Earth's axis of rotation. Paleo-lake Olduvai fluctuated widely in extent and depth through the early Pleistocene, eventually drying up around 1.15 Ma. The fossil site at Olorgesailie, in southern Kenya, was sometimes beneath a lake and at other times quite dry, as it is at present.[12] Lake Malawi, currently very deep, shrank to a series of puddles between 135 and 75 ka.[13] Water levels in various eastern African lakes were high during the period between 145 and 120 ka, then again from 80 to 65 ka.[14] The water level in Lake Tanganyika dropped by over 400 m around 90 ka, more than took place during the LGM, but had risen back to its former level by 75 ka.[15]

While responding to climatic and tectonic variability, lakes in and near the Eastern Rift accumulated the sediments that preserved fossils, of human ancestors along with numerous other animals. Those in eastern Africa associated with the Western Rift were too deep to provide fossil accumulations.

Overview

Because of Africa's relatively dry climate, its rivers carry less water than those in most other continents (except for Australia), and many cease flowing seasonally. Annual variation in peak river flow is also wider than in other continents.[16] All of Africa's easterly flowing rivers can readily be crossed in their upper reaches during the dry season and thus have not formed barriers to animal dispersals between the northern and southern sections of the continent. Some of its rift valley lakes are exceptionally deep while others are shallow and saline. Pools retained in seasonal rivers, pan depressions and wetlands play an important role in drawing concentrations of large herbivores away from major rivers and lakes. In South America, the enormous Amazon River plus its tributaries and the Orinoco River further north form major obstacles to north–south movements by animals not able to swim competently. Although Africa has several large wetlands, those formed in the Llanos of Venezuela and Pantanal of Brazil are far vaster. Europe and North America have numerous deep-flowing rivers navigable far upstream from the coast, plus many large lakes. Most of interior Australia is largely waterless and devoid even of long-lasting pools.

Water movements redistributing rainfall have further influences on soil fertility, in association with bedrock geology, as will be explained in the next two chapters.

Suggested Further Reading

Fynn, RWS, et al. (2015) African wetlands and their seasonal use by wild and domestic herbivores. *Wetlands Ecology and Management* 23:559–581.

Maslin, M. (2017) *The Cradle of Humanity*. Oxford University Press, Oxford.

Walling, DE. (1996). Hydrology and rivers. In Adams, WM, et al. (eds) *The Physical Geography of Africa*. Oxford University Press, Oxford, pp. 103–121.

References

1. McCarthy, T; Rubidge, BS. (2005) *The Story of Earth and Life. A Southern African Perspective on a 4.6-Billion-Year Journey*. Struik, Cape Town.
2. Moore, A, et al. (2017) A geomorphic and geological framework for the interpretation of species diversity and endemism in the Manica Highlands. *Kirkia* 19:54–69.
3. Burrough, SL, et al. (2009) Mega-lake in the Kalahari: a Late Pleistocene record of the Palaeolake Makgadikgadi system. *Quaternary Science Reviews* 28:1392–1411.
4. Beverly, EJ, et al. (2020) Rapid Pleistocene desiccation and the future of Africa's Lake Victoria. *Earth and Planetary Science Letters* 530:115883.

5. Maslin, M. (2017) *The Cradle of Humanity*. Oxford University Press, Oxford.
6. Lyons, RP, et al. (2015) Continuous 1.3-million-year record of East African hydroclimate, and implications for patterns of evolution and biodiversity. *Proceedings of the National Academy of Sciences of the United States of America* 112:15568–15573.
7. Mendelsohn, JM, et al. (2010) *Okavango Delta: Floods of Life*. IUCN Gland (Suiza) Harry Oppenheimer Okavango Research Centre, Maun (Botswana).
8. Shaw, PA; Thomas, DSG. (1996) The quaternary palaeoenvironmental history of the Kalahari, Southern Africa. *Journal of Arid Environments* 32:9–22.
9. Naidoo, R, et al. (2020) Mapping and assessing the impact of small-scale ephemeral water sources on wildlife in an African seasonal savannah. *Ecological Applications* 30:e02203.
10. Parris, R. (1984) Pans, rivers and artificial waterholes in the protected areas of the south-western Kalahari. *Koedoe: African Protected Area Conservation and Science* 27:63–82.
11. Lupien, RL, et al. (2018) A leaf wax biomarker record of early Pleistocene hydroclimate from West Turkana, Kenya. *Quaternary Science Reviews* 186:225–235.
12. Potts, R, et al. (2018) Environmental dynamics during the onset of the Middle Stone Age in eastern Africa. *Science* 360:86–90.
13. Scholz, CA, et al. (2007) East African megadroughts between 135 and 75 thousand years ago and bearing on early-modern human origins. *Proceedings of the National Academy of Sciences of the United States of America* 104:16416–16421.
14. Blome, MW, et al. (2012) The environmental context for the origins of modern human diversity: a synthesis of regional variability in African climate 150,000–30,000 years ago. *Journal of Human Evolution* 62:563–592.
15. Burnett, AP, et al. (2011) Tropical East African climate change and its relation to global climate: a record from Lake Tanganyika, Tropical East Africa, over the past 90+ kyr. *Palaeogeography Palaeoclimatology Palaeoecology* 303:155–167.
16. Walling, DE (1996). Hydrology and rivers. In Adams, WM et al. (eds) *The Physical Geography of Africa*. Oxford University Press, Oxford, pp. 103–121.

Chapter 4: Bedrock Geology: Volcanic Influences

Because much of Africa's high interior constitutes an eroding surface, the bedrock geology must be taken into account in interpreting its ecology. This brings into play distinctions between granitic ('felsic') and volcanic ('mafic') formations on soil features. Sedimentary deposits are limited in their extent mostly to South Africa and the extension of Kalahari sands in the west. The Congo Basin has accumulated fluvial (river-washed) sediments, but is peripheral to the story.

The important geochemical feature of the bedrock is its silica content, contributing to the granular texture of the soils formed (Box 4.1). Silica-rich granite and sandstone yields sandy soils of low intrinsic fertility. Volcanically derived basalts as well as mudstones produce finely textured soils that are inherently richer in mineral nutrients. Furthermore, the clay content retains the mineral nutrients supporting plant growth against the forces of leaching. The consequent soil fertility underlies a functional subdivision within Africa's savannas between places with lots of herbivores, and regions where herbivores are locally concentrated in current or former wetlands. Geology can become somewhat complex. In relation to ecology, the summary outline presented in Box 4.1 is adequate. Silica-rich rocks are typically pale pink or brown in colour while volcanic rocks are generally dark brown or reddish (Figure 4.1).

I became aware of geological influences on ecology during my doctoral research within the Hluhluwe-iMfolozi Park located in northern KwaZulu-Natal of South Africa. The escarpment foothills in this region are mostly underlain by shale and sandstone layers derived from sediments deposited while Africa was part of Gondwana, representing the Karoo Supergroup. Locally, dolerite sills intrude, derived from the feeder pipes conveying molten basalt to the surface via the fissure eruptions that initiated the break up of Gondwana. The underlying geology affected not only the kinds of trees and grasses that grew, but also where white rhinos were most likely to be found.

The Basement Shield

Granites, gneisses and allied igneous or metamorphosed rocks form the basement shield upon which more recent sedimentary rocks and volcanic material

Box 4.1 Rock Types

The products of bedrock weathering forming soils are dependent on how the rock was formed as well as its geochemical composition. Igneous rocks originate from the cooling of molten magma, either deep beneath the land surface or following surface eruptions. The mineral content and rate of cooling of the molten material influence the grain or crystal size. *Felsic* rocks have high contents of quartz (silica oxide) and feldspar (alumino-silicates) and are thus coarsely crystalline, especially if formed deep underground ('plutonic'). They include forms of granite, typically light grey or pinkish in colour because of the low iron content (Table 4B.1). Gneiss is a metamorphosed (melted and recrystallised) form of granite or other rocks, sometimes showing banding because specific minerals have separated. Felsic rocks are labelled 'acidic', because the soils they form typically show a low pH.

Mafic rocks have high contents of magnesium and iron silicates (pyroxene) and are labelled 'basic'. Mafic rocks are intrinsically dark grey in colour due to their high iron oxide content, but weather to reddish brown on their surfaces. Basalt is formed when mafic lava is extruded on the land surface and cools rapidly, producing fine-grained crystals. Dolerite (or diabase) is similar in mineral composition to basalt, but is formed intrusively in fissures or cracks beneath the land surface, generating somewhat coarser crystals. Gabbro, produced by slow cooling of mafic lava deep underground, is still more coarse-textured. Volcanic substrates containing greater amounts of silica give rise to rhyolite or granodiorite. Phonolite, also of volcanic origin, is intermediate in its chemical composition between felsic and mafic. Very ancient ('Archean') volcanic material that has subsequently been metamorphosed yields greenstone. This is especially rich in magnesium and hence is labelled 'ultramafic'. Banded ironstone is derived from alternating precipitates of iron oxide and silica formed in ancient oceans, before there was much oxygen in the atmosphere. Granite can vary quite widely in its mineral composition, affecting the texture of the soils formed.

Sedimentary rocks are generated where erosion deposits have been compressed under pressure. The texture of the sedimentary source distinguishes pebbly conglomerates, coarse sandstones, finer-grained siltstones and mudstones or shales. Limestone and dolomite (also called dolostone) are derived from marine sediments that are rich in calcium from shell fragments. Sedimentary deposits may become metamorphosed by melting under extreme pressure followed by recrystallisation. Sandstone is thereby transformed into quartzite. Granites, gneisses and allied igneous rocks constitute the basement shield upon which more recent sedimentary rocks and volcanic material have been deposited.

Box 4.1 (cont)

Table 4B.1 Classification of rock types based on their texture or grain size and mineral composition, particularly of silica relative to iron and magnesium minerals

	Label	Felsic	Intermediate	Mafic	Ultramafic
	Colour	Light grey to pink	Medium grey	Dark grey	Dark grey to black
	Mineral composition	Quartz Alkali feldspar ± Amphibole	Alkali feldspar Calcium feldspar Amphibole	Calcium feldspar Pyroxene	Olivine Pyroxene
	Silica content	>69%	52–69%	45–52%	<45%
Texture					
	Igneous				
Fine	Volcanic – surface	Rhyolite	Andesite	Basalt	Komatitite
Intermediate	Volcanic – fissures	Microgranite	Microdiorite	Dolerite	Microperidotite
Coarse	Plutonic – deep	Granite	Diorite Phonolite	Gabbro	Peridotite
	Metamorphic				
Fine					Greenstone
Intermediate			Schist		
Coarse		Gneiss			
	Sedimentary				
Fine			Shale Mudstone	Limestone Banded ironstone	
Intermediate			Siltstone		
Coarse		Sandstone conglomerate Quartzite			

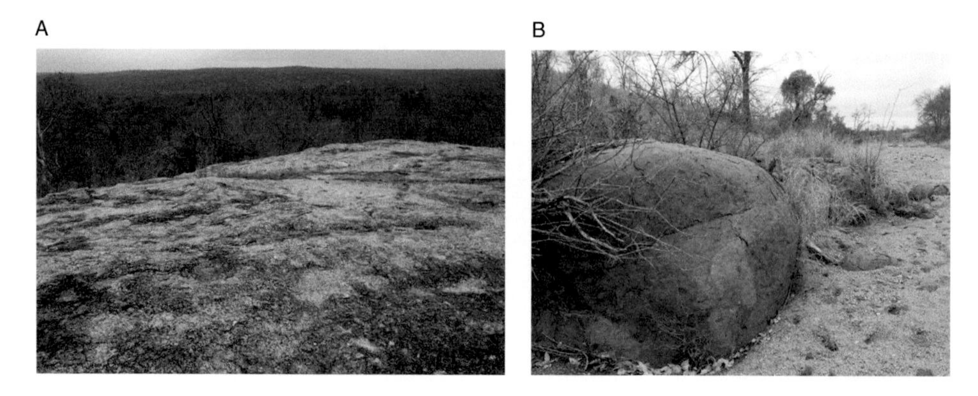

Figure 4.1 Basic contrasts in geological substrates. (A) Felsic granitic–gneiss pinky-beige in colour; (B) mafic dolerite dark brown in colour (both photos from South African Lowveld region near Kruger NP).

have been deposited. These plutonic rocks formed originally deep beneath the Earth's surface have become exposed by erosion over many millions of years, initiated by the break up of Gondwana and renewed by subsequent uplift during the Pliocene and Pleistocene.[1] The rocks themselves were formed between 3.6 and 2 billion years ago, initially taking the form of Archean greenstones and banded ironstones formed before there was much oxygen in the Earth's atmosphere. Basement granitic rocks constitute the prevalent geological substrate from northern parts of South Africa through Zimbabwe, Zambia and southern Tanzania into sections of Kenya and Uganda, as well as through much of western Africa (Figure 4.2). Archean greenstone and ironstone outcrop in the Witwatersrand ridges, Barberton mountain-land adjoining Swaziland, Great Dyke running through Zimbabwe and near Lake Victoria in eastern Africa. The Bushveld Igneous Complex, formed about 2 billion years ago, fills an extensive basin situated north of the Witwatersrand. It is mined for platinum and chrome. It is rimmed in the south by quartzite forming the Magaliesberg range of hills, which separates Highveld grassland from the bushveld region to the north. Dipping deep beneath the Witwatersrand watershed are the conglomerate layers, formed in ancient stream channels 2.8 billion years ago, which have yielded most of South Africa's gold.

Sedimentary Formations

Most of South Africa is underlain by rocks formed from sediments laid down in the interior basin north of the Cape Fold Mountains, during the time around 250 Ma when Africa was situated centrally within Gondwana (Figure 4.2). The various layers form the Karoo Supergroup. Tillite of glacial origin lies at the

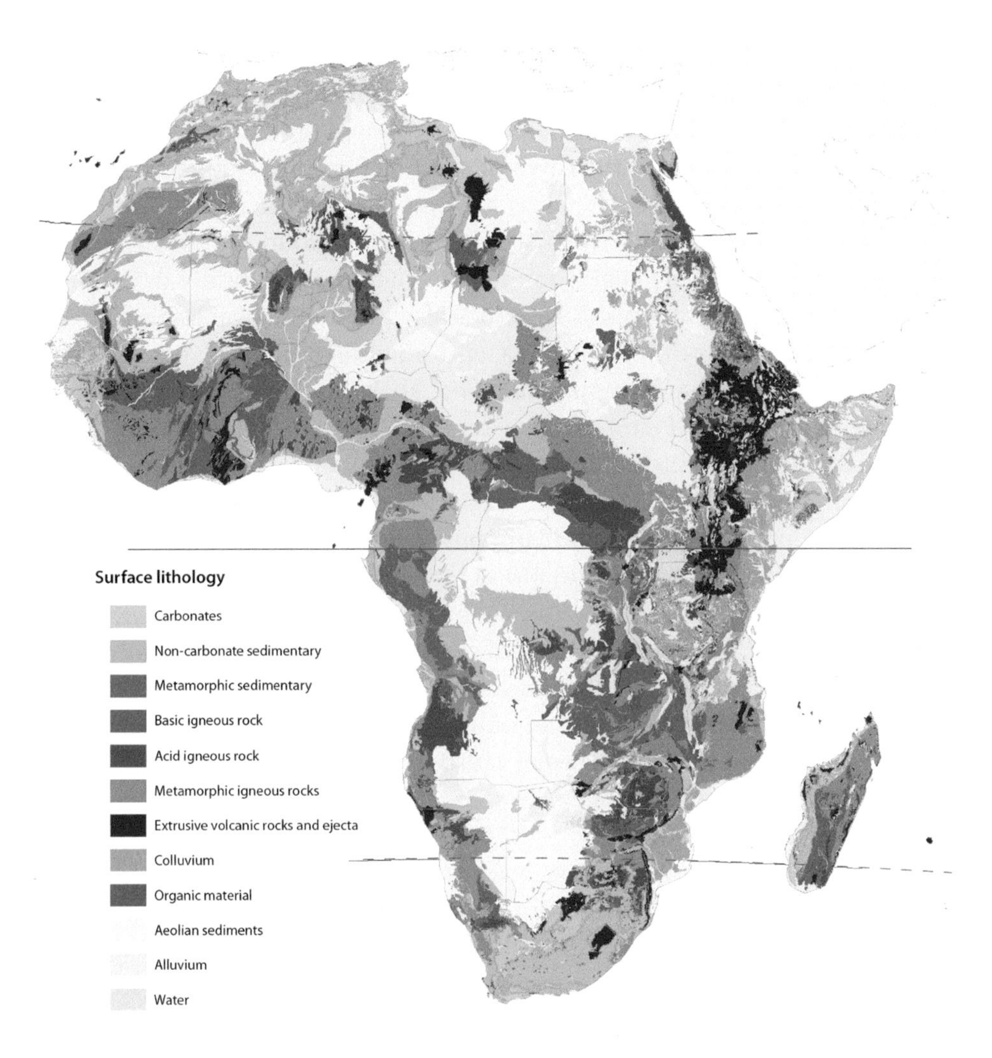

Figure 4.2 Surface geology map of Africa. Notable features are (i) widespread presence of basement ('acid' or 'metamorphic') igneous rocks, (ii) local prevalence of volcanic intrusions in the rift valley region extending from Ethiopia into northern Tanzania, and again in pockets in southern Africa, (iii) geological complexity within eastern Africa, (iv) sedimentary deposits forming the Karoo Supergroup surrounding the volcanic Drakensberg basalt in South Africa, and (e) deposits of wind-blown Kalahari sand extending over a vast area in the west (from Jones et al. (2013) *Soil Atlas of Africa*).

base, testifying to the former polar location of the subcontinent. Above it are the layers deposited during the Permian period when the mammal-like reptiles (or therapsids) were the predominant big animals. The coal beds it contains, derived from swampy vegetation, are mined. The sequence upwards from mudstones through shales and sandstones reflects the progressive dryness that prevailed through the Triassic into the early Jurassic. The sequence was terminated by the eruption of the Drakensberg lavas. However, around Johannesburg more ancient formations have been exposed as a result of the continental uplift. They include dolomites deposited on a sea floor 2.5 billion years ago, which generated the caves or sinkholes accumulating fossils in the 'Cradle of Humankind' proclaimed as a World Heritage Site.

Permian sediments allied with the Karoo Supergroup are also found in western Zimbabwe where coal is mined. They are present locally in Zambia's Luangwa Valley and from north-eastern Malawi to the Rukwa Basin in southern Tanzania. In north-west Zambia and adjoining parts of Congo DRC, copper is mined where ancient sandstone, shale and limestone formations outcrop. Sediments of Permian age extend again from the coastal region of Kenya near Mombasa into adjoining Ethiopia. Sedimentary deposits of Jurassic age containing dinosaur fossils occur in the Tendaguru formation in southern Tanzania and in the Zambezi and Limpopo valleys.

Unconsolidated aeolian (wind-blown) or colluvial (water-washed) sands that were eroded from the high-lying interior accumulated in the Kalahari Basin in the west. They cover a vast area extending from northern South Africa through Angola, western Zimbabwe and Zambia as far north as the Bateke Plateau in Congo-Brazzavillle and adjoining Gabon (Figure 4.2). Fluvial (water-born) sediments fill the Congo Basin. The coastal plain extending from southern Mozambique into northern KwaZulu-Natal in South Africa has been formed by sand of Cretaceous or later origin deposited by the Limpopo and Zambezi rivers. Sediments of Miocene or later age, containing fossils covering the crucial periods in human and mammalian evolution, are found mostly within or adjoining the East African Rift System. The fossil-bearing sediments of Olduvai Gorge and Laetoli in Tanzania to the south of the Serengeti plains were laid down in an ancient lake basin to the west of the highlands bordering the Eastern Rift.[2]

Volcanic Intrusions

Volcanic deposits associated with rift valley formation are prominent within an area extending from Ethiopia as far south as Tanzania to the east of Lake Victoria (Figure 4.2). In Ethiopia, volcanic eruptions commenced around 30 Ma during the Oligocene, building up the broad basalt platform underlying the Simien Mountains.[3,4] The thickness of the volcanic deposit there originally approached 3000 m, but has been reduced by erosion.[5] To the south-east, the Bale Mountains were formed by lava flows over a base of sedimentary sandstone and limestone. The Afar Depression to the east is covered by volcanic lavas of Quaternary age up to 1500 m thick.[6] Lava flows had spread into the Samburu region of central Kenya by 15 Ma in the early Miocene. The Aberdare

Range adjoining the Eastern Rift in central Kenya was formed by basaltic lava erupted during the late Miocene, followed by further eruptions in the Pliocene and late Pleistocene.[2] Further south in Kenya, phonolite lavas dated to ~13 Ma occur around Nairobi and border the Yatta Plateau in Tsavo East NP. Basaltic eruptions that took place merely 150 years ago feature in the Chyulu Hills and adjoining regions of Tsavo West NP (Figure 4.3D). Mount Longonot, situated within Kenya's rift valley, spewed volcanic ash only 100 years ago. In northern Tanzania, volcanic deposits span an east–west distance exceeding 200 km near the southern extremity of the Eastern Rift, from mounts Kilimanjaro and Meru in the east to west of the Ngorongoro highlands. Mount Oldoinyo Lengai still periodically spews volcanic ash towards the nearby Serengeti Plains. Lavas associated with both rift valley arms are notable chemically for their high sodium and carbonatite contents plus unusually low silica, making them exceptionally alkaline.[2]

Figure 4.3 Volcanic overlays. (A) Basalt ramparts of the Drakensberg escarpment in South Africa; (B) basalt traps forming the Simien Mountains, Ethiopia (photo: Craig R. Sholley); (C) outcropping columnar dolerite in Karoo region, South Africa (photo: Trishya Owen-Smith); (D) recent basalt lava flow in Tsavo West NP, Kenya, with Mount Kilimanjaro vaguely discernible behind.

Along the Western Rift, Mount Nyamuragira has erupted 40 times since 1885, while Mount Nyiragongo erupted 34 times during the same period. Remnant cones exemplify recent volcanic activity on the Ugandan side of the Congo border within the Queen Elizabeth (formerly Rwenzori) NP. Rwanda is mostly covered by volcanic material emanating from Western Rift volcanoes. In western Africa, a region of ongoing volcanic activity extends from the Cameroon highlands to islands in the Gulf of Guinea. Northern Nigeria has remnants of ancient lava flows intruding through basement granite on the Jos and Bui plateaus. In Guinea, dolerite and gabbro sills penetrate the predominant sandstone on the Fouta-Djallon highlands.

In South Africa, basalt derived from fissure eruptions initiated 183 Ma covers quite a small area, because most has been eroded away (Figure 4.2). Nevertheless, the massive outpouring of lava left a spectacular feature along the border between South Africa and Lesotho. Cliffs forming the Drakensberg/Maloti escarpment rise almost 1000 m vertically (Figure 4.3A). Further remnants of this volcanic mantle persist in the Lebombo range of hills running along South Africa's border with Mozambique, in the Springbok flats north of Pretoria, and from south of Victoria Falls in Zimbabwe into western Zambia. In Namibia, the Etendeka lavas that erupted preceding the split with South America occur in the north-west. Basalt lies beneath the Kalahari sand cover through much of Botswana and adjoining parts of northern Namibia, Angola and Zambia.[7] Widely prevalent, but not shown on the continental map, are numerous dolerite dykes and sills, remnants of feeder pipes supplying the surface-erupted lavas. They are especially prominent in the Karoo region of South Africa (Figure 4.3C). Intrusions of gabbro, much earlier than the basalt in their origin, occur in the eastern highlands of Zimbabwe and in a strip of the Lowveld extending into the Kruger NP in South Africa. Even more ancient Ventersdorp lavas dating back more than 2.7 billion years outcrop in the Witwatersrand region.

The influence of the rift valley volcanoes extends well beyond the surface area that they cover. Those in the Ngorongoro highlands contributed to the formation of the Serengeti plains, and their influence on soils can be detected over 100 km to the west. In southern Africa, most of the basalt may be gone, but its legacy in the form of dolerite dykes remains widespread and contributes to local heterogeneity in soils and vegetation. Volcanic material is absent from Malawi through southern Tanzania in south-central Africa.

Overview

Basement granitic substrates form the bedrock over much of Africa. Sedimentary deposits of Kalahari sand occur extensively through the south-west, while much of southern Africa is underlain by sedimentary layers formed in the Karoo basin. Volcanic intrusions are most widespread in the north-east, from Ethiopia southwards in association with rift valley faults. South Africa retains a more ancient legacy of volcanism in the form of the dolerite intrusions representing feeder pipes to the flood basalts associated with the breakup of Gondwana. Volcanic eruptions continue along the Western Rift and occurred quite recently elsewhere in eastern Africa.

Compared with the Pacific Rim and southern Europe, the extent of the volcanic influence in Africa is quite trivial. Nevertheless, it is more widespread than in the other two southern continents. Basalt contemporaneous with the Etendeka lavas in Namibia covers parts of the Parana plateau of Brazil and adjoining Uruguay, but most of the Brazilian plateau is underlain by sandstone. To the south, the land cover in the Argentinian pampas is mainly windblown silt (or loess) originating from the Andean volcanics. Parts of Patagonia are covered by volcanic lavas, tufts and ashes of late Cretaceous age. Lavas produced by volcanoes that erupted during the Plio–Pleistocene cover sections of the high interior plateau of central Mexico. An active mantle plume underlies the high plateau of Yellowstone NP with its geysers and hot springs. In Australia, basalt eruptions dated to 60 Ma underlie parts of the eastern tablelands. The Deccan traps ('stairs') in west-central India were formed by multiple layers of flood basalts more than 2000 m deep covering a vast area. They date back to the end of the Cretaceous 66 Ma, intriguingly synchronous with the terminal extinctions of the dinosaurs. Much of central Eurasia and North America is covered by windblown deposits of loess generated by continental glaciation during the Pleistocene, highly fertile but divorced from the underlying geology.

Although volcanic intrusions are limited in their extent in Africa, their consequences for large herbivores and hence human evolution were profound through their effects on soil fertility, as will be described in the next chapter.

Suggested Further Reading

McCarthy, T; Rubidge, BS. (2005) *The Story of Earth and Life. A Southern African Perspective on a 4.6-Billion-Year Journey*. Struik, Cape Town.

References

1. Furon, R. (1963) *Geology of Africa*. Oliver and Boyd, London.
2. Scoon, RN, et al. (2018) *Geology of National Parks of Central/Southern Kenya and Northern Tanzania*. Springer, Cham.

3. Macgregor, D. (2015) History of the development of the East African Rift System: a series of interpreted maps through time. *Journal of African Earth Sciences* 101:232–252.
4. Partridge, TC. (2010) Tectonics and geomorphology of Africa during the Phanerozoic. In Werdelin, L; Sanders, WJ (eds) *Cenozoic Mammals of Africa*. University of California Press, Berkeley, pp. 3–17.
5. Asrat, A. (2016) The Ethiopian highlands. In Viljoen, R (ed.) *Africa's Top Geological Sites*. Struik, Cape Town, pp. 197–205.
6. Asrat, A. (2016) The Danakil depression of Ethiopia. In Viljoen, R (ed.) *Africa's Top Geological Sites*. Struik, Cape Town, pp. 189–196.
7. Haddon, IG. (2004) *The Sub-Kalahari Geology and Tectonic Evolution of The Kalahari Basin, Southern Africa*. University of the Witwatersand, Johannesburg.

Chapter 5: Soils: Foundations of Fertility

Soils are formed from the merger of inorganic material supplied by the disintegration of rocks along with decomposing organic matter. They provide the mineral nutrients plants need to grow. The most basic mineral nutrients are those we apply in fertiliser for our crops or gardens, symbolised 'NPK': nitrogen (N), phosphorus (P) and potassium (K). Out in nature these are recycled by decomposition of plant matter. The extent to which these nutrients are retained against leaching by rainwater percolating through soils depends largely on the clay content. This is governed by the bedrock composition and modified by rainfall.

Soils typically exhibit distinct layers (Figure 5.1A). A topsoil layer called the *A* horizon incorporates decomposing vegetation called humus. Below it a subsoil or *B* horizon accumulates the downwash of clay particles and other leached products. Deeper still lies the *C* horizon, constituted by decaying bedrock. A hardpan (or duricrust) may form at the base of the *B* horizon above the bedrock. However, this ideal profile may not be shown by soils that are rather shallow or very sandy (Figure 5.1B,C). On eroding uplands, the mineral composition largely reflects the underlying bedrock. Lowlands accumulate the products of erosion transported from upslope or elsewhere by water, wind or animals. Specific soil features depend on (1) the underlying bedrock providing the parent material, (2) prevailing climatic conditions affecting weathering (mineral decomposition), (3) local topography determining water movements, (4) vegetation cover controlling organic inputs and, not to be overlooked, (5) time, affecting soil development. Soils may retain features derived from rainfall and temperature conditions back in the distant past. The pertinent outcome for ecology is the soil fertility, determined by the soil texture, in turn governed by the content of fine clay particles able to retain nutrients (Box 5.1).

The soil texture influences both fertility and water penetration or hydrology. While clay is beneficial for holding cations, soils that are too rich in clay become sticky and difficult to cultivate when wet. The ideal texture for agriculture is loam constituted by a mix of clay and sand. Soil fertility is represented most simply by the cation exchange capacity (CEC), or total exchangeable bases (TEB), taking into account the extent to which this

Figure 5.1 Soil profiles. (A) Idealised soil profile showing distinct *A* horizon where organic matter (humus) is concentrated, *B* horizon where clay and nutrients are leached downwards and *C* horizon where decaying rock becomes incorporated (from Jones et al., 2013), licensed under a Creative Commons Attribution 4.0 International License. (B) Soil on granitic bedrock under grassland in Nyika Plateau of Malawi, showing lack of humus in the surface layer. (C) Soil formed from deep deposit of Kalahari sand in south-eastern Gabon.

capacity is filled, called the base saturation. The units are expressed either in milli-equivalents per 100 g of soil, or as centimoles of charge per kilogram, which are numerically identical. Soils are rated as relatively fertile if the sum of exchangeable bases exceeds about 20 cmol/kg (Figure 5.2). Soil fertility tends to be greatest where the moisture input from rainfall closely balances potential evaporation, so that there is little leaching of base cations.[2] Some of the phosphorus present (in the form of phosphate) also gets removed by

Box 5.1 Soil Features Affecting Fertility

Soil fertility is governed most fundamentally by the granular texture of the soil particles, in particular the clay content. Clay particles are made up of alternating layers of silica and aluminium oxides, forming a lattice structure.[1] One silica sheet coupled with one alumina sheet represents a 1:1 lattice, whereas two silica sheets per alumina sheet form a 2:1 lattice. Negative charges formed on the lattice attract cations, most notably potassium, sodium, calcium and magnesium. Smectite or illite clays with a 2:1 lattice structure have a high cation-holding capacity, in contrast to kaolinite clays with a 1:1 lattice. If cations are leached, they get replaced by hydrogen ions, making soils more acidic. Over time, the mineral component of soils can become weathered into unstructured iron or aluminium sesquioxides with little or no capacity to retain cations, especially in warm tropical regions with high rainfall. Iron oxides contribute to the redness typical of tropical soils. Under wet conditions, prolonged weathering can give rise to duricrusts constituted by iron oxides, variously called ferricrete, plinthite or laterite. Under drier conditions, calcium carbonate precipitates as nodules or larger chunks of calcrete, while sand and gravel can become cemented by dissolved silica to form silcrete. In hot and dry climates, the evaporation of moisture can cause sodium salts to accumulate at the soil surface. This salinisation may lead to the compaction of clay-rich soils, thereby resisting water and air infiltration.

The underlying bedrock contributes to soil fertility by being the source of base cations as well as through its contribution to the phosphorus taken up by plants mainly through the agency of soil fungi. Soils derived from mafic volcanic rocks, like basalt, both retain cations on clay particles and contribute relatively high amounts of phosphorus. Felsic igneous or metamorphic rocks that have high silica contents, like granite or gneiss, yield sandy soils with less capacity to hold mineral nutrients. Very sandy soils derived from sandstone depend almost entirely on organic matter for their cation-holding capacity. Fine-grained sediments formed from wind-blown loess or lacustrine (lake bed) deposits, widely prevalent in Europe, are especially fertile. Limestone and dolomite, derived from marine deposits, produce fine-textured soils that are moderately fertile, as also are soils derived from shale or mudstone.

The humus content can make a further contribution to soil fertility because it forms the basis for nutrient recycling, from the soil via plants, and perhaps also herbivores, back into the soil. Mineral nitrogen originates from the decomposition of organic matter and is taken up by plants in the form of

continues

Box 5.1 (cont)

nitrate (NO_3^-) or ammonia (NH_4^+). Most of the available phosphorus is also recycled through the organic matter component. Under warm and moist climates, the organic matter content is decomposed rapidly by soil bacteria, reducing the effective soil fertility. Recurrent fires lower the organic matter input, and cause losses particularly of nitrogen. Many legumes have the capacity to alleviate nitrogen shortages by fixing atmospheric nitrogen in root nodules with the aid of symbiotic bacteria.

Plants with the capacity to fix nitrogen may do so only when it is most needed, for example during seedling growth, because of the costs involved. Because phosphate is highly insoluble, its availability for plant growth can depend on how it is rendered to plant roots by soil fungi called mycorrhizae. In heavily weathered soils, phosphorus can become bound to iron and effectively unavailable. Phosphates may be released by volcanic lavas and ash, but high calcium contents (carbonitites) may restrict the effective availability of the phosphorus to herbivores feeding on the vegetation.

weathering due to water passage. However, in warm, tropical climates fertility is actually highest where the water balance is somewhat more negative than is the case for temperate regions; i.e. in situations where evaporative losses somewhat exceed rainfall inputs. Soils in tropical and subtropical latitudes in the southern hemisphere are especially low in organic carbon content compared with those at counterpart latitudes in the northern hemisphere.[3] This contributes further to the importance of the bedrock in governing the effective fertility in savanna regions of Africa.

Water infiltration is slower in clay soils than in sand, although very heavy clays that swell when wet and crack open when dry allow more water to penetrate. However, clay soils bind whatever moisture remains intensely when they become dry, making this less available to plants. Sandy soils allow water to percolate to greater depths so that it remains available for longer, but in very deep sands, water may sink beyond the reach of all but the most deeply rooted trees. Overall, clay soils retain less water into the dry season than loamy or sandy soils, because more runs off while less of the moisture that penetrates can be extracted.

There is no globally agreed system for naming soil types. In the US taxonomy, savanna soils are typically mollisols, ultisols, oxisols or alfisols. In the World Reference Scheme, they are called luvisols, lixisols or acrisols. In the

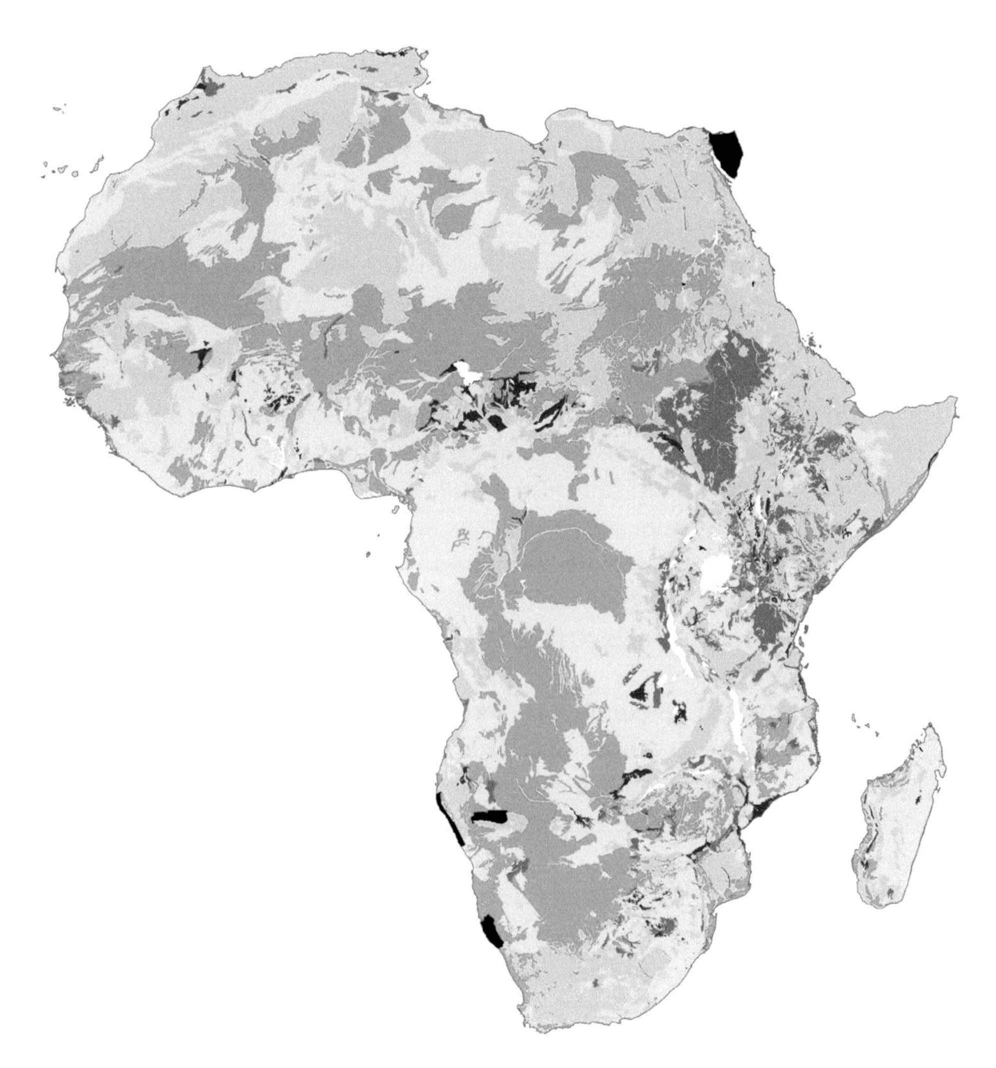

Figure 5.2 Soil fertility map of Africa based on cation exchange capacity of soils. High fertility is shown by dark purple (20-40 cmol/kg) and light purple (10–20 cmol/kg) and low cation exchange capacity by shades of green (<10 cmol/kg). Note the region of high soil fertility extending from Ethiopia through the rift valley region of eastern Africa where volcanic substrates are widespread. The most nutrient-deficient soils occur in the Kalahari Sand region, Congo Basin, the dry northern region of western Africa to the south of the Sahara and in the moist region of south-central Africa with mainly granitic soils (from Jones et al. (2013) *Soil Atlas of Africa*), licensed under a Creative Commons Attribution 4.0 International License.

South African classification, they are described as melanic, plinthic, calcic or duplex. Sticky clays are called vertisols, or vertic in the South African classification. None of these systems clearly exposes the ecologically relevant distinctions for African savanna regions where geology plays such an important role.

Geological Influences on Fertility

I was inducted into the relationships between soil features and the underlying geology during my white rhino study in the Hluhluwe-iMfolozi Park. Sticky clays were associated with Ecca shales of the Karoo Supergroup, interspersed with sandy soils derived from sandstone layers. Dolerite dykes penetrating the Karoo sediments and capping some of the hills formed clay soils that were less sticky and allowed greater water infiltration on account of their more granular texture.

More broadly, felsic granite or gneiss and sandstone produce sandy soils deficient in calcium, magnesium and iron.[1] Mafic basalt, dolerite and gabbro, as well as fine-grained sediments, yield clay-rich soils with higher CECs and base saturations and are typically also higher in available phosphorus content (Table 5.1). Schist, syenite and biotite as well as limestone and dolomite generate soils with intermediate properties. Phosphorus appears to be the mineral nutrient most generally restricting soil fertility across Africa.[4] This is partly a result of the prolonged erosion to which Africa's ancient land surfaces have been subjected.

Infertile sandy soils predominate through much of southern Tanzania, Zambia and Zimbabwe, which are widely underlain by basement granitic bedrock (Figure 5.2). In central and western Africa where rainfall is high, granitic bedrock has become weathered into red soils containing iron oxide concretions, with low intrinsic fertility. Infertile sandy soils also extend through the Kalahari basin from southern Africa via interior Angola and western Zambia as far as south-eastern Gabon.[1,5] They retain a legacy of past weathering in the form of pebbles, clay layers and calcrete or silcrete crusts deep beneath the soil surface. In South Africa's Highveld region, duricrusts of silcrete or calcrete can be several metres thick.[6] Highly weathered profiles with mainly kaolinite (commercially known as white clay or talcum) persist there and elsewhere as a legacy of the more humid conditions that prevailed in the past.

Soils of volcanic origin with high inherent fertility are prevalent in and around the East African Rift System (Figure 5.2). Where rainfall is high enough, they support dense crop production, most notably in Rwanda, situated adjoining the Western Rift only a short distance south of the Equator.[1] Soils on the Serengeti Plains in Tanzania have been enriched by volcanic ash deposited by nearby volcanoes, which generated a calcrete crust at shallow depths.[7,8] The ash is rich in calcium and sodium, while leaching is blocked by the hardpan of calcrete overlying the granite beneath. Although these soils are too shallow and saline to support crop production, they produce grasses that are especially nutritious for large herbivores. A volcanic influence can detected

Table 5.1 Soil features in relation to bedrock geology

Country	Place	Bedrock geology	Clay content (%)	Cation exchange capacity (cmol/kg)	Calcium content (cmol/kg)	Available phosphorus (mg/kg)	Reference
Tanzania	Serengeti Plains	Volcanic ash	–	44	33		1
South Africa	Hluhluwe	Dolerite	–	–	39.2	10.9	2
Zimbabwe	Country	Basalt and dolerite	50	39.0			3
Tanzania	Country	Volcanics	35	29.5	11.6		4
South Africa	Kruger NP	Basalt	30	28.0	17.0	8.6	
Tanzania	Serengeti	Ultrabasic	30	22.0	13.0		5
Tanzania	Serengeti	Granite	22	16.2	12.0		5
Malawi	Kasungu	Biotite gneiss	–	15.0	9.8	11	6
Tanzania	Country	Sediments	26	12.5	5.8		4
Tanzania	Serengeti	Gneiss	14	11.7	8.4		4
Tanzania	Country	Granites	20	6.9	1.7		4
Zimbabwe	Country	Granite	10	6.4			3
Zimbabwe	Country	Sandy sediments	10	5.4			3

continues

Table 5.1 (cont)

Country	Place	Bedrock geology	Clay content (%)	Cation exchange capacity (cmol/kg)	Calcium content (cmol/kg)	Available phosphorus (mg/kg)	Reference
Zambia	Kasama	Granite	–	5.6	1.5	–	6
South Africa	Kruger NP	Granitic gneiss	10	2.6	2.0	2.5	7

References: 1: de Wit, HA (1978) PhD thesis, Wageningen University, Wageningen. 2: Howison, RA, et al. (2017) In *Conserving Africa's Mega-Diversity in the Anthropocene*, ed. Cromsigt et al., pp. 33–55, Cambridge University Press. 3: Thompson, JG and Purves, WD (1978) *A Guide to the Soils of Rhodesia*. Ministry of Agriculture, Rhodesia. 4: Funakawa, S, et al. (2012) In *Soil Health and Land Use Management*, ed. MC Hernandez Soriano, Intech. 5: Jager, T (1982) Soils of the Serengeti woodlands, Tanzania. PhD dissertation, Netherlands. 6: Frost, P (1996) The ecology of miombo woodlands. In *The Miombo in Transition*, ed. Campbell B, Centre for International Forestry Research, Bogor, India, pp. 11–57. 7: Venter, FJ, et al. (2003) In *The Kruger Experience*, ed. JT du Toit et al., pp. 83–129, Island Press.

as far as 200 km west of the volcanoes located in the Crater Highlands.[9] In parts of the west-central region of Serengeti NP, ultramafic greenstone and banded ironstone have become exposed, generating soils that are also relatively clay-rich and fertile. In central and northern parts of this park, soils derived from granite with relatively low silicate and high biotite contents are sufficiently fertile to be classified as mollisols. A subtle shift in the substrate in the north-west from granite to gneiss is associated with a vegetation transition from fine-leaved to broad-leaved savanna. South Africa's Kruger NP presents a striking geological contrast between basaltic soils in the east and granitic soils in the west, intruded locally by dolerite and gabbro sills. A narrow strip of Karoo sediments divides the granite from the basalt.

Hydrological Redistribution

Topography modifies soil features via upland erosion and lowland deposition, especially on Africa's interior plateau. A topo-sequence or 'catena' develops, with shallow eroded soils on crests, while soils on mid or lower slopes and pediments are most leached of cations by water percolating through them. Clay particles and mobile cations get concentrated on foot slopes or bottomlands. Seep zones form where percolating water is forced to the surface by underlying bedrock or ferricrete. Consequently, lowland soils near drainage lines or dambos can be relatively fertile even in landscapes underlain by granitic bedrock.

Soil texture influences how readily rainfall penetrates to deeper depths. The intensity of the rain received during thunderstorms can exceed the infiltration capacity of soils, causing much of this water to run off down slope.[8] This threshold capacity lies between 10 and 30 mm/h of rainfall, depending on soil features such as texture, depth and structure.[10] Accordingly, on uplands the soil moisture effectively available to support plant growth can be less than half of the total amount of rain received as measured in a rain gauge. Deeper soils in bottomlands gain additional moisture from run on and retain moisture longer into the dry season, especially if their surface texture is sandy. More extensive wetlands develop where drainage is blocked. This redistribution of soil moisture contributes fundamentally to the spatial heterogeneity that is a striking feature of African savannas.

Nutrient Recycling

The recycling of nutrients to support plant growth depends on the mineralisation of organic matter into nitrates and other inorganic compounds that can be taken up by roots. The breakdown of plant detritus ceases when soils become too dry for microbial decomposers to function. This results in a pulsed release of mineral nutrients following the first adequate rains at the start of the wet season.[11] Sufficient soil moisture to support decomposer activity may persist quite late into the dry season, extending the period available for nutrient uptake by plants.

The mound-building activities of termites also influence soil features. Through most of the dry tropics and subtropics of Africa, soil has been worked and reworked by termites since these insects evolved in the Triassic period over 200 Ma.[12,13] Mound-building species in the subfamily Macrotermitinae convey plant residues into their large mounds where the cellulose content gets degraded by fungi, which in turn are consumed by the termites. Organic breakdown products accumulate within the mounds, along with termite faeces and bodies, and get transported out by animals feeding on the termites. In constructing their mounds, termites raise clay particles towards the surface, generating hotspots of soil fertility within savanna landscapes (see Chapter 13).[11] Both CEC and available phosphorus are elevated in the vicinity of *Macrotermes* mounds. Upward soil movements contribute to the generation of stone-lines in subsoils, which persist long after the mounds have disintegrated.

Human presence has modified soil features in places, especially where dwellings and livestock enclosures were located. Mineral nutrients accumulated where dung from cattle was concentrated can persist for decades or even centuries, most especially phosphorus.[14,15,16] These sites stand out as open glades with short grass, attracting concentrations of grazing ungulates. In the Nylsvley Nature Reserve in South Africa, sites of former human habitation exhibited a 10-fold increase in total phosphorus, without much change in the sandy soil texture, several centuries after the inhabitants had moved on.[17] Organic matter contributions from human settlements and associated livestock counteract the tendency for phosphorus to become locked in refractory compounds as soils age over time. In eastern Botswana, grass species associated with high soil phosphorus levels indicate sites where Iron Age settlements were located.[18]

Overview

Across much of Africa, soil fertility is dependent largely on the underlying bedrock. This is because most of its eastern and southern regions form an eroding plateau surface exposing bedrock influences. Duricrusts formed by prolonged weathering restrict water infiltration. The relatively low rainfall over the interior plateau contributes additionally. Where rainfall is high, the nature of the underlying geology matters little because leaching overrides it. If rainfall is low, water does not penetrate far and mineral nutrients are retained, especially sodium of high importance for large herbivores. The intermediate range between 500 and 1000 mm in MAR, typical of most of High Africa, is most conducive to bedrock influences on soil fertility. In South American savannas, rainfall is generally double that prevailing in Africa, producing intensely weathered soils with little capacity to retain nutrients. Soils over the southern tropics and subtropics are generally low in organic carbon content compared with those at counterpart latitudes in the northern hemisphere.[18] This probably reflects the rapid decomposition that takes place following rainfall events.

Within Africa, volcanic influences on soils are widespread in the rift valley region extending from Ethiopia into northern Tanzania. Sedimentary deposits are less extensive than in other continents, covering the vast Amazon Basin of South America, former lakebeds and intermontane valleys in central Europe, the periglacial region of North America, and temperate China where windblown loess formed deep deposits. Soils regarded as adequately fertile cover merely 10 percent of the land surface of South America, restricted mostly to valleys where residual minerals accumulated, or localities where calcareous bedrock or volcanic intrusions occur.[19] West African soils are mostly infertile due to high leaching. Within Australia, soils are generally less subject to leaching, but lack much phosphorus, except in the eastern seaboard.[20] Fungus-growing termites are absent from Australia and the Americas, so their contribution towards enhancing soil fertility is lacking.

Additional influences on soil functioning beyond the geology and climate come from recurrent fires and through the recycling of vegetation biomass via large herbivores and termites. These will be addressed in Part III of this book.

Suggested Further Reading

Jones, A, et al. (2013) *Soil Atlas of Africa*. European Commission, Publications of the European Union, Luxembourg.

References

1. Areola, O. (1996) Soils. In Adams, WM, et al. (eds) *The Physical Geography of Africa*. Oxford University Press, Oxford, pp. 134–147.
2. Huston, MA. (2012) Precipitation, soils, NPP, and biodiversity: resurrection of Albrecht's curve. *Ecological Monographs* 82:277–296.
3. Crowther, TW, et al. (2019) The global soil community and its influence on biogeochemistry. *Science* 365:772.
4. Pellegrini, AFA. (2016) Nutrient limitation in tropical savannas across multiple scales and mechanisms. *Ecology* 97:313–324.

5. Jones, A, et al. (2013) *Soil Atlas of Africa*. European Commission, Publications of the European Union: Luxembourg.
6. Runge, J. (2016) Soils and duricrusts. In Knight J; Grab SW (eds) *Quaternary Environmental Change in Southern Africa*. Cambridge University Press, Cambridge, pp. 234–249.
7. Anderson, GD; Talbot, LM. (1965) Soil factors affecting the distribution of the grassland types and their utilization by wild animals on the Serengeti Plains, Tanganyika. *The Journal of Ecology* 53:33–56.
8. de Wit, HA. (1978) Soils and grassland types of the Serengeti Plains (Tanzania). PhD thesis, Wageningen University, Wageningen.
9. Jager, T. (1982) Soils of the Serengeti woodlands, Tanzania. PhD thesis, Wageningen University, Wageningen.
10. Anderson, TM. (2008) Plant compositional change over time increases with rainfall in Serengeti grasslands. *Oikos* 117:675–682.
11. Grant, CC; Scholes, MC. (2006) The importance of nutrient hot-spots in the conservation and management of large wild mammalian herbivores in semi-arid savannas. *Biological Conservation* 130:426–437.
12. Goudie, AS. (1988) The geomorphological role of earthworms and termites in the tropics. In Viles, H (ed.) *Biogeomorphology*. Blackwell, Oxford, pp. 43–82.
13. Jouquet, P, et al. (2011) Influence of termites on ecosystem functioning. Ecosystem services provided by termites. *European Journal of Soil Biology* 47:215–222.
14. Augustine, DJ. (2003) Long-term, livestock-mediated redistribution of nitrogen and phosphorus in an East African savanna. *Journal of Applied Ecology* 40:137–149.
15. Muchiru, AN, et al. (2009) The impact of abandoned pastoral settlements on plant and nutrient succession in an African savanna ecosystem. *Journal of Arid Environments* 73:322–331.
16. van der Waal, C, et al. (2011) Large herbivores may alter vegetation structure of semi-arid savannas through soil nutrient mediation. *Oecologia* 165:1095–1107.
17. Blackmore, AC, et al. (1990) The origin and extent of nutrient-enriched patches within a nutrient-poor savanna in South Africa. *Journal of Biogeography* 17:463–470.
18. Denbow, JR. (1979) *Cenchrus ciliaris*: an ecological indicator of Iron Age middens using aerial photography in eastern Botswana. *South African Journal of Science* 75:405–408.
19. Cole, MM. (1986) *The Savannas. Biogeography and Geobotany*. Academic Press, New York.
20. Lavelle, P; Spain, A. (2001) *Soil Ecology*. Kluwer, Dordrecht.

Part I: Synthesis: Structure of the Physical Cradle

The preceding chapters have described the physical features distinguishing Africa from other continents. Africa is mostly high-lying in its eastern and southern interior with almost equal sections north and south of the equator. Its high interior elevation is due initially to its central location within Gondwana, amplified by tectonic uplift continuing from the Miocene into the Pleistocene. Eastern escarpments generated rain shadows, lowering precipitation leeward in tropical and subtropical latitudes and especially via season restrictions. Both rivers and lakes have fluctuated widely in flow regimes and water levels. All east-flowing rivers may be crossed without swimming in their upper reaches during dry seasons. Ramifying rift valleys disrupted surface topography through the east and generated the local depositional basins that have retained fossils. The major depositional basin in the west accumulated loose sands while in the east sediments extended coastlines. Africa's eroding plateau surface exposed soils to bedrock influences, fostered additionally by reduced leaching due to the lowered rainfall. Volcanic activity associated with rifting contributed mafic geological substrates generating relatively fertile soils for the tropics. Southern Africa retains volcanic contributions from feeder pipes to the basalt eruptions that accompanied the breakup of Gondwana. Within regions underlain by basement granitic substrates, local hotspots of fertility are associated with wetlands, former lake beds and smaller-scale drainage basins.

Water redistribution within landscapes and geological influences on soils contribute to the spatial heterogeneity that is a notable feature of African savannas. Fluctuations in rainfall contribute to temporal variation driven by the Earth's shifting orbital geometry. As the Earth has progressively cooled since the late Miocene, conditions within Africa have become increasingly dry, intensifying the seasonal aridity.

Summarising, the distinguishing features of Africa are:

1. contributions of the high interior to rain shadows and bedrock influences on soils;

2. continuing tectonic influences on local precipitation and lake levels;

3. pervasiveness of volcanic enrichment of soil fertility;

4. reductions in locally effective rainfall due to rain shadows and surface run-off;

5. erratic temporal variation in rainfall seasonally and between years;

6. lack of river barriers to block dispersal between north and south.

In combination, these features have contributed to the prevalence of savanna grasslands and the consequent radiation of large grazing ungulates. They also shaped the evolutionary transitions that led from forest-dwelling apes to the sophistication of modern humans, the story to be developed in the chapters that follow.

Figure I.2 Formerly perennial Letaba River reduced to scattered pools in the dry season, Kruger NP, South Africa.

Part II: The Savanna Garden: Grassy Vegetation and Plant Dynamics

Figure II.0 Savanna landscape, Serengeti NP.

The tropical savanna biome, defined most simply by the coexistence of trees and grasses, covers nearly half of Africa's surface south of the Sahara (Figure II.1). Much attention has been given to explaining why the tree cover doesn't close up. But while the shade cast by tree canopies can suppress grasses, if sufficiently dense, sunlight is not the main limitation where savanna vegetation formations prevail. It is water in the soil, supplied seasonally and somewhat erratically within seasons by rainfall, and redistributed spatially. Competition among trees and grasses operates primarily underground in the rooting zone and thus out of sight. It takes place amid the mat of grass roots and the roots of woody plant seedlings probing for soil moisture and the mineral nutrient resources that this water conveys. The competitive

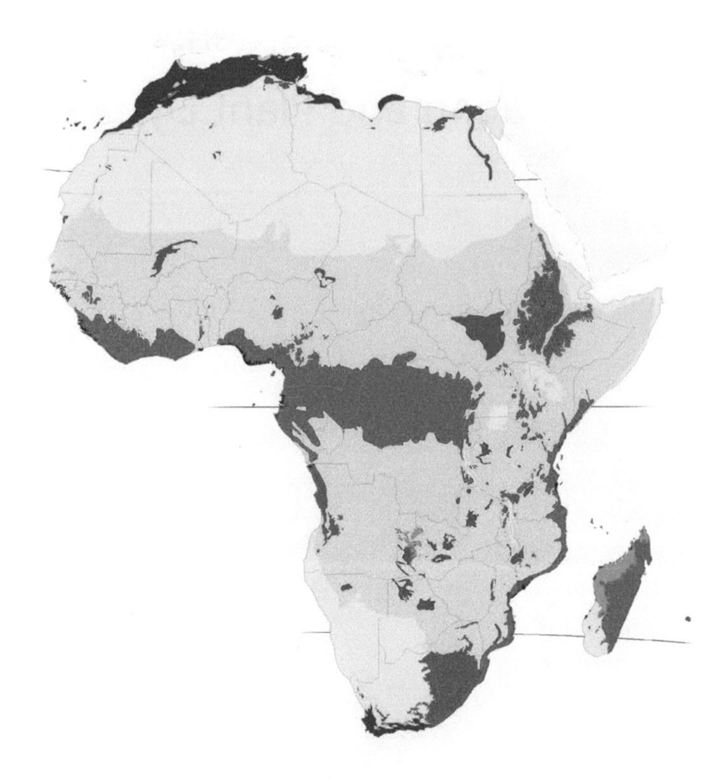

Figure II.1 African biomes, distinguishing evergreen forest in green, savanna in tan, grassland in brown, ericaceous shrubland in red, arid shrubland and desert in yellow, plus lakes or wetlands in blue (from World Wildlife Fund Ecoregions, reproduced in Jones et al. (2013) *Soil Atlas of Africa*).

interaction enters a second stage once established tree saplings emerge from the grass layer, only to be burnt back periodically by the recurrent fires sustained by the grasses. While awaiting a sufficient interval between fires to raise their foliage above the flame zone, juvenile trees are exposed to further tissue losses and damage from browsing herbivores. Grasses are superbly adapted to accommodate variable rainfall, withstand fires and tolerate herbivory, as the chapters forming this section will reveal. The feature defining savanna formations is specifically the presence of a grass layer sufficiently well-developed to support fires. Hence grasslands lacking trees are functionally allied in a broadened category of tropical grassy biomes.[1] What needs explanation is where and how woody plants manage somehow to establish and persist in regions where grasses dominate.

Textbooks commonly place savannas climatically where precipitation is too low to support closed woodland or forest; effectively as a seasonally dry form of forest (Figure II.2). Grasslands are located where precipitation is too low to support trees. Both placements are misleading. The feature distinguishing

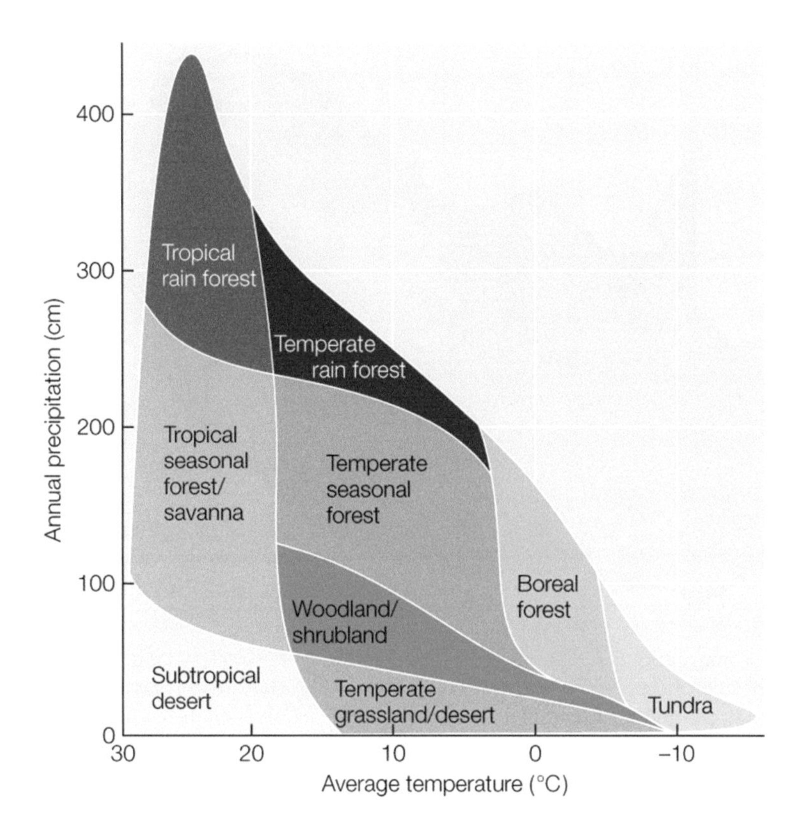

Figure II.2 Ordination of world biomes in relation to temperature and precipitation. Savanna is misleadingly located as a form of seasonally dry forest (source: Wikipedia, modified from Whittaker (1975) *Communities and Ecosystems*).

savanna climates is the recurrence of a dry season receiving too little rainfall to support much plant growth. The amount of rain falling during the wet season months may be no different from that associated with forests. South Africa's open grasslands extend into regions that are wetter than those associated with savanna, not drier (Figure II.3). Where rainfall is generally low but more evenly distributed through the year, savanna grades into low thicket or shrub steppe. Where winter rather than summer months are wet, ericaceous shrubland prevails. Closed-canopy forests develop where soil moisture remains ample for tree growth year-round.

The dependence of the vegetation cover on soil moisture generates a further characteristic feature of savannas: spatial heterogeneity in the tree canopy cover. Forest patches form in moist hollows or flanking rivers. Impermeable layers in the soil may keep trees out. Soil texture and depth along with topography modify the tree cover. Recognising this heterogeneity, it has been proposed that 80 percent or more of the landscape should have an adequately robust grass layer in order to be mapped within the savanna biome.[2]

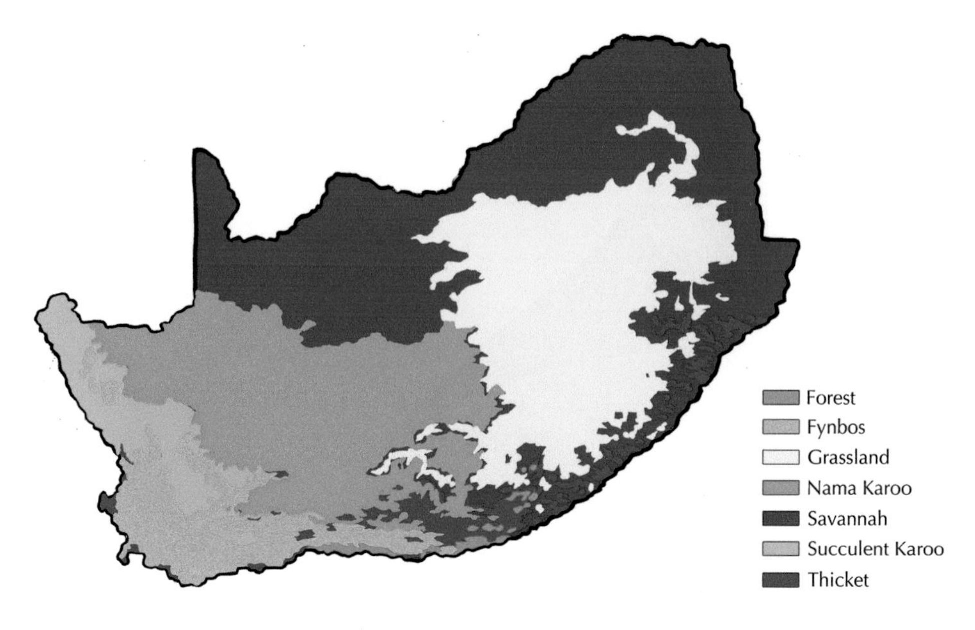

Figure II.3 Biome distribution in South Africa (from Mucina, L; Rutherford, M. (2006) *Strelitzia 19: The Vegetation of South Africa, Lesotho and Swaziland*). Note the broad extent of grassland in the north-eastern interior, the savanna extension into the arid western interior and the savanna mosaic interspersed with thicket patches spreading along the coast.

My experience of African savannas was honed within the Hluhluwe-iMfolozi Park, situated in foothills below the interior escarpment in the northern region of KwaZulu-Natal formerly known as Zululand. It encloses a mean annual rainfall (MAR) gradient from 600 to 950 mm within its 900-km^2 bounds. Thorn savanna prevails in the dry thornveld in the low-lying south-west and a grassland–forest–thicket mosaic in the moist north-east. The local composition of the trees and grasses, and thus where I was most likely to encounter white rhinos, depended further on the underlying soils, in places modified by past human settlements. Similar contributions to spatial heterogeneity occur in savannas elsewhere, but gradients were especially compressed because of the topographic diversity associated with the transitional location of this park.

Savanna ecology is thus intimately entwined with the ecology of grasses. In the first chapter introducing Part II, I will describe how climate, soil features, hydrology and fire regimes generate heterogeneity in tree:grass ratios. This concludes with an appendix introducing some of the prominent tree and grass species. The next chapter details the physiological processes governing competition between grasses and trees, taking place largely underground. The third chapter shifts the focus above-ground to population interactions mediated by repeated fires. The final chapter looks back on how vegetation features have changed through time, particularly during the period when hominins evolved. Not covered in this section are the profound impacts that large mammalian herbivores, in interaction with fire, can have in modifying vegetation and hence ecosystem function. This is deferred to Part III of the book, after the large herbivore fauna has been introduced.

References

1. Parr, CL, et al. (2014) Tropical grassy biomes: misunderstood, neglected, and under threat. *Trends in Ecology & Evolution* 29:205–213.
2. Scholes, RJ; Walker, BH. (1993) *An African Savanna. Synthesis of the Nylsvley Study*. Cambridge University Press, Cambridge.
3. Mucina, L; Rutherford, M. (2006) *Strelitzia 19: The Vegetation of South Africa, Lesotho and Swaziland*. South African National Biodiversity Institute, Pretoria.

Chapter 6: Forms of Savanna: From Woodland to Grassland

How might we classify the variability in structure and composition inherent within savannas? Reviewing savanna vegetation formations world-wide, Cole[1] recognised the following structural divisions: (1) open parkland with widely scattered trees, (2) denser woodland with fairly tall and more closely spaced trees, (3) low tree and shrub savanna with a fairly dense woody canopy, (4) grassland lacking trees, and (5) thicket with little grass (Figure 6.1). The rangeland classification developed for eastern African savannas distinguishes woodland, bushland, wooded grassland, bushed grassland, dwarf shrubland and pure grassland, based on tree cover and height.[2,3] The structural classification of vegetation developed by UNESCO applies the following definitions: (1) forest – a continuous stand of trees taller than 10 m with interlocking crowns; (2) woodland – an open stand of trees at least 8 m tall with woody cover >40 percent and a field layer of grasses; (3) bushland – an open stand of trees or shrubs 3–8 m tall with woody cover >40 percent; (4) thicket – a closed stand of trees or shrubs usually 3–8 m tall; (5) shrubland – an open or closed stand of shrubs up to 2 m tall; (6) wooded grassland – tree cover however tall of 10–40 percent; and (7) grassland – woody cover <10 percent. Almost all of these divisions focus attention on the woody plant cover rather than features of the grass layer. None are readily mapped at continental scale because patches of forest, varying tree cover and open grassland are commonly interspersed within local landscapes. Variation in grass structure and composition exists at even finer scales within these structural formations defined by tree cover.

Within southern Africa's savanna biome, six 'bioregions' were distinguished.[4] These were named (1) Central Bushveld, located in the north; (2) Lowveld, situated below the escarpment in the east; (3) Sub-Escarpment Savanna, present in parts of KwaZulu-Natal in the south-east; (4) Mopane Woodland, located in lower-lying regions of the north; (5) Eastern Kalahari Bushveld; and (6) Kalahari Duneveld. Note that the Afrikaans word 'veld' recognises the presence of a field layer, i.e. grass. Furthermore, the Kalahari region is regarded as arid savanna, not desert, because its MAR exceeds 250 mm, which is adequate to support occasional fires in wet years. Its aridity

Figure 6.1 Structural savanna formations in Africa. (A) Broad-leaved miombo (*Brachystegia*) woodland with fairly tall trees, Malawi; (B) bushwillow (*Combretum*)–clusterleaf (*Terminalia*) woodland of medium height, Kruger NP; (C) open parkland of paperbark thorn (*Acacia sieberiana*) and sausage trees (*Kigelia africana*) in Kidepo NP, Uganda; (D) low tree savanna mainly of fine-leaved acacias, Samburu, Kenya; (E) open grassland on shallow soils, Serengeti Plains, Tanzania; (F) subtropical thicket, Addo NP, South Africa, distinct from savanna because of the lack of much grass.

for animals, and people, is due especially to the lack of surface water because of the deep sand cover. Subtropical thicket is interpreted as a distinct biome and not as a form of savanna, due to its rather patchy grass cover. The grassland biome, separated from savanna, encompasses (1) Mesic Highveld, in the east; (2) Dry Highveld, in the west; (3) Montane Grassland, below the Drakensberg escarpment; and (4) Sub-Escarpment Grassland, elsewhere in KwaZulu-Natal

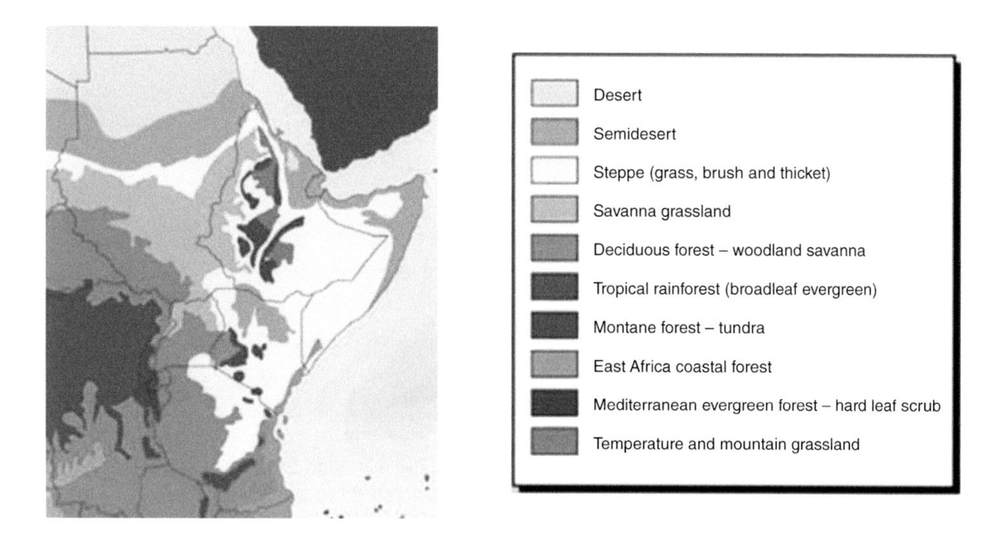

Figure 6.2 Vegetation mosaic in eastern Africa. Units labelled deciduous forest, woodland, savanna grassland and steppe represent forms of savanna (http://exploringafrica.matrix.msu.edu/wp-content/uploads/2016/05/Vegetation-of-East-Africa-Map.png).

and the Eastern Cape, including coastal regions. Temperate grassland including heath-like shrubs prevails at elevations exceeding 2800 m on top of the Drakensberg/Maloti escarpment in Lesotho, and similar grassy vegetation is represented on hillcrests within the Nama-Karoo biome.

For eastern Africa, the broadest mapped vegetation zone was 'steppe', represented by a mixture of grass, 'brush' and thicket (Figure 6.2). It grades into 'savanna grassland' towards the west where the rainfall is higher and into 'woodland savanna' in the south-west and south. In western Africa, a broad zonation in the woody vegetation cover reflects the rainfall gradient diminishing inland from the coast. The Guinean zone takes the form of a forest–savanna mosaic in the south, replaced by broad-leaved savanna woodlands in the Sudanian zone and by an open cover of low trees, mainly acacias, in the Sahelian region, bordering the Sahara Desert (Figure 6.3). Sudanian savanna formations extend westward into Uganda and southern Ethiopia, mostly as open woodland with broad-leaved trees along with tall thatch grasses.

Climatic Controls: Total Annual Rainfall Versus Dry Season Duration

Climatically, tropical savanna formations span an extreme range in MAR from 1750 mm bordering the transition into rainforest down to as little as 200 mm in the driest parts of the Kalahari and where the Sahel grades into the Sahara

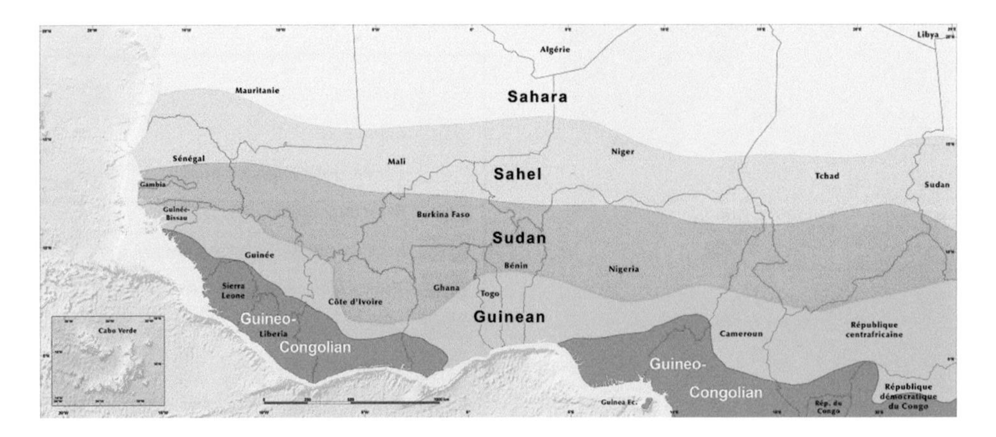

Figure 6.3 Western African savanna formations. Guinean, Sudanian and Sahelian savanna types form bands across the diminishing rainfall gradient from the coastal forest to the Sahara desert (eros.usgs .gov/westafrica/sites/default/files/inline-images/160823_Climate_zones.jpg)

Desert.[5] The defining feature is the duration and intensity of the dry season. Savanna prevails where the dry season lasts five months or longer spanning the cooler months of the year.[5,6] In south-central Africa, dry season rainfall diminishes to as little as 5 percent of the MAR and the dry season extends over seven months (see Table 1.1). In eastern Africa near the equator, the dry season is less intensely dry and as much as 25 percent of the annual rainfall total may be received during five dry season months.

The tree canopy cover tends to increase with rainfall, but there is much variation among savanna sites in the actual woody plant cover exhibited locally (Figure 6.4).[6,7] Below a rainfall threshold of around 650 mm, the maximum tree cover recorded diminishes with decreasing MAR. Above this threshold, the local tree canopy can potentially reach 80 percent, but generally is maintained below 40 percent by fire and other influences.[8] Once the MAR exceeds 1000 mm, two alternative states can prevail: either (a) a densely wooded savanna, or (b) open grassland interspersed with forest patches.[9] The latter tends to prevail at altitudes above 1800 m where temperatures get cooler: for example, below the Drakensberg escarpment, on the Nyika Plateau in Malawi, in the eastern highlands of Zimbabwe, and along the Eastern Arc Mountains in Tanzania (Figure 6.5). The broad-leaved deciduous woodlands known as miombo prevail through south-central Africa where hot conditions occur from August through November before the rainy season begins. Sudanian savanna formations replace it north of the equator under similar climatic regimes. Once forest trees become established, the dense shade they cast restricts the herbaceous cover and hence fire penetration.[10] If the forest canopy gets opened, by people or elephants, the evergreen forest trees have little capacity to resist being burnt and may get eliminated.

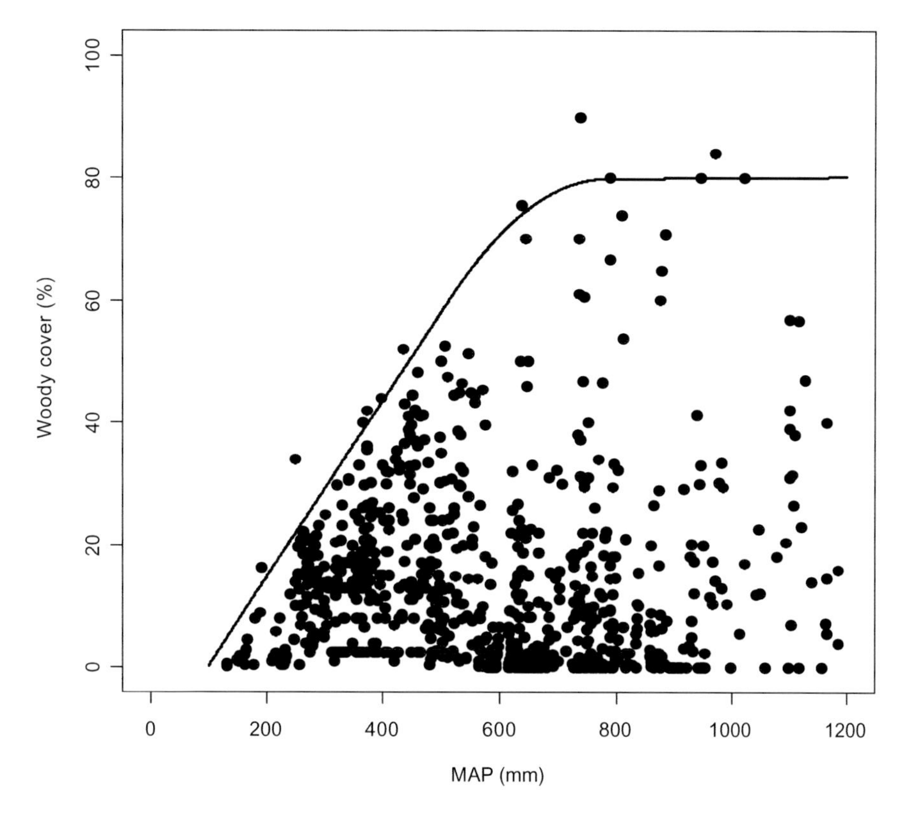

Figure 6.4 Woody vegetation cover in relation to mean annual precipitation (MAP) recorded at various sites within savanna regions of Africa. Diminishing rainfall restricts the tree cover below 650 mm, while above this threshold the tree cover is widely variable (from Sankaran et al. (2005) *Nature* 438: 846–849).

A potential division thus exists between moist savannas, with MAR exceeding 650 mm, where broad-leaved woodlands prevail, and dry savannas with MAR less than 650 mm, where rainfall limits the cover of mainly fine-leaved trees. But the woody plant cover is controlled not only by seasonal moisture – the nature of the soil also influences the kinds of trees and grasses that predominate, as I will explain next.

Geological Substrate and Soil Fertility

Soil fertility is governed by the capacity of soil particles to retain the mineral nutrients needed for plant growth. This is determined largely by the clay fraction, as outlined in Chapter 5. The clay component is derived from the geological substrate, with mafic volcanic rocks contributing much clay and felsic granitic rocks forming rather sandy soils. The clay contributed by sedimentary rocks depends on whether they were consolidated from mud, silt or sand. The bedrock contributes also to the mineral nutrient pool, while the

Figure 6.5 Grassland–forest mosaics. (A) Lope NP, Gabon, with forest pockets in lowlands (MAR 1750 mm); (B) Nyika Plateau, Malawi, with forest patches on mid-slopes (MAR 1200 mm); (C) Drakensberg foothills, South Africa, with forest patches on slopes and lowlands (MAR 1050 mm); (D) Maasai-Mara, Kenya, with forest flanking Mara River (MAR 1000 mm); (E) Eastern Highveld grassland, MAR 700 mm; (F) mostly treeless grassland in the Bateke plateau in south-eastern Gabon, underlain by deep sandy soils derived from Kalahari sand (MAR 1500 mm).

humus component helps retain nutrients, especially nitrogen, derived from organic matter, not rocks.

The effective fertility is influenced also by rainfall. Water percolating through the soil leaches nutrients and weathers the clay content to kaolinite with lowered cation-holding capacity. Where rainfall is low, water penetrates less deeply into soils and mineral nutrients get retained near the surface. With high amounts of rainfall, soil fertility depends largely on the organic matter input, whatever the geological substrate. This leads to a 'double-barrelled' subdivision between dry and/or 'eutrophic' and moist and/or 'dystrophic' savanna forms, reflecting the interaction between rainfall and the bedrock geology.[11] Clay-rich soils are typically 'well-nourished' ('*eu*-trophic') for plants, while sandy soils provide poor nourishment ('*dys*-trophic'). South African cattle ranchers recognise the implications for the performance of their livestock, distinguishing 'sweet bushveld', capable of supporting cattle year-round, from 'sour bushveld', where livestock lose condition during the dry season.[12] It is in the middling range of MAR between 500 and 1000 mm, prevalent through much of Africa's savanna biome, that bedrock geology is most influential. With higher rainfall, nutrients get intensely leached, whatever the soil parent material. With lower rainfall, water does not penetrate far and nutrients get retained (Box 6.1).

The subdivision between dry/eutrophic and moist/dystrophic savanna forms is associated with distinctions in the predominant tree species. Dry/eutrophic savannas are typified by the prevalence of fine-leaved legumes formerly grouped in the genus *Acacia* within the subfamily Mimosoideae (see Box 6.2). Moist/dystrophic savannas prevalent in south-central Africa are dominated by broad-leaved legumes in the subfamily Caesalpinioideae (recently revised to Detariodeae), especially the genera *Brachystegia, Julbernadia* and *Isoberlinia* (Appendix 6.1 lists common and scientific names of species mentioned in the text along with their families). North of the equator in the Sudanian zone, broad-leaved trees in the family Combretaceae (*Combretum* and *Terminalia* spp.) tend to predominate and grow also in the understorey of miombo woodlands. Then there is the anomalous mopane (*Colophospermum mopane*), also a member of the Caesalpinioideae, which attains almost monospecific dominance in low-lying hot and dry areas in southern Africa, especially but not exclusively on clay soils. While the predominant trees are deciduous in both savanna subdivisions, they retain leaves longer into the dry season in moist/dystrophic savannas than in dry/eutrophic savannas.

The rainfall influence alone was explored along a transect consistently on infertile Kalahari sand stretching from the Northern Cape of South Africa through Botswana into Zambia.[13] There was a transition from acacia trees to bushwillow or mopane once MAR exceeded 400 mm, and a variety of broad-leaved leguminous trees became predominant above 600 mm MAR. The prevalent grass types also changed, from aristoid (three-awn or needle grasses, *Aristida* spp.) at the dry end to panicoid (buffalo grass *Panicum* spp. and finger grass *Digitaria* spp.) in the middle, and andropogonoid species (thatch grasses *Hyparrhenia* spp.) in the wettest north.

Box 6.1 What Governs the Distinction in Soil Fertility Between Dry/Eutrophic and Moist/Dystrophic Savannas?

Soil fertility controls how fast plants can potentially grow when water is sufficient. It is governed by nutrients like nitrogen recycled through decaying organic matter, basic elements like potassium, magnesium and calcium held on clay particles and phosphorus contributed from degrading rock material. Fertiliser applied to crops and gardens consists basically of nitrate, phosphate and potassium salts (NPK).

The bedrock geology supplies the mineral elements and phosphate (PO_4^-), but more importantly provides the clay particles that hold cations, including nitrogen in the form of ammonia (NH_4^+), against leaching by water percolating through the soil. Once generated, the pool of both mineral and organic nutrients gets recycled mostly through the soil organic matter or humus content via growth and decay of plants, below- as well as above-ground, plus animal contributions in the form of dung, urine and dead bodies. However, the humus content of African savanna soils is generally low,[14] because primary productivity is restricted by low rainfall and plant litter is decomposed rapidly by soil microbes in hot environments. Moreover, there are leakages. Much of the nitrogen is volatilised in smoke during fires, and some nitrogen in the form of nitrate (NO_3^-) is lost through leaching. However, phosphates are highly insoluble and can contribute to maintaining high fertility for a long time in sites enriched by settlements where animal bones accumulate.[15]

The capacity of clay particles to hold cations like K^+ and Ca^+ depends on the mineral form of the clay. Clay is constituted by alternating layers of silica and aluminium oxides, which develop negative surface charges attracting cations. These charges are best developed where the layers form a 2:1 lattice, called smectite or illite. After they get weathered to a 1:1 lattice (kaolinite, the substance of talcum powder) this capacity is greatly reduced, and mineral nutrients are held more precariously in the organic matter.

Legumes in the subfamilies Mimosoideae (including the acacias) and Papillionoideae (including beans) can reclaim nitrogen from the atmosphere through the agency of bacteria housed in root nodules. However, this biological nitrogen fixation is energetically costly, and beneficial only if phosphorus is sufficiently available to complement the nitrogen gained.[16] Where phosphorus is deficient, in sandy soils derived from granitic rocks or sandstone that typically underlie miombo woodlands, no benefit would be derived from nitrogen fixation.

In African savannas, more nutrients get recycled via clay particles and less through soil organic matter than in temperate latitudes, which is why the

Box 6.1 (cont)

bedrock geology has such a great influence on fertility. Where rainfall is high, all soils become reduced to deeply weathered 'oxisols' with little capacity to retain nutrients. Accordingly, in tropical forests most recycling take place via the mulch of leaves and other plant litter on the surface before it enters the soil. The link between soil fertility and geology in African savannas is thus via (1) the clay amount and form generated by the bedrock type, which determines the cation exchange capacity (ability to retain nutrients), and (2) the content of mineral nutrients, especially P, in the parent rock material.

Box 6.2 Status of the Genus *Acacia*

The fine-leaved legumes in the Mimosoideae that were grouped in the genus *Acacia* have recently become partitioned between newly erected genera *Vachellia* and *Senegalia* for the African representatives, with the generic label *Acacia* confined to the Australian wattles. Species now subsumed within *Vachellia* have fairly straight spines developed from stipules located at the base of leaves, while the species placed in *Senegalia* have recurved prickles located at nodes along stems. The former produce yellow or cream pom-pom flowers, while the latter produce white or cream catkins. Species labelled *Vachellia* tend to flower throughout the wet season, while those in *Senegalia* generally flower early, before leaf flush. The former produce either indehiscent pods containing very hard seeds that fall to the ground intact, or dehiscent pods that split to shed their seeds, while the latter produce only dehiscent pods.

While there are these valid distinctions between these subgroups, the ecological unity among the African acacias has thereby been obscured. It would have been more meaningful ecologically had the distinctions that exist among the African acacias been recognised only at subgeneric level. Furthermore, the new generic assignments were erected irregularly by reassigning the type species, and so remain contentious. In this book, I will retain the original assignment of Africa's fine-leaved thorn trees to the genus *Acacia*. Common names together with scientific names and other features of all plant species mentioned are listed in the Appendix at the end of this book.

The soil effect is strikingly apparent in South Africa's Kruger NP. The eastern region underlain by basalt supports knob thorn (*Acacia nigrescens*) parkland, under MAR regimes between 550 and 650 mm, while the western region on granitic gneiss carries a broad-leaved bushwillow woodland under similar rainfall. Where gabbro or dolerite sills intrude within the granite, acacias once again predominate. In the Serengeti ecosystem in Tanzania where soils are widely volcanically enriched, acacia savanna extends into regions with MAR exceeding 1000 mm in the north. Red grass (*Themeda triandra*) features in the herbaceous layer on clay-rich soils, while thatch grasses and various love grasses are prevalent on soils with a higher sand content.

Broad-leaved savanna woodlands known as miombo are typically associated with sandy soils derived from basement granite, sandstone or Kalahari sand, mostly in regions with MAR exceeding 800 mm. Where soils are highly infertile, broad-leaved savanna formations occur even under MAR regimes as low as 600 mm. Examples include woodlands dominated by wild seringa (*Burkea africana*) on sandstone-derived soils in the Nylsvley Nature Reserve and elsewhere in the 'sour' bushveld of South Africa; and places with Zimbabwe teak (*Baikiaea plurijuga*) on deep Kalahari sand in northern Botswana and adjacent parts of Zimbabwe where MAR is similarly low.[17] Sudanian savannas are associated with regions where soils became intensely weathered down to iron concretions.[18] This is derived from when they were located close to the climatic equator while Africa was drifting northward after the breakup of Gondwana.

Phosphorus seems to be the key soil nutrient governing the prevalence of fine-leaved acacias. Nitrogen availability becomes most limiting where the available phosphorus is adequate. Legumes in the Mimosoideae overcome the nitrogen limitation via a symbiotic relationship with bacteria housed in nodules on their roots.[19] These bacteria are capable of fixing atmospheric nitrogen, subsidised energetically by carbohydrates from the host plant. The phosphorus influence was evident in the Nylsvley Nature Reserve where the prevalent broad-leaved woodland gave way to patches dominated by acacia trees on sites of former human occupation.[20] In these places, the phosphorus content in the sandy soil had been enriched by wood-burning and bone accumulations. The grassland composition also changed from dominance by tall and highly fibrous broom love-grass (*Eragrostis pallens*) to a prevalence of love grasses (other *Eragrostis* spp.) and blue buffalo grass (*Cenchrus ciliaris*) in the acacia patches.

Thickets of low-growing trees, commonly evergreen, occur patchily within savanna regions, either on deeper soils that retain moisture longer into the dry season or on rocky hills. Only in South Africa are thickets sufficiently extensive to be regarded as forming a distinct biome, occurring in a region of the Eastern Cape where rainfall is low but distributed year-round and deep sandy soils retain moisture. Once formed, thickets resist fires because of the sparse grass cover. These climatically generated thickets are compositionally distinct from thickets that have developed due to fire exclusion or lack of grass cover brought about by heavy grazing.[21]

Soil Water: Topo-hydrology

The amount of moisture effectively available for plant growth depends not simply on the rainfall, but also on how rainwater is redistributed within landscapes. The deluge during thunderstorms can exceed the infiltration capacity of the soil, so that a substantial fraction of the water gravitates down slope, eroding and leaching in the process.[22] How much seeps in or runs off depends on the soil texture as well as the soil depth. Uplands tend to be driest, while lowlands accumulate water and soil, especially finer clay particles. The resultant topo-sequence linking vegetation formations is known as a catena. In general, tree cover tends to be least and grasses shortest on uplands where soils are shallowest, while woodland or forest flanks river channels[23] (Figure 6.6A). However, this pattern gets reversed in broad-leaved savannas, especially on sandy soils. Tree cover becomes densest on crests, while zones of treeless grassland, known locally as dambos, mbugas or vleis, occupy the broad valleys where soils are seasonally waterlogged (Figure 6.6B–D). Seep zones may also promote open grassland on lower slopes where water gets forced to the surface over bedrock or a hardpan. Sandier soils allow greater infiltration and retain moisture for longer than compacted clay soils, while the latter dry more intensely. Consequently, sandy soils tend to support a denser tree cover than clay-rich soils. Soil depth may be restricted by hardpan layers in the subsoil.

The open short-grass plain in south-eastern Serengeti is edaphically controlled by a calcrete hardpan developed from volcanic ash and tuff deposited by nearby volcanoes.[24] With increasing rainfall westward, soils get deeper and the vegetation shifts from treeless grassland into acacia savanna with a taller grass cover. Zones of open grassland more generally are associated with reduced rates of water infiltration into soils.[22,25] Tree cover decreases with increasing intensity of rainfall on clay soils.[26] Various forms of hardpan can restrict the rooting depths of trees elsewhere in Africa. On ancient land surfaces in South Africa's Highveld region, as well as further north in Africa, erosion over many millions of years has formed ferricrete (or plinthite), silcrete or calcrete hardpans at shallow depths, inhibiting tree establishment.[27,28] Shrubs occur on rocky outcrops where soil moisture is concentrated between the rocks and fires are deflected.

Quite extensive grasslands are found on floodplains and former lake beds where clay seals restrict water penetration, as shown in the Savuti and Mababe depressions in northern Botswana.[29] Besides the Okavango Delta, extensive floodplains border sections of the Kafue River and occur in the Bangweulu

Figure 6.6 Topographic sequences of vegetation formations within local landscapes. (A) Open grassy upland grading into shrub cover near drainage lines, Serengeti NP, Tanzania; (B) grassy seep zone with tall grassland in mid-slope and clusterleaf woodland on crests, south-west Kruger NP; (C) grassy dambo intersecting miombo woodland, Zambia; (D) wide dambo grassland flanked by broad-leaved woodland, north-west Serengeti NP.

Swamp in Zambia, below Gorongoza Mountain on the edge of the rift in Mozambique and in the Katavi region of western Tanzania. These sump grasslands represent hotspots of nutrient enrichment within predominantly nutrient-deficient woodlands. Termite mounds raised within the floodplain allow trees to escape waterlogging in local clumps. Under considerably drier conditions in the southern Kalahari, pan depressions and infilled valleys have likewise remained treeless due to clay seals. Tree establishment may also be inhibited on very deep sands, as seems to be the case on the Bateke Plateau in Congo Brazzaville and adjoining Gabon underlain by ancient deposits of Kalahari sand (Figure 6.5F). The influential factor in such localities may be the difficulty for woody seedlings to access water at depth.[30]

Combined influences from the redistribution of rainwater, bedrock geology and hydrology generate a complex mosaic of vegetation patterns, most extremely between treeless grassland and closed forest. The composition of the grass layer also responds to soil depth and texture, with shortest grasses prevailing on uplands and tallest grass in lowlands and wetlands. These effects are modified additionally by the recurrent fires promoted by the seasonal dryness.

Fire Regimes: Recurrent Incineration

Recurrent fires are an intrinsic feature of savannas due to the seasonal drying of the grass layer. The impact of these fires on the vegetation cover depends on the frequency of the burns, the intensity of the fires and the stage within the seasonal cycle when the fires occur.[31] Currently, the fire return interval is typically 2–3 years across a range in MAR from 700 to 1200 mm. In regions with lower rainfall than this, there is less fuel to support the spread of fires and burns occur mostly during unusually wet years, perhaps at quite long intervals.

However, intervals between fires tend to vary quite widely locally. Some places may escape being burnt for many years, due to local soil moisture or grass removal by herbivores. Within Serengeti NP, some localities got burnt twice annually, while others escaped being burnt for 10 years or longer, mostly on the heavily grazed plains.[32,33] Regular annual fires promote an open tree canopy, without necessarily affecting the density of woody plants, because small trees and shrubs can remain hidden in the grass layer. With protection from fire, the increased tree cover can shift the grass layer towards shade-loving grasses less supportive of fires, reinforcing the woody thickening trend.[34] Variable intervals between burns may enable juvenile trees to grow beyond the flame zone. With long intervals, moribund grass accumulated over several years can promote fires hot enough to kill tree saplings several metres tall plus some canopy trees. Fires occurring in drier savannas, although less intense, can still be quite destructive to woody plants that lack adequate protective features.

Higher rainfall promotes greater grass growth and hence hotter fires (Figure 6.7). Weather conditions at the time of the fire are also influential.[35] Extremely hot 'firestorms' are generated by 30:30:30 conditions: air temperature >30°C, relative humidity <30 percent and wind speed >30 km/h.[36] Head fires fanned by wind are hot but brief, and temperatures at the soil surface may not rise much. Back burns into the wind can be more destructive to grass tufts as well as tree saplings, because flames remain longer. Intense 'firestorms' can kill even established trees by burning through their protective bark, especially if some bark had been removed by elephants, porcupines or other animals.

Fires can get ignited by lightning when the first thunderstorms occur while the grass still remains dry. If the onset of the rains is delayed, burns may extend into the wet season months, and thus be especially detrimental to trees that had flushed new leaves. However, humans have been major agents of fire ignition for at least 300,000 years in Africa, perhaps even much earlier

Figure 6.7 Savanna fires. (A) Hot fire burning tall savanna grassland, Kruger NP (photo: E. Le Roux); (B) gentle fire burning grass layer in miombo woodland, Luangwa Valley; (C) fire burning into forest margin, Gabon; (D) herbivores grazing green-up following burning, Serengeti NP.

during the Pleistocene.[37] Hunter-gatherers lighting fires to clear impeding grass and attract animals to the post-burn flush of green grass would have set them earlier in the dry season than those caused by lightning. After humans acquired livestock, early dry season burns may have been used to improve the nutritive value of the grass cover for these domestic herbivores during the dry season. In moist miombo woodlands in Zambia, early season burns promote an open understorey coupled with pockets of resistant evergreen thicket, while annual late burns cause the death of some canopy trees, replaced by coppice regeneration.[38,39] Wetland grasslands ranging from dambos to floodplains maintained by waterlogging can support fires after floodwaters have retreated and grasses have dried out.

Regular burns help maintain a sharp boundary between grassland and forest patches in grassland–forest mosaics.[40] Shrubs bordering forest margins may form partial barriers to fire penetration into the forest. Recurrent annual or semi-annual fires contribute to the mosaic of open grassland and stunted woodland interspersed with forest patches in regions of montane grassland.[41,42]

Recurrent fires thus help maintain an open woody canopy within the savanna biome, especially in moister regions. The suppression of fires by European settlers, combined with heavy stocking with domestic grazers, has contributed to the widespread expansion in woody vegetation cover within savanna regions of southern Africa[43] and in parts of eastern Africa[44] over recent decades. A further factor may have been the post-industrial rise in atmospheric carbon dioxide levels, favouring woody plants over grasses.[45,46] However, the susceptibility of the grass cover to being burnt is modified by the activities of large herbivores, both grazers and browsers. These interactions will be considered within an ecosystem context in Part III of the book.

River Corridors

River corridors contribute importantly to vegetation patterns within savannas. Regions bordering river channels, from headwater gullies to broad river basins, have more fertile soils and retain soil moisture longer into the dry season than the surrounding savanna matrix, favouring woodland or forest development. Fires get suppressed not only because of wetter soils and greater shading, but also because heavy grazing and trampling by herbivores concentrating near water reduces the fuel load.[47] Tall trees characteristic of river margins include sycamore figs (*Ficus sycomorus*), ana trees (*Faidherbia albida*), jackalberry (*Diospyros mespilliformis*), leadwood (*Combretum imberbe*) and mahogany (*Trichilia emetica*) (see Appendix 6.1). Fever trees (*Acacia xanthophloea*) may form groves in regions subject to seasonal flooding. Stands of reeds (*Phragmites* spp.) develop in sandy river beds.

Periodic floods can remove even quite large trees bordering the river channel and redistribute the areas covered by water, sand, reeds or forest.[48] Hence the current state of the vegetation alongside or within the river channel depends largely on the time elapsed since the last big flood. Pools in rivers support hippos and attract concentrations of grazers and browsers to feed on the locally greener vegetation as well as to access water for drinking, increasing the spatial heterogeneity within the riparian zone.

Open Grasslands: Why No Trees?

Regions with predominantly open grassland are mapped as a distinct biome in South Africa, but not generally elsewhere in the continent. Various factors have been proposed to explain the lack of trees, but no consensus has been reached on their relative importance. Temperatures sufficiently cool to generate winter frosts have been invoked, but grasslands without trees are not restricted to highlands where frosts are frequent. Highveld climates do not prevent shrubs from growing on rocky outcrops, where they gain some protection from fire. Grassland and forest are interspersed in a mosaic bordering the rainforest biome and even in coastal regions, where frosts never occur. Fires burn intensely hot through open grasslands, especially when fanned by winds, but the presence of a few trees with fire-resistant bark does little to reduce the intensity, which depends mainly on the grass fuel load. Soil depth on eroding uplands and elsewhere where hardpan layers have formed may block deeper penetration by tree roots[27]; but open grasslands exist on deep sands in the Bateke Plateau in Congo-Brazzaville and on the Mozambican coastal plain. The lack of trees may even be a legacy of the last glacial maximum when atmospheric levels of CO_2 dropped too low to support much woody plant growth, but there should have been adequate time since then for trees to spread back.

Overview

The savanna biome is defined by the presence of a grass layer sufficiently dense to support recurrent fires over most of the landscape. The local tree cover is climatically allied with a dry season period when plant growth ceases. Geological influences on soil fertility, modified by rainfall, underlie a functional subdivision between fine-leaved dry/eutrophic savannas, characterised by thorny acacias, and broad-leaved moist/dystrophic savannas where other leguminous trees predominate. Treeless grasslands are found on cooler uplands and elsewhere where shallow soils restrict rooting depth, although not restricted to such conditions. A forest–grassland mosaic prevails adjoining the rainforest zone. The spatiotemporal redistribution of soil moisture contributes to spatial heterogeneity in the woody plant canopy cover. Variable fire return intervals further modify the tree and grass cover.

The tropical savanna biome is less extensive in other continents. South American savanna formations, locally known as 'cerrado', prevail in a fairly large block south of the equator on the Brazilian highlands and in a smaller area just north of the equator in Venezuela, known as the llanos, separated by the forested Amazon basin. Australia has an arc of savanna in its north, running from Western Australia through Northern Territory into Queensland. In parts of India and south-east Asia, vegetation formations mapped as 'tropical scrub forest' and 'tropical deciduous forest' have been allied with savanna forms because of their substantial grass cover.[49] Savanna vegetation extends into regions with higher rainfall in other continents than in Africa: up to 2500 mm in South America and 2000 mm in Australia, compared with 1750 mm in Africa.[5] Fire incidence has less influence on the woody plant cover in both of these southern continents than in Africa.[50]

The dry/eutrophic savanna subdivision seems to be exclusive to Africa. South American cerrado found on Brazil's Parana plateau is associated with a range in MAR from 1100 to 1900 mm.[51,52,53] The tree cover there gives way to open grassland towards higher elevations, disrupted by shrubby trees on rocky hills, but the underlying bedrock, whether sedimentary sandstone or mafic basalt/diabase, has little effect on vegetation structure or composition. Soils are generally highly weathered, degrading the clay content to kaolinite with little cation-holding capacity.[54] Trees retaining evergreen foliage are intermingled among those shedding their leaves during the dry season. Grasses are generally low in nutritional value and include a substantial component of species with the C_3 photosynthetic pathway (see Chapter 7). The grassy llanos occupies a low-lying plain in Venezuela and adjoining parts of Colombia where MAR ranges from 1200 mm up to 2750 mm. Over much of its extent the open aspect is maintained by seasonal flooding, despite the 5–6 months dry season. Smaller blocks of grassland–forest mosaic emerge from the Amazon rainforest near the border between Brazil, Venezuela and Guyana. The closest approach to African savannas is the mix of dry woodland, thicket and grassland called 'chaco' found in Paraguay and adjoining countries. In north-eastern Brazil, a dry spiny thicket or low forest with stem-succulents prominent called 'caatinga' replaces cerrado where MAR lies between 450 and 1000 mm with a less distinctive dry season.[55] The Pantanal in south-western Brazil is a vast wetland grassland, although some areas do burn during the dry season. Well south of the tropics, savanna grades into temperate 'pampas' grassland and steppe in Argentina.

Australia exhibits a lower tree cover relative to MAR than Africa, perhaps due to tree mortality during prolonged extreme droughts.[56] Australian savanna formations are dominated mostly by evergreen or semi-evergreen gum trees (*Eucalyptus* spp.).[57] Although wattle trees (various *Acacia* spp.) tend to predominate where soils are less infertile, there is no clear division between moist savanna and dry savanna. Trees assigned to the genus *Acacia* there lack thorns and many have phyllodes (expanded stalks) functioning as leaves. Treeless grassland occurs in the Barkly Tableland under MAR as low as 350 mm on vertisols derived from limestone, which exclude trees by swelling and shrinking in response to moisture variation. Other areas of open grassland occur locally on basalt substrates. The form of red grass (*Themeda* sp.) growing in Australia lacks the ability of the African species to support grazing and African grasses have been introduced to promote the livestock industry.

Within India and parts of south-east Asia, vegetation formations resembling the moist savannas of Africa are present in uplands of the Western Ghats on basalt and more locally elsewhere.[49] Dipterocarp woodlands with a grassy understorey, which does burn periodically, occur amid closed-canopy forest under MAR levels approaching 2000 mm in continental south-east Asia.[1] Dry areas with thorn trees prominent prevail in parts of north-west India, but lack sufficient grass cover to be called savanna. Dense human occupation plus a policy of actively suppressing fires has perhaps obscured savanna affinities with Africa in these regions.

Thus, tropical savannas outside Africa are generally rather wet and mostly extremely dystrophic. Within Australia recurrent fires burn the woody canopy as well as the grass layer due to the flammability of the prevalent eucalypt trees. Fires seem less influential in tropical Asia, perhaps because they are generally suppressed, while dry seasons there are less dry than within Africa and Australia.

To establish how grasses exclude trees, we need to explore the mechanisms of competition between these plant forms, taking place below-ground as well as above-ground. This is the topic to be addressed in the next chapter. Before pitting trees against grasses, you need to be reminded that individual species of each of these plant types differ in their ecologies. Profiles of some of the prominent species, plus pictures, are presented in Appendix 6.1.

Suggested Further Reading

Bond, WJ. (2019) *Open Ecosystems. Ecology and Evolution Beyond the Forest Edge.* Oxford University Press, Oxford.

Cole, M. (1986) *The Savannas. Biogeography and Geobotany.* Academic Press, New York.

Huntley, BJ; Walker, BH. (1982) *Ecology of Tropical Savannas.* Springer, Berlin.

Lehmann, CER, et al. (2011) Deciphering the distribution of the savanna biome. *New Phytologist* 191:1970209.

McClenahan, TR; Young, TP. (1996) *East African Ecosystems and their Conservation.* Oxford University Press, Oxford.

References

1. Cole, MM. (1986) *The Savannas. Biogeography and Geobotany.* Academic Press, New York.
2. Pratt, DJ; Gwynne, M. (1977) *Rangeland Management and Ecology in East Africa.* Hodder and Stoughton, London.
3. Grunblatt, J, et al. (1989) A hierarchical approach to vegetation classification in Kenya. *African Journal of Ecology* 27:45–51.
4. Mucina, L; Rutherford, M. (2006) *The Vegetation of South Africa, Lesotho and Swaziland.* South African National Biodiversity Institute, Pretoria.
5. Lehmann, CER, et al. (2011) Deciphering the distribution of the savanna biome. *New Phytologist* 191:197–209.

6. Good, SP; Caylor, KK. (2011) Climatological determinants of woody cover in Africa. *Proceedings of the National Academy of Sciences of the United States of America* 108:4902–4907.
7. Sankaran, M, et al. (2005) Determinants of woody cover in African savannas. *Nature* 438:846–849.
8. Sankaran, M, et al. (2008) Woody cover in African savannas: the role of resources, fire and herbivory. *Global Ecology and Biogeography* 17:236–245.
9. Staver, AC, et al. (2011) History matters: tree establishment variability and species turnover in an African savanna. *Ecosphere* 2:1–12.
10. Charles-Dominique, T, et al. (2018) Steal the light: shade vs fire adapted vegetation in forest-savanna mosaics. *New Phytologist* 218:1419–1429.
11. Huntley, BJ. (1982) Southern African savannas. In Huntley, BJ; Walker, B (eds) *Ecology of Tropical Savannas*. Springer, Berlin, pp. 101–119.
12. Ellery, FN. (1995) The distribution of sweetveld and sourveld in South Africa's grassland biome in relation to environmental factors. *African Journal of Range and Forage Science* 12:38–45.
13. Scholes, R, et al. (2002) Trends in savanna structure and composition along an aridity gradient in the Kalahari. *Journal of Vegetation Science* 13:419–428.
14. Crowther, TW, et al. (2019) The global soil community and its influence on biogeochemistry. *Science* 365:eaav0550.
15. Augustine, DJ. (2003) Long-term, livestock-mediated redistribution of nitrogen and phosphorus in an East African savanna. *Journal of Applied Ecology* 40:137–149.
16. Hogberg, P. (1986) Nitrogen-fixation and nutrient relations in savanna woodland trees (Tanzania). *Journal of Applied Ecology* 23:675–688.
17. Childes, SL; Walker, BH. (1987) Ecology and dynamics of the woody vegetation on the Kalahari sands in Hwange National Park, Zimbabwe. *Vegetatio* 72:111–128.
18. Spinage, CA. (1988) First steps in the ecology of the Bamingui-Bangoran National Park, Central African Republic. *African Journal of Ecology* 26:73–88.
19. February, EC, et al. (2019) Physiological traits of savanna woody species: adaptations to resource availability. In Scogings, PF; Sankaran, M (eds) *Savanna Woody Plants and Large Herbivores*. Wiley, Oxford, pp. 309–329.
20. Blackmore, AC, et al. (1990) The origin and extent of nutrient-enriched patches within a nutrient-poor savanna in South-Africa. *Journal of Biogeography* 17:463–470.
21. Bond, WJ, et al.(2017) Demographic bottlenecks and savanna tree abundance. In Cromsigt, JPG, et al. (eds) *Conserving Africa's Mega-Diversity in the Anthropocene*. Cambridge University Press, Cambridge, pp. 161–188.
22. Jager, T. (1982) Soils of the Serengeti woodlands, Tanzania. PhD thesis, Wageningen University, Wageningen.
23. Colgan, MS, et al. (2012) Topo-edaphic controls over woody plant biomass in South African savannas. *Biogeosciences* 9:1809–1821.
24. de Wit, HA. (1978) Soils and grassland types of the Serengeti Plains (Tanzania). PhD thesis, Wageningen University, Wageningen.

25. Holdo, RM, et al. (2020) Spatial transitions in tree cover are associated with soil hydrology, but not with grass biomass, fire frequency, or herbivore biomass in Serengeti savannahs. *Journal of Ecology* 108:586–597.
26. Case, MF; Staver, AC. (2018) Soil texture mediates tree responses to rainfall intensity in African savannas. *New Phytologist* 219:1363–1372.
27. Tinley, K. (1982) The influence of soil moisture balance on ecosystem patterns in southern Africa. In Huntley, BJ; Walker, BH (eds) *Ecology of Tropical Savannas*. Springer, Berlin, pp. 175–192.
28. O'Connor, TG; Bredenkamp, GJ. (1997) Grassland. In Cowling, RM, et al. (eds) *Vegetation of Southern Africa*. Cambridge University Press, Cambridge, pp. 215–257.
29. Sianga, K; Fynn, R. (2017) The vegetation and wildlife habitats of the Savuti–Mababe–Linyanti ecosystem, northern Botswana. *Koedoe* 59:1–16.
30. Knoop, WT; Walker, BH. (1985) Interactions of woody and herbaceous vegetation in a southern African savanna. *The Journal of Ecology* 73:235–253.
31. Archibald, S, et al. (2013) Defining pyromes and global syndromes of fire regimes. *Proceedings of the National Academy of Sciences of the United States of America* 110:6442–6447.
32. Eby, S, et al. (2015) Fire in the Serengeti ecosystem: history, drivers, and consequences. In Sinclair, ARE, et al. (eds) *Serengeti IV: Sustaining Biodiversity in a Coupled Human–Natural System*. University of Chicago Press, Chicago, pp. 73–103.
33. Probert, JR, et al. (2019) Anthropogenic modifications to fire regimes in the wider Serengeti–Mara ecosystem. *Global Change Biology* 25:3406–3423.
34. Higgins, SI, et al. (2007) Effects of four decades of fire manipulation on woody vegetation structure in savanna. *Ecology* 88:1119–1125.
35. Govender, N, et al. (2006) The effect of fire season, fire frequency, rainfall and management on fire intensity in savanna vegetation in South Africa. *Journal of Applied Ecology* 43:748–758.
36. Browne, C; Bond, W. (2011) Firestorms in savanna and forest ecosytems: curse or cure? *Veld & Flora* 97:62–63.
37. Archibald, S, et al. (2012) Evolution of human-driven fire regimes in Africa. *Proceedings of the National Academy of Sciences of the United States of America* 109:847–852.
38. Trapnell, CG. (1959) Ecological results of woodland and burning experiments in Northern Rhodesia. *The Journal of Ecology* 47:129–168.
39. Smith, P; Trapnell, C. (2002) Chipya in Zambia: a review. *Kirkia* 18:16–34.
40. Titshali, LW, et al. (2000) Effect of long-term exclusion of fire and herbivory on the soils and vegetation of sour grassland. *African Journal of Range and Forage Science* 17:70–80.
41. Oliveras, I; Malhi, Y. (2016) Many shades of green: the dynamic tropical forest–savannah transition zones. *Philosophical Transactions of the Royal Society B – Biological Sciences* 371.
42. Walters, G. (2012) Customary fire regimes and vegetation structure in Gabon's Bateke Plateaux. *Human Ecology* 40:943–955.
43. Stevens, N, et al. (2017) Savanna woody encroachment is widespread across three continents. *Global Change Biology* 23:235–244.

44. Sinclair, ARE, et al. (2007) Long-term ecosystem dynamics in the Serengeti: lessons for conservation. *Conservation Biology* 21:580–590.
45. Bond, WJ; Midgley, GF. (2000) A proposed CO_2-controlled mechanism of woody plant invasion in grasslands and savannas. *Global Change Biology* 6:865–869.
46. Buitenwerf, R, et al. (2012) Increased tree densities in South African savannas: >50 years of data suggests CO_2 as a driver. *Global Change Biology* 18:675–684.
47. Smit, IP; Archibald, S. (2019) Herbivore culling influences spatio-temporal patterns of fire in a semiarid savanna. *Journal of Applied Ecology* 56:711–721.
48. Rountree, M, et al. (2000) Landscape state change in the semi-arid Sabie River, Kruger National Park, in response to flood and drought. *South African Geographical Journal* 82:173–181.
49. Ratnam, J, et al. (2016) Savannahs of Asia: antiquity, biogeography, and an uncertain future. *Philosophical Transactions of the Royal Society B – Biological Sciences* 371.
50. Lehmann, CER, et al. (2014) Savanna vegetation–fire–climate relationships differ among continents. *Science* 343:548–552.
51. Eiten, G. (1982) Brazilian 'savannas'. In Huntley, BJ; Walker, BH (eds), *Ecology of Tropical Savannas*. Springer, Berlin, pp. 25–47.
52. Sarmiento, G. (1984) *The Ecology of Neotropical Savannas*. Harvard University Press, Cambridge, MA.
53. Borghetti, F. (2020) South American savannas. In Scogings, PF; Sankaran, M (eds) *Savanna Woody Plants and Large Herbivores*. Wiley, Oxford.
54. Medina, E; Silva, JF. (1990) Savannas of northern South America: a steady state regulated by water–fire interactions on a background of low nutrient availability. *Journal of Biogeography* 17:403–413.
55. Bucher, EH. (1982) Chaco and Caatinga – South American arid savannas, woodlands and thickets. In Huntley, BJ; Walker, BH (eds) *Ecology of Tropical Savannas*. Springer, Berlin, pp. 48–79.
56. Fensham, RJ, et al. (2005) Rainfall, land use and woody vegetation cover change in semi-arid Australian savanna. *Journal of Ecology* 93:596–606.
57. Williams, RJ, et al. (1997) Leaf phenology of woody species in a North Australian tropical savanna. *Ecology* 78:2542–2558.

Appendix 6.1 Some Tree and Grass Species Typical of African Savannas

Plant species matter ecologically, not just the broad types labelled trees and grasses. Let me introduce you to some of their notable characteristics, in text and in pictures.

All of the acacias are thorny as well as possessing finely divided leaves (Figure 6A.1). Umbrella thorn (*Acacia tortilis*) is the most widely distributed, found throughout Africa and broadly tolerant of different soil substrates. Other acacias have distinct soil associations. Knob thorn (*Acacia nigrescens*) is typical of basaltic or doleritic clay soils, but is also present in bottomlands of granitic landscapes where clay accumulates. Giraffe thorn (*Acacia erioloba*) is characteristic of arid savannas on Kalahari sand and flanks dry river beds, accessing soil moisture deep below. Whistling or ant-gall thorn (*Acacia drepanolobium*) occurs in shrubby form on heavy clays, while black monkey thorn (*Acacia burkei*) favours quite sandy soils and common hook-thorn (*Acacia caffra*) grows on rocky hillsides. White thorn (*Acacia polyacantha*) is usually found on alluvial soils near rivers. Paperbark thorn (*Acacia sieberiana*) grows both in alluvial lowlands and in moister savannas elsewhere. Sweet thorn (*Acacia karroo*) is restricted to South Africa, growing either as a spreading tree or as a pole-like or cage-like shrub. Black thorn (*Acacia mellifera*) grows as a multi-stemmed shrub forming thickets in dry regions of southern and eastern Africa. Most acacias shed their leaves quite early in the dry season, but splendid (or stinkbark) thorn (*Acacia robusta*) retains leaves until the end of the dry season, while ana trees (*Faidherbia albida*) carry leaves throughout the dry season and shed them at the start of the wet season. Both of the latter species commonly flank rivers. Acacia species also differ in their adaptations to withstand fires.[1] Thus, each of the acacia species that I know has distinctive features of its ecology – there are no neutrally equivalent niches.

Broad-leaved savanna woodlands feature numerous species placed within the legume subfamily Caesalpinioideae (recently revised to Detariodeae).[2] Three species of *Brachystegia* and one species of *Julbernadia* represent miombo woodland at its southern limit in Zimbabwe. Zebrawood or msasa (*Brachystegia spiciformis*) is ubiquitous on comparatively shallow but well-drained sands from Zimbabwe to the Congo (Figure 6A.2). Prince of Wales feathers or mfuti

Figure 6A.1 Some representative acacias. (A) Umbrella thorn widespread through Africa, Luangwa, Zambia; (B) giraffe thorn, typically found in arid savanna, central Kalahari, Botswana; (C) paperbark thorn, especially resistant to fires, Kidepo, Uganda; (D) splendid thorn, typically in river margins, Luangwa Valley, Zambia; (E) knob thorn, favouring clay soils, Hluhluwe-iMfolozi Park, South Africa; (F) fever trees, typical of swampy soils, Kruger NP, South Africa.

Figure 6A.2 Some miombo woodland and allied trees. (A) Msasa, the most widespread species, Zambia; (B) *Brachystegia* spp., early leaf flush, Zambia; (C) munondo, found on shallower soils, Zimbabwe; (D) Zimbabwe teak, typical of deep Kalahari sands, Zimbabwe; (E) wild seringa, found in drier sandy soils than typical miombo, Magaliesberg in South Africa; (F) tall mopane woodland, predominant in hot dry lowlands of south-central Africa, Luangwa Valley, Zambia.

(*Brachystegia boehmi*) is also widely distributed, favouring shallow infertile soils. Mountain acacia (*Brachystegia tamarindoides*) is associated with rocky hills. Munondo (*Julbernadia globiflora*) is associated with relatively dry conditions at low altitudes, often on rocky slopes with thin soils. Representatives of other genera within this subfamily include wild seringa (*Burkea africana*) and Zimbabwe teak (*Baikiaea plurijuga*), which are locally common on sandy soils in drier regions. Mutondo (*Isoberlinea angolensis*) is characteristic of wetter forms of miombo woodland from Zambia into western Africa. In this heartland, 12 or more species in the miombo group may be intermingled without obvious differences in their ecology. All of these broad-leaved legumes evidently contain chemical deterrents against being munched by browsing ungulates. Miombo woodland trees characteristically produce new leaves several weeks before the early rains, coloured red or yellow by anthocyanin precursors of tannins.

Then there is, once again, the enigmatic mopane (*Colophospermum mopane*; Figure 6A.3). Despite being a broad-leaved member of the Caesalpinioideae, it occurs in hot and dry low-lying regions in an east–west swathe from northern South Africa through Botswana into Namibia and Angola, but no further north than the Luangwa Valley in Zambia, achieving monospecific dominance over much of this range. Mopane can grow either as a woodland of tall trees or a low

Figure 6A.3 Some widely distributed broad-leaved trees. (A) Widely spread marula tree, Kruger NP; (B) widely distributed sausage tree, Luangwa Valley, Zambia; (C) semi-evergreen jackalberry tree on termite mound, Northern Botswana; (D) clusterleaf tree (*Terminalia trichopoda*), Serengeti, Tanzania; (E) enormous baobab succulent, Luangwa Valley, Zambia; (F) huge sycamore fig tree, Ndumo GR, South Africa.

tree shrubland, dependent on soil depth governed by calcrete hardpans. The grass cover within mopane woodlands is commonly sparse with annuals predominating, reducing the effects of fires. Mopane woodlands in south-central Africa currently support most of Africa's elephants.

Bushwillows in the Combretaceae are commonly found in the understorey of miombo woodlands as well as predominating on sandy soils in regions too dry to support the broad-leaved legumes. Typical genera are *Combretum* (bushwillows) and *Terminalia* (clusterleafs), some growing as low trees and others as quite tall trees. A giant among them is leadwood (*Combretum imberbe*), growing in locations with deep soil water. Its wood is especially heavy and thus resistant to decay, and specimens can live for over a thousand years.

Other broad-leaved trees are not restricted to either savanna form. Marula (*Sclerocarya birrea*) trees (Figure 6A.3) are widely distributed in both dry/eutrophic and moist/dystrophic savanna and dominate the tree biomass in some regions. Baobab trees (*Adansonia digitata*) have enormous succulent trunks and are typical of drier regions. Species commonly represented in riverine woodlands, but not restricted to them, include jackalberry (*Diospiros mespiliformis*) and sausage tree (*Kigelia africana*). Species of wild myrrh (*Commiphora* spp.) are a common constituent of dry savannas in eastern Africa, while the evergreen shepherd's tree (*Boscia albitrunca*) is revered for shade as well as forage in arid savanna regions.

African savannas support an impressive diversity of grass species forming fine-scale spatial mosaics.[3,4] Tribal divisions are associated with ecological distinctions. Grasses in the tribe Andropogoneae tend to grow tall and stemmy and thereby promote hot fires (Figure 6A.4).[5,6] Typical representatives are thatch grasses falling within the genera *Hyparrhenia* and *Hyperthelia*. They prevail under relatively high rainfall conditions and resist being heavily grazed on account of their high fibre contents. Red grass, sole representative of the genus *Themeda*, is widely dominant on clay-rich soils. It retains adequate nutritional value for livestock into the dry season, at least in drier areas, making it characteristic of sweetveld. It is promoted by frequent fires. The tribe Paniceae includes grasses highly palatable to grazers, but less supportive of fires, notably species in the genera *Panicum* (Guinea grass and others), *Digitaria* (finger grasses), *Urochloa* (bushveld signal grasses) and *Setaria* (bristle grasses). Guinea grass (*Panicum maximum*) is commonly prominent under tree canopies where soils are enriched in nitrogen from leaf litter deposited by the trees. The tribe Chlorideae includes lawn-forming grasses in the genera *Cynodon* (couch grasses) and *Sporobolus* (dropseed grasses), as well as numerous species of *Eragrostis* (love grasses). Pan dropseed (*Sporobolus ioclados*) attracts grazing by accumulating sodium in its leaves. Couch grass (*Cynodon dactylon*) is widely cultivated as lawn cover as well as for animal fodder, but can become toxic

Figure 6A.4 Grassland types. (A) Red grass grassland, Serengeti NP, Tanzania; (B) red grass grassland, Kruger NP, South Africa; (C) thatch grass (*Hyparrhenia rufa*) grassland, Kidepo, Uganda; (D) tall thatch grass, Lope NP, Gabon; (E) Guinea grass, Kruger NP, South Africa; (F) finger grass grassland, Botswana; (G) pan dropseed grass, Kalahari, Botswana; (H) mixed short grass lawn with flowering *Ammocharis* lily, Mfolozi GR; (I) finger grass (*Digitaria macroblephara*) showing connecting stolons (runners), Serengeti NP, Tanzania.

through a high cyanide content. Three-awn or needle grasses in the tribe Aristideae, including the genera *Aristida* and *Stipagrostis*, are most commonly found in arid savannas or even semi-deserts. They have narrow or rolled leaves. Then there are the wetland grasses tolerant of seasonal inundation, including the genera *Leersia* and *Oryza* within the tribe Oryzeae (wild rices), associated with tall sedges (*Cyperus papyrus*), rushes (*Typha* spp.) and reeds (*Phragmites* spp.). Genera with aromatic oils inhibiting their consumption by grazers include stinking grasses (*Bothriochloa* spp.) and lemon grasses (*Cymbopogon* spp.). All grasses contain silica bodies in their leaves, believed to inhibit grazing by increasing tooth wear of ungulate herbivores. Various grass genera and even species are shared between eastern and southern Africa[7,8] and some African grasses have become distributed worldwide.[9]

Figure 6A.5 Underground trees or geoxyles. (A) *Pygmaeothamnus zeyheri*, Magaliesberg; (B) *Lannea edulis*, Magaliesberg; (C) *Parinari capensis*, Magaliesberg; (D) highly toxic gifblaar *Dichapetalum cymosum*, Nylsvley, South Africa; (E,F) unidentified geoxyles, Gabon.

Other plant forms found in savannas show various adaptations to cope with recurrent fires as well as herbivory. Shrubs, i.e. multi-stemmed woody plants not growing much taller than 3 m, retain sufficient underground reserves in roots and tubers to regenerate stems and leaves lost during fires.[10] Geoxyles growing with their woody stems mostly underground and merely their branch tips bearing leaves protruding are a feature of sandy soils (Figure 6A.5).[11] They evade the recurrent loss of woody growth to fires, but need to have their exposed leaves defended against large herbivores by deterrent chemicals. Most genera with geoxyles also have species that grow into tall trees, including *Brachystegia, Erythrina, Lannea, Parinari, Dichapetalum* and *Combretum*.

Flowering herbs apart from grasses and sedges commonly get lumped as forbs. These include dwarf shrubs remaining under 0.5 m in height along with herbaceous shoots growing from woody roots, tubers or bulbs below ground. Creepers and woody vines ascend from the herbaceous layer using the support provided by trees and grasses. Forbs and shrublets in the pea family (e.g. *Indigofera* spp.) make an important contribution to soil fertility through nitrogen fixation in their root nodules, although this capacity seems to diminish as plants mature.[12] Forbs of various forms typically contribute most of the taxonomic diversity in the herbaceous layer of savannas. Although forming only a minor component of the vegetation, plants with underground storage organs contributed importantly to the survival of hominins through the dry season.

Then there are plants with succulent stems or leaves, commonly associated with but not exclusive to arid environments, such as aloes and certain euphorbias. Some of these succulents attain tree stature, notably the candelabra euphorbia (*Euphorbia ingens*), which is distributed widely from Uganda into southern Africa. Baobab trees are effectively enormous succulents with their swollen stems storing moisture in the arid regions they occupy.

References

1. Staver, AC, et al. (2012) Top-down determinants of niche structure and adaptation among African acacias. *Ecology Letters* 15:673–679.
2. Smith, P; Allen, Q. (2004) *Field Guide to the Trees and Shrubs of the Miombo Woodlands.* Royal Botanic Gardens Kew, Richmond, pp. 132–133. Includes a picture.
3. McNaughton, SJ. (1983) Serengeti grassland ecology – the role of composite environmental factors and contingency in community organization. *Ecological Monographs* 53:291–320.
4. Augustine, DJ. (2003) Spatial heterogeneity in the herbaceous layer of a semi-arid savanna ecosystem. *Plant Ecology* 167:319–332.
5. Hempson, GP, et al. (2019) Alternate grassy ecosystem states are determined by palatability–flammability trade-offs. *Trends in Ecology & Evolution* 34:286–290.
6. Archibald, S, et al. (2019) A unified framework for plant life-history strategies shaped by fire and herbivory. *New Phytologist* 224:1490–1503.
7. Fynn, R, et al. (2011) Trait–environment relations for dominant grasses in South African mesic grassland support a general leaf economic model. *Journal of Vegetation Science* 22:528–540.
8. Cromsigt, J, et al. (2017) The functional ecology of grazing lawns – how grazers, termites, people, and fire shape HiP's savanna grassland mosaic. In Cromsigt JPG, et al. (eds) *Conserving Africa's Mega-diversity in the Anthropocene: the Hluhluwe-iMfolozi Park Story*. Cambridge University Press, Cambridge, pp. 135–160.
9. Linder, HP, et al. (2018) Global grass (Poaceae) success underpinned by traits facilitating colonization, persistence and habitat transformation. *Biological Reviews* 93:1125–1144.
10. Wigley, BJ, et al. (2009) Sapling survival in a frequently burnt savanna: mobilisation of carbon reserves in *Acacia karroo. Plant Ecology* 203:1.
11. Maurin, O, et al. (2014) Savanna fire and the origins of the 'underground forests' of Africa. *New Phytologist* 204:201–214.
12. February, EC, et al. (2019) Physiological traits of savanna woody species: adaptations to resource availability. In Scogings, PF; Sankaran, M (eds) *Savanna Woody Plants and Large Herbivores*. John Wiley & Sons, Oxford, pp. 309–329.

Chapter 7: How Trees and Grasses Grow and Compete

Trees potentially out-compete grasses by growing taller and thereby shading grasses growing beneath their canopies, but soil moisture is the limiting resource in savannas, not sunlight. Water enables both the capture of carbon dioxide from the atmosphere and the uptake of mineral nutrients from the soil. Competition for access to water and nutrients operates mostly underground, out of sight and is thus easily overlooked. Once established, trees can build up supporting stems to overtop grasses and raise their leaf canopies beyond the flame zone. However, grasses can activate faster than trees in accessing soil moisture when conditions permit. Competition is effective mainly during the establishment phase for tree seedlings, which must somehow claim rooting space amid the grass roots.

Plants derive carbon, the structural constituent of their biomass, from atmospheric carbon dioxide. Mineral nutrient resources are obtained primarily from the products of decaying organic matter in the soil. Nitrogen, taken up in the form of nitrate or ammonia, is an essential constituent of proteins. Potassium is needed for ionic regulation. Magnesium is a constituent of chlorophyll. Calcium is a component of cell walls. Phosphorus, in the form of phosphates, fuels cell metabolism. Sulphate is needed for sulphur-containing amino acids. Legumes in the subfamilies Mimosoideae and Papillionoideae overcome nitrogen deficiencies by gaining atmospheric nitrogen fixed by bacteria housed in root nodules. Phosphorus uptake is facilitated by associations between roots and mycorrhizal fungi in the soil.[1] These processes are energised by solar radiation captured by chlorophyll molecules located in chloroplasts in leaves.

Organic matter decomposition is brought about by soil organisms. Termites and other invertebrates play the leading role in breaking down plant litter. Bacteria and fungi complete the mineralisation of the organic matter. Their activity is dependent on sufficient moisture and ceases at some stage during the dry season. This results in a pulse of nutrient release following the first adequate rains initiating the wet season.[2,3] Most of the available nutrients are generated in the top 20 cm of the soil.[4] Water availability determines when growth can occur, while soil nutrient availability can restrict the rate of growth achieved while moisture is sufficient. The low concentration of carbon dioxide in the atmosphere constrains the growth rate achieved when soil nutrient supplies are adequate.[5]

In this chapter, I will outline how trees and grasses grow and compete in the context of the seasonal restriction in rainfall and its erratic distribution even within the wet season. Large herbivores set back plant growth through their consumption of leaves, stems and sometimes roots, but also benefit plants by dispersing seeds contained in fruits and grass inflorescences (flowering heads).

Growth of Woody Plants

Woody plants increase in standing biomass by annual increments to stems and branches bearing leaves. Savanna trees typically produce their new leaves shortly before the start of the rains by drawing on moisture stored in their stems or accessed from deeper soil levels.[6,7] Miombo woodland trees flush their new leaves especially early, 4–8 weeks before the first rain showers initiate the wet season.[8] These leaves are tinted with red or orange colours indicating high anthocyanin contents, which may act as a sunscreen (Figure 7.1). The leaf display is followed by shoot extension, then stem expansion and lastly root growth, during the course of the wet season.[9] The initial growth of new leaves at the start of the wet season is generated from stored carbohydrate and mineral reserves. Heights reached by savanna trees rarely exceed 20 m, and most remain under 10 m. Shrubs typically do not exceed 3–5 m in height and are characteristically multi-stemmed from their base. Savanna trees typically exhibit spreading canopies, in contrast to the upward growth towards light shown by forest trees.

Most of the woody plants in African savannas are deciduous, shedding their leaves at some stage during the dry season.[10,11] This forestalls the moisture loss that they would have incurred had leaves been retained.[12] The thorny acacias

Figure 7.1 Pre-rain flush of red leaves on (A) *Brachystegia* in Zambia and (B) *Julbernadia* in Zimbabwe.

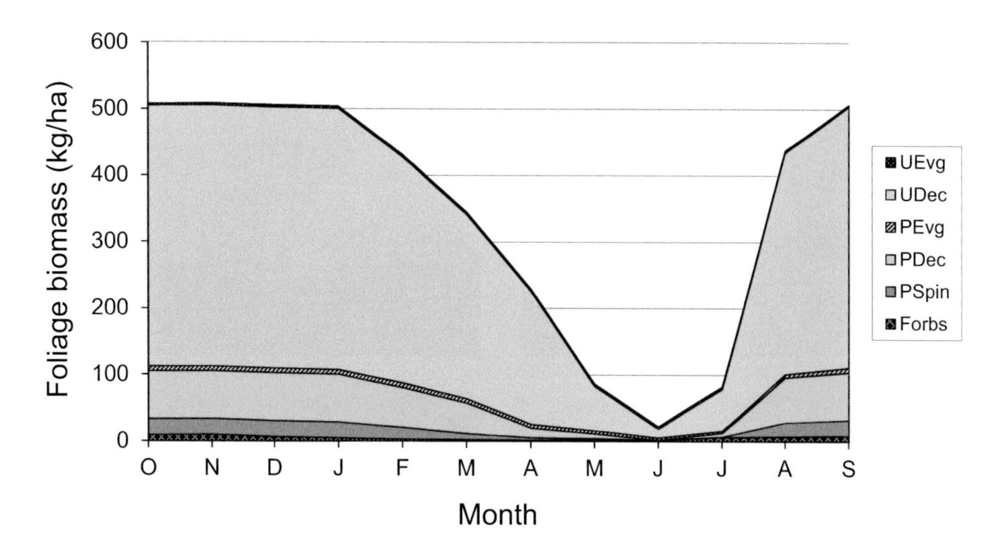

Figure 7.2 Seasonal variation in tree foliage retained within browsing height in mixed broad-leaved woodland in the Nylsvley Nature Reserve, subdivided by plant types. PSpin, spinescent acacias; PDec, deciduous bushwillows and wild raisins; PEvg, evergreen monkey oranges and karee; UDec, wild seringa and other deciduous species with foliage defended by tannins; UEvg, evergreen shrubs with high tannin contents in their leaves. Note that most acacias lose their leaves quite early in the dry season, while the broad-leaved legumes (unpalatable deciduous) produce new leaves earliest preceding the start of the wet season.

typical of dry/eutrophic savannas generally drop their leaves quite early in the dry season, while the broad-leaved trees found in moist/dystrophic woodlands on sandy soils tend to retain their leaves somewhat later into the dry season. Nutrients are mostly withdrawn before the leaves are shed. Evergreen or semi-evergreen trees and shrubs that retain their leaves through the dry season tend to be found in bottomlands where soil moisture persists longer and less fire penetrates as a result of the trampled grass cover. Termite mounds also provide refuge sites from fires. Evergreen leaves are tougher than leaves of deciduous species in order to resist wilting when water is deficient. Only a tiny remnant of the foliage produced by savanna trees remains available to herbivores through the dry season (Figure 7.2). By the end of the dry season, savannas become bare of tree leaves and bereft of shade (Figure 7.3). Conditions may become baking hot and dry before the first rains.

Because trees commence their seasonal growth of leaves and shoots ahead of the rains, their annual production of these tissues is not determined by the current season's rainfall.[13] However, leaves may be shed earlier in drier years, and there may be carryover effects on the commencement of growth in the following wet season, mediated by soil water reserves but buffered by moisture stored within stems.

Figure 7.3 Dry season aspects. (A) Bushwillow trees bare of leaves, Kruger NP; (B) mopane trees bare of leaves, Luangwa Valley, Zambia; (C) bared soil in central Kruger by end of dry season, September 2016; (D) eastern short grass plains in the dry season, Serengeti, Tanzania; (E) thatch grass grassland grazed down to stubble in Maasai Mara NR, February 2017; (F) bared soil and leafless trees, Chobe riverfront, northern Botswana; (G) bare ground and trees during the 1983 drought, Kruger NP; (H) leafless mopane and bare soil during drought, near Kruger NP.

Rooting Patterns

Rooting patterns are a fundamental indication of niche distinctions among tree species related to resource capture, but are exposed only after considerable effort in digging.[14,15,16] Some trees extend shallow roots outwards well beyond their leaf canopies, competing with grass roots for both moisture and mineral nutrients. Others extend taproots downwards several metres to access water that persists longer at greater depths, particularly in sandy soils. Most acacias have well-developed lateral root systems, coupled with taproots in some cases. Umbrella thorn has contorted surface roots enabling it to draw water from a wide area. Leadwood trees have exceptionally large taproots and relatively small lateral roots. Red bushwillow (*Combretum apiculatum*), a small tree widely prevalent on shallow sandy or rocky soils in southern Africa, has mainly an extensive lateral root system. Russet bushwillow (*C. hereroense*), commonly found low down on catenas where soils are more clayey, has several descending roots as well as abundant lateral roots. Silver clusterleaf (*Terminalia sericea*), common in sandy soils both on crests and further down bordering seep zones, has roots extending outwards from the central rootstock. Marula trees are comparatively deep-rooted. Shepherd's tree, evergreen despite being prevalent in dry environments, has a large taproot but lacks lateral roots. It exhibits the deepest root depth recorded for any tree: 68 m.[17] Second in line for rooting depth is giraffe thorn, with a taproot penetrating down to 60 m. Trees associated with deep sandy soils, such as Zimbabwe teak and msasa, also exhibit a comparatively deep root distribution.[18] A teak seedling can produce a taproot extending 1.5 m deep within one year while the above-ground shoot at that stage remains under 15 cm tall.[19] In contrast, wild seringa has predominantly shallow roots. Shrubs growing on sandy sediments such as sand camwood (*Baphia massaiensis*) and Zambezi jessebush (*Combretum celastroides*) are also relatively shallow-rooted, with 95 percent of their roots less than a metre deep. Nevertheless, they have proportionately more biomass below ground than most trees, and some have swollen lignotubers facilitating stem regeneration after fires. Trees with shallow roots are subject to greater mortality during severe droughts.[16] On firmer, more clay-rich soils, roots of both trees and shrubs are restricted to shallower rooting depths.[20,21] Mopane trees have mostly shallow roots because they grow in arid regions where little moisture penetrates deeper.[22] Trees with deep taproots still depend mostly on water in the topsoil layer for nutrient capture and thus do not evade competition with grasses during the wet season.[23]

Fruits

Savanna trees generally start producing fruits when around 20 percent of their potential height and reach maximum production once they attain 67 percent or more of maximum size.[24] The predominance of leguminous trees means that fruits mainly take the form of dry pods in both fine-leaved and broad-leaved savannas. Nevertheless, certain acacias drop their ripened pods early in the dry season and these are consumed by various ungulates, including elephants,[25] dispersing the seeds contained far and wide. Seeds that have not passed through the gut of an ungulate germinate less well. Marula produce abundant crops of plum-like drupes dropped on the ground and so available for consumption late in the wet season (Figure 7.4A). Antelope spit out their hard seeds while ruminating in the shade some distance away. Elephants travel long distances to feed on marula fruits and play a major role in seed dispersal. Various species of monkey orange (*Strychnos* spp.) produce large hard-shelled fruits with fleshy interiors late in the dry season when few other fruits are available (Figure 7.4E). These are sought out by primates (including humans) as well as ungulates. However, elephants seem not to eat them, perhaps because their powerful jaws would crush the seeds, releasing toxic alkaloids. Baobab trees (Figure 7.4F) produce huge pods containing seeds buried in a dry pulp, which are shed during the dry season. Humans are again among the primates consuming them. Small berries are produced by jackalberry, bird plum (*Berchemia* spp.), raisin bush (*Grewia* spp.), gwarrie (*Euclea* spp.) and snow-berry (*Flueggia virosa*) shrubs and consumed by birds, primates and, along with leaves, by ungulates.

The pods produced by members of the Caesalpinoideae break open to scatter their seeds, which are dispersed only a few metres beyond the parent trees. Nevertheless, miombo woodlands also contain trees producing a rich variety of succulent fruits. These include drupes of mobola plum (*Parinari curatellifolia*), mahobohobo or sugar plum (*Uapaca* spp.) and large waterberry (*Syzigium guineaense*), which all typically ripen late in the dry season (Figure 7.4B,C).

Species in the Combretaceae produce dry winged pods that are not attractive to ungulates, although they are occasionally eaten. They are evidently adapted for wind dispersal over short distances. Other large fruits remain puzzling with regard to the role of herbivores in seed dispersal. Sausage trees produce huge woody pods sometimes eaten by hippos, giraffes and black rhinos, but not by much else. Wild gardenias (*Gardenia volkensii*) bear hard round fruits the size of a cricket ball supposedly eaten occasionally by baboons and monkeys as well as antelope.

Resistance to Fire

Juvenile woody plants are vulnerable to having all of their top growth burnt by recurrent fires while they remain below 2–3 m in height.[26] Because most savanna species can resprout from buds in the root bole or stem, they are

Figure 7.4 Savanna fruits eaten by humans as well as baboons and antelope. (A) Drupes from widespread marula tree, South Africa, dropped when ripe; (B) mobola plums, Zambia, featuring in miombo woodlands; (C) mahobohobo fruits, Zambia, also featuring in miombo woodlands; (D) sour plums (*Ximenia caffra*), South Africa; (E) monkey orange fruits with a hard shell and soft pulp inside, South Africa, ripening during the dry season; (F) baobab pods, gathered by people as well as eaten by baboons (photo: Sarah Venter).

rarely killed by fires once they have built up sufficient root reserves to restore the incinerated parts. Saplings of savanna trees progressively develop a thickened bark providing protection for their woody stems, and resistance to fire depends also on stem thickness relative to height and bark moisture content.[27] Vulnerability to top-kill decreases sharply with increasing height in the range 1–4 m. Shrubs gain some protection against losing all of their above-ground biomass by having multiple stems. Woody seedlings are vulnerable to being killed before they have established sufficient underground reserves. Beyond this stage, fires only temporarily set back the growth of trees and shrubs unless burns are both frequent and especially hot. This occurs mainly where there is little grazing pressure on the grass layer. Nevertheless, even quite large trees can be killed if bark damage allows fire to burn into their trunks. Trees typical of miombo woodlands regenerate via profuse stem coppicing after being burnt back.

Trees with evergreen or semi-evergreen foliage suffer a greater setback than deciduous species when their potentially long-lasting leaves get burnt. These species occupy sites where they are less exposed to fire because of the lack of much grass cover, flanking rivers and on termite mounds. The shade cast by the spreading canopies of savanna trees favours grass types less likely to burn, conferring some protection against fires once established. Woody plant species typical of forests or thickets are much more vulnerable to mortality from fire than fire-adapted savanna trees and shrubs.[28]

Plant Anti-herbivore Defences

Plants can have much of their energy-gathering potential removed through consumption of their leaves by herbivores. They can inhibit consumption by incorporating secondary chemicals not involved in leaf metabolism in their foliage. For trees, these anti-herbivore defences commonly take the form of carbon-based secondary metabolites like tannins or other phenolics, although terpenoids, alkaloids and toxic amino acids may also come into play.[29] Trees growing on nutrient-deficient soils potentially have a surplus supply of carbon, which continues to be captured while stomata remain open to draw up scarce mineral nutrients.[30] Little extra cost may be entailed in elaborating the surplus carbon into chemical deterrents. Biochemically, tannins form complexes with proteins, thereby blocking the enzymatic action of these proteins. The prime adaptive value of plant tannins may be to protect leaves from fungal or bacterial pathogens. They also restrict rates of decomposition of fallen leaves, thereby helping retain nutrients against leaching. However, tannins also interfere with the digestion of leaves by large herbivores, both by complexing with plant proteins and by interfering with the activity of cellulolytic enzymes secreted by gut microbes. Herbaceous plants growing under shade canopies of trees or grasses where carbon capture is restricted tend to be defended against herbivory by nitrogen-based compounds like alkaloids or non-protein amino acids.

The anthocyanins responsible for the red spring colours of the leaves of miombo woodland trees may be precursors to the formation of tannins in these leaves. Woody plants associated with dystrophic savannas generally have higher tannin or phenolic contents in their leaves than those growing in more fertile soils.[29,31] Plants growing in fertile soils have less surplus carbon relative to nutrients so that producing carbon-based chemical deterrents would be at the cost of growth. Thus, few of the acacias prevalent in dry/eutrophic savannas contain much in the way of tannins. They rely mainly on structural defences to restrict leaf losses (Figure 7.5).[32] However, structural defences in the form of thorns, spines or prickles on stems are not effective against insect herbivores.

Smaller antelopes are less restricted by thorns than larger ones because they can nibble leaves between the thorns. Giraffes overcome the effects of spines by stripping leaves from branch tips of acacias with their lower incisor teeth. Juvenile trees may deter browsing by developing a cage structure coupling prominent spines with short internodes, protecting inner leaves (Figure 7.5H).[33,34] Spinescent trees and shrubs experience less leaf loss than plants lacking such defences.[35] Many acacia species produce longer, thicker or denser spines when exposed to browsing than when protected from browsers.[36] Shrubs with evergreen leaves attracting browsing during the dry season produce especially formidable spines (Figure 7.5E).

The chemicals that serve as deterrents against large mammals may not be effective against insect herbivores, and vice versa. At Nylsvley, some of the tree species with high tannin contents in their leaves were subject to defoliating outbreaks of caterpillars. Those containing aromatic oils showed little leaf damage from insects, but were readily consumed by ungulates.[29] None of the acacias incurred much leaf loss to insect herbivores, despite being most nutritious in terms of leaf protein content.

The chemical and structural defences presented by trees and shrubs generally do not prevent browsing. Nevertheless, they adequately restrict rates of leaf loss and impose metabolic costs on herbivores that influence the diet choice of the latter.[37] At Nylsvley, all of the 60 woody species present in the study enclosure were eaten by kudus at some stage. Even those containing high tannin levels were nibbled when little else remained to eat. However, certain forbs and geoxyles were never eaten, suggesting that they contain more potent toxins, like the monofluoracetatic acid present in gifblaar ('poison-leaf' *Dichapetalum cymosum*; Figure 6A.5D), which acts as a heart poison by disrupting energy metabolism. Cattle and goats die after eating just a few leaves, but wild ungulates seem somehow to know to avoid it, even hand-reared impalas.

Figure 7.5 Plant structural defences against browsers. (A) Conspicuously white straight spines of sweet thorn; (B) prominent straight spines of ant-gall thorn (*A. drepanolobium*); (C) mixed straight and

Grass Growth

Grasses retain living biomass underground in the form of root tissues and buds. Thus, while their above-ground stems and leaves are removed by dry season fires, these parts grow back during the start of the following wet season. Hence their growth patterns are adapted to cope with seasonality in rainfall and recurrent fires. The primary buds (meristems) that generate above-ground growth are low down in the crown, protected from fire and herbivory (Figure 7.6). Single or multiple stems or tillers extend upwards from the crown with attached leaves. Tufted or bunch grasses are formed by clusters of tillers, each functioning largely independently. Eventually stems rise beyond the last leaf to produce inflorescences in the form of clusters of tiny flowers and later seeds. Some grasses extend their stems laterally along the ground in the form of stolons, with roots developing where nodes make contact with the soil. Roots underground typically constitute over 60 percent of the peak biomass of grasses, with the finest roots concentrated in the top 10 cm of the soil. During the course of the year, most of the grass leaves above-ground not consumed by herbivores get transformed into standing dead tissues or 'necromass' (Figure 7.7).

Grasses generate above-ground biomass during the course of the wet season either by extending tillers or by producing new tillers. Individual leaves rarely last longer than a few weeks, while tillers typically persist for 1–2 years. This means that there is ongoing turnover of the above-ground parts, so that the amount of biomass accumulated by the end of the wet season is less than the total biomass actually produced. Stems remaining from the preceding growing season hamper the growth of new tillers by shading them. Accordingly, bunch grasses depend on regular fires or grazing to enable their annual growth. More than 80 percent of the root biomass of grasses can turn over annually in tropical grasslands as some fine roots extend their growth towards sources of mineral nutrients, while others die back.[38]

Annual growth of the above-ground parts of grasses is directly controlled by rainfall in savannas. Grasses grow while the soil remains sufficiently moist so that growth may be pulsed in response to rainfall events during the course of the wet season. Grasses growing on sandy soils derived from granite retain green leaves longer into the dry season than grasses growing on clayey basaltic soils.[39] There is an almost linear relationship between the annual accumulation of above-ground biomass by grasses and the wet season rainfall total below a MAR of around

Figure 7.5 (*cont.*) hooked spines of umbrella thorn; (D) hooked entangling spines formed by knob thorn; (E) formidable branched thorns of num-num (*Carissa bispinosa*), an evergreen shrub; (F) sharp branchlet tips of sickle-bush (*Dichrostachys cinerea*); (G) spiny knobs formed on stem of young knob thorn; (H) prickly cage formed by sweet thorn spines.

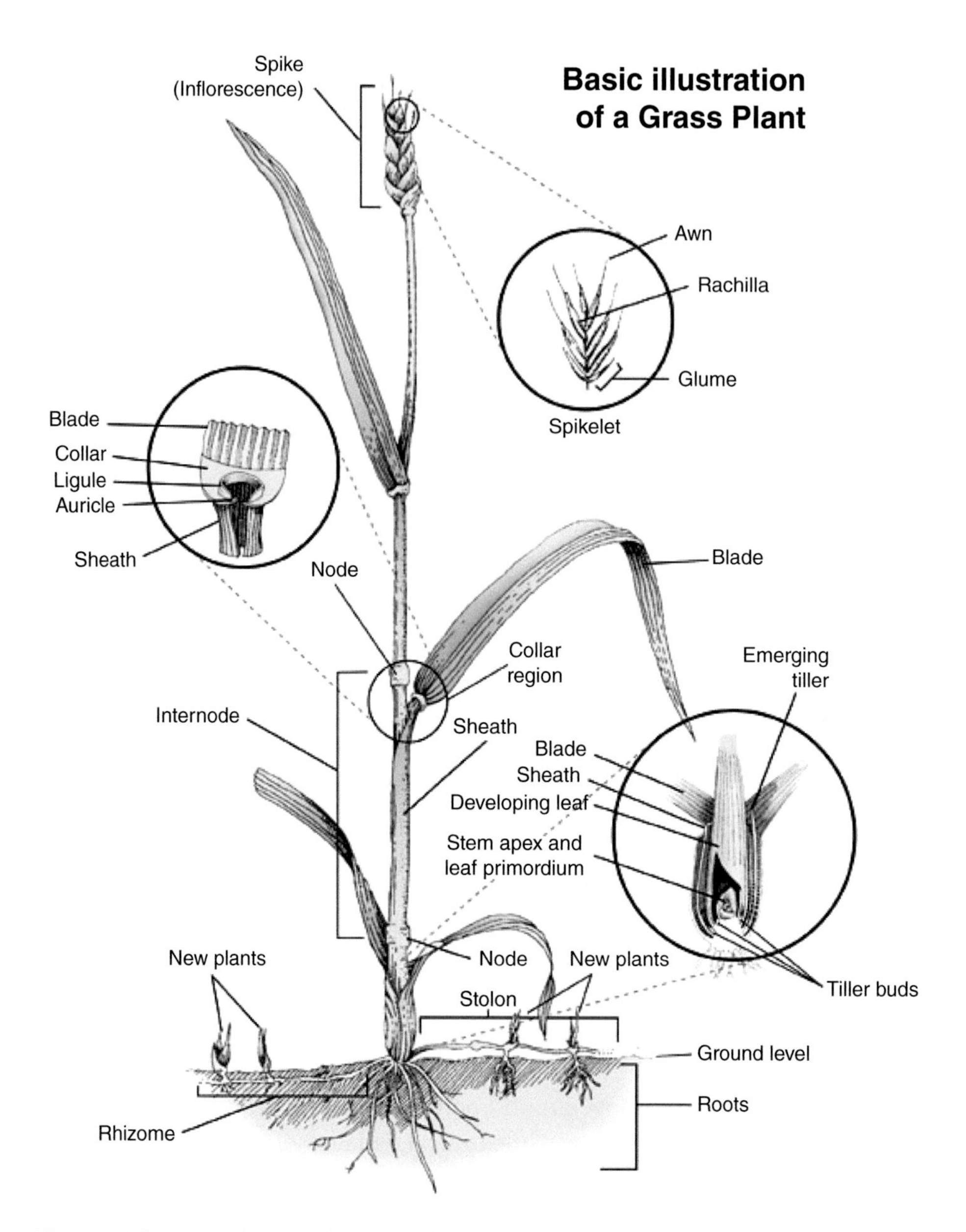

Figure 7.6 Structure of a grass plant. Reproduced with permission from the University of Missouri.

1000 mm.[40] Carbohydrates and other nutrients stored in the crown and roots enable the above-ground tissues to be regenerated at the start of each wet season.[41]

C_4 Versus C_3 Photosynthesis

Photosynthesis takes place in the chloroplasts present within leaves, which absorb infrared radiation and reflect green wavelengths to energise

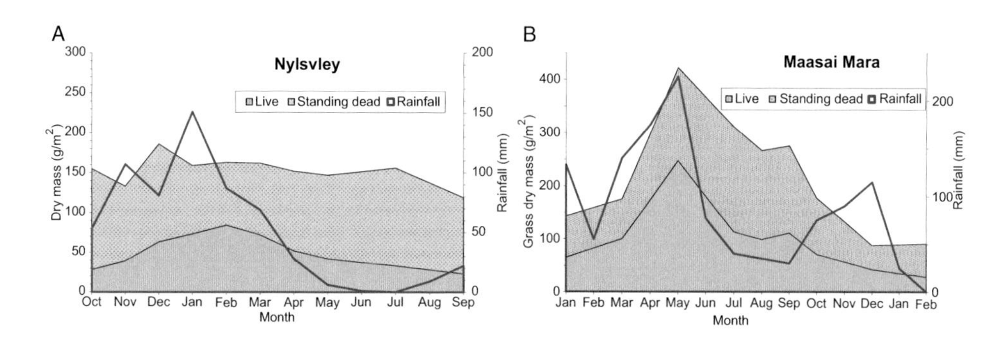

Figure 7.7 (A) Seasonal dynamics of grass parts above-ground partitioned between green and brown (derived from Grunow et al. (1980) *Journal of Ecology* 68:877–889); (B) grassland on fertile clay soils in Maasai Mara Reserve, Kenya, subject to grazing by abundant wild herbivores (derived from Boutton et al. (1988a) *African Journal of Ecology* 26:89–101).

carbon fixation. Tropical grasses have evolved a special mechanism to concentrate carbon dioxide at the sites where rubisco, the enzyme catalysing photosynthesis, is lodged. Carbon capture is mediated via a four-carbon molecule rather than simply by the three-carbon molecule used by temperate grasses and most other plants.[42] The C_4 mechanism effectively acts as a pump drawing in carbon dioxide to the sites in bundle sheath cells where photosynthesis takes place. This enhances the efficiency of carbon gained relative to the water and nutrient resources needed, compared with grasses employing the C_3 pathway, despite the extra energy required. It also helps overcome the limitation on growth rates posed by low atmospheric concentrations of carbon dioxide. Accordingly, tropical C_4 grasses are able to grow rapidly in hot and dry climates during periods while soil moisture remains adequately available. While C_4 grasses can grow faster than C_3 trees and shrubs while water remains available, they close their stomata sooner than woody plants when soils become dry. The evolution of this biochemical mechanism among tropical grasses was a major factor in the expansion of savanna grasslands during the Miocene, as will be explained in Chapter 9.

Some succulents have evolved the C_4 photosynthesis pathway, with a difference. They open their stomata at night, incorporate CO_2 in a four-carbon molecule, then continue with photosynthesis during daylight using the captured carbon.[43] This is called the CAM (crassulacean acid metabolism) pathway, because it was originally identified in the family Crassulaceae. Rather than increasing growth rate, this biochemical adaptation, along with moisture retained in leaves or stems, enables succulents to continue growing during prolonged periods without rainfall.

Grass Adaptations to Grazing

Grasses may either deter herbivory by their high fibre contents, governed largely by stem:leaf ratios, or restrict tissue losses by keeping their leaves close to the ground. If ungrazed, or unburnt, erect bunch grasses shade out shorter grasses. Locally concentrated grazing can enable the spread of low-growing grasses to form grazing lawns (Figure 6A.4F–I).[44] Some grass species have the capacity to grow either erect or decumbent.

Mega-grazers, in the form of white rhinos and hippos, are the archetypal lawn-mowers, using their broad lips to crop grasses evenly.[45] Wildebeest equipped with broader muzzles than other ruminants can also maintain grazing lawns.[46] Smaller grazers like impalas and gazelles selectively nibble individual leaves, thereby exerting a more detrimental impact on grass plants than larger herbivores. However, they can contribute to maintaining grazing lawns once these have been formed.[47] Lawn-forming grasses tend to predominate on uplands where soils are shallower and around termitaria.[48] Grazing concentrated on more palatable grass species can lead to the replacement of these grasses by less nutritious types of grasses, which are called 'increasers' in the range management literature.

The extent of the grazing lawns cultivated by white rhinos facilitated my wanderings through the acacia parkland of the Hluhluwe-iMfolozi Park. Traversing the taller grasslands in the wetter north of the park, sopping wet with dew on summer mornings, loaded with ticks and potentially concealing dangerous animals, was much more inhibiting. Similar experiences must have been shared by ancestral humans, considered in Part IV of this book.

How Grasses Overrule Trees

Grass roots are vastly more abundant than tree roots in the upper 20 cm of savanna soils, except under tree canopies.[49] Rainfall seldom penetrates much deeper than this, especially in clay-rich soils,[50] and nutrient release occurs mostly in the upper topsoil. Numerous experiments show how woody seedlings grow more slowly and survive less well in the presence of grasses than they do following grass removal. [28,51,52,53] This shows how grasses outcompete trees for access to water in the topsoil zone. Tree seedlings need to gain access to sufficient soil moisture both during the season of their germination and through the next growing season in order to become established.[51] To do so, they must occupy a gap within the dense mat of shallow roots established by grasses. Gaps can arise where grass tufts have died during droughts or following severe fires, or where some animal has dug up the soil to get at a bulb. Savanna grasslands have more gaps than montane or Highveld grasslands, as a consequence of their more erratic rainfall.[54] Within Highveld grasslands, deeper penetration by tree roots can be blocked by the hardpan layer typically present beneath the topsoil.[55,56] Tree roots are relatively more abundant than grass roots only beyond a depth of around 80 cm, i.e. in fairly deep soils. Savanna tree seedlings must succeed in extending their taproots to deeper levels in the soil, where moisture reserves remain, before growing much in height, in order to evade intense competition from grass roots.

Juvenile trees can potentially access soil water not taken up by grasses by exploiting windows in time. By putting out their new leaves before the first rains, woody plants can draw on soil moisture provided by the early rains, producing a pulse of nutrient release, before the grasses commence growth.[57,58] Juvenile trees can also retain their leaves longer into the dry season, thereby extending their growth after grasses have mostly become dormant.[59,60] By coupling spatial partitioning in rooting depth with temporal separation in growth, woody plants may bypass the superior rates of extraction of soil water by C_4 grasses, especially in drier savannas, when circumstances allow.

However, the difference in rooting depth between trees and grasses diminishes with increasing rainfall.[61] Grasses generally have shallower roots than trees, but trees become less deep-rooted with increasing rainfall.[62] In wetter savannas, water saturation in the topsoil can be more than sufficient for the needs of both trees and grasses through much of the wet season. In drier savannas, established trees can gain water from deeper soil levels by hydraulic lift, although this mechanism also benefits grasses.[63,64] Earlier leaf flush may benefit trees primarily via access to the pulse of mineral nutrients released by the first soil wetting, especially in infertile savannas where nutrients are especially precious.

The grass cover restricts the establishment of juvenile trees further by promoting repeated fires, which are more intense in moist savannas because of the greater grass biomass produced, and thus more likely to incinerate small trees. A multi-year interval in time between fires is needed for juvenile trees to escape the flame zone.

Tree establishment can thus depend on chance opportunities in space for water abstraction coupled with windows in time, governed by the seasonal hydrology of soil moisture. Water remains available for longer at greater depths, especially in places where soils are sandy and in lowlands where soils are deeper, unless water logging occurs. The tree cover thus depends on landscape position and on soil texture and depth in addition to the local rainfall.

Despite variation in the annual growth produced by trees and grasses, one fundamental reality remains. Within the savanna biome, tree canopies generally cover less than 40 percent of the land surface, so that the grasses growing between and beneath them capture more solar radiation and hence potentially produce more plant biomass than all of the trees. Trees get prime attention because of the capital accumulation presented by their prominent stems and branches, but these are constituted mostly by accumulated deadwood rather than metabolically active biomass. The grasses operate more nimbly, particularly underground, in response to erratic rainfall and pulsed releases of mineral nutrients. They are better adapted to resist the destructive effects of fires as well as defoliation by large herbivores.

Overview

Grasses restrict the presence of trees in savannas through being superior competitors for soil moisture, thereby retarding the growth of juvenile trees and holding the latter in the flame zone for longer. Grasses with the C_4 photosynthetic pathway can grow rapidly in hot and periodically dry environments while water remains available, overcoming limitations on growth posed by low atmospheric levels of carbon dioxide. Trees may coexist by tapping into water at deeper depths, thereby extending their growth periods. Earlier seasonal growth by trees enables them to capture the pulse of mineral nutrients released by the first rains before grasses start growing. Juvenile trees need sufficient time to extend taproots and establish space for their surface roots amid the mat of grass roots. At a later stage, tree growth switches from gaining height towards spreading the leaf canopy, which ameliorates fire intensity in the shade cast. Between-tree competition contributes to maintaining an open woody canopy in drier savannas, while allowing ample space for grasses to coexist beneath and between trees. Trees tend to get excluded where soils are shallow and hardpans restrict water infiltration. This limitation contributes to the treeless grasslands prevalent in the South African Highveld and on upland surfaces elsewhere in Africa. Woodlands or even forests develop in lowlands and elsewhere where soils are deeper and retain moisture longer, except where seasonal waterlogging restricts rooting depth. The annual production of grass biomass above-ground is controlled largely by the wet season rainfall total, while the annual production of leaves and shoots by woody plants is affected little by the current year's rainfall.

The trees that predominate in South American savannas are mainly evergreen,[12] and the eucalypts that typify Australian savannas are also mostly evergreen. The evergreen habitat is believed to aid nutrient conservation in regions with extremely infertile soils, avoiding nutrients lost when leaves are shed. It could also be advantageous in dry regions where rainfall and hence opportunities for plant growth are highly unpredictable, as in much of Australia. In tropical Asia, most of the trees growing in savanna-like woodlands are deciduous, but shed their leaves quite late during a brief dry season. Fires generally have a lesser influence on the tree cover than in Africa, perhaps because more rainfall is received during the dry season.

Several of the grasses prevalent in South America cerrado exhibit the C_3 pathway. Relative to African grasses, South American grasses have slower growth rates and appear less efficient in dry matter produced per unit amount of nitrogen or phosphorus absorbed.[65] They are generally less tolerant of defoliation than African grasses and less digestible.[66] Some introduced African grasses have become aggressive invaders in South America, including signal grasses (*Brachiaria* spp.), red thatching grass (*Hyparrhenia rufa*) and Guinea grass, probably due to the fertilisation provided by agriculture. Blue buffalo grass has spread widely in Australia, but needs fertiliser applications to be maintained. Native Australian grasses tend to be particularly low in nitrogen content, although readily digestible.[65] Tussock-forming Mitchell grasses (*Astrebla* spp.) form the nearest approach to sweetveld.

However, we need to look beyond immediate or annual growth and consider changes in the populations of the individual plant species contributing to the vegetation cover over multi-annual periods. The comparative life-history patterns of trees and grasses in the context of recurrent fires will be the subject of the next chapter.

Suggested Further Reading

Gibson, DJ. (2009) *Grasses and Grassland Ecology*. Oxford University Press, Oxford.

McNaughton, SJ. (1983) Serengeti grassland ecology: the role of composite environmental factors and contingency in community organization. *Ecological Monographs* 53:291–320.

O'Connor, TG; Bredenkamp, GJ. (1997) Grassland. In Cowling, RM, et al. (eds) *Vegetation of Southern Africa*. Cambridge University Press, Cambridge, pp. 215–257.

References

1. Stevens, BM, et al. (2018) Mycorrhizal symbioses influence the trophic structure of the Serengeti. *Journal of Ecology* 106:536–546.
2. Augustine, DJ; McNaughton, SJ. (2004) Temporal asynchrony in soil nutrient dynamics and plant production in a semiarid ecosystem. *Ecosystems* 7:829–840.
3. Higgins, SI, et al. (2015) Feedback of trees on nitrogen mineralization to restrict the advance of trees in C4 savannahs. *Biology Letters* 11.
4. Wigley, BJ, et al. (2013) What do ecologists miss by not digging deep enough? Insights and methodological guidelines for assessing soil fertility status in ecological studies. *Acta Oecologica – International Journal of Ecology* 51:17–27.

5. Kgope, BS, et al. (2010) Growth responses of African savanna trees implicate atmospheric [CO_2] as a driver of past and current changes in savanna tree cover. *Austral Ecology* 35:451–463.
6. Higgins, SI, et al. (2011) Is there a temporal niche separation in the leaf phenology of savanna trees and grasses? *Journal of Biogeography* 38:2165–2175.
7. February, EC; Higgins, SI. (2016) Rapid leaf deployment strategies in a deciduous savanna. *PLoS One* 11:e0157833.
8. Frost, P. (1996) The ecology of miombo woodlands. In Campbell, B (ed.) *The Miombo in Transition: Woodlands and Welfare in Africa*. Centre for International Forestry Research, Bogor, India, pp. 11–57.
9. Rutherford, MC. (1983) Growth rates, biomass and distribution of selected woody plant roots in *Burkea africana–Ochna pulchra* savanna. *Vegetatio* 52:45–63.
10. Boaler, S. (1966) Ecology of a miombo site, Lupa North Forest Reserve, Tanzania: II. Plant communities and seasonal variation in the vegetation. *The Journal of Ecology* 54:465–479.
11. Ryan, CM, et al. (2017) Pre-rain green-up is ubiquitous across southern tropical Africa: implications for temporal niche separation and model representation. *New Phytologist* 213:625–633.
12. Eamus, D. (1999) Ecophysiological traits of deciduous and evergreen woody species in the seasonally dry tropics. *Trends in Ecology & Evolution* 14:11–16.
13. Rutherford, MC; Panagos, MD. (1982) Seasonal woody plant shoot growth in *Burkea africana–Ochna pulchra* savanna. *South African Journal of Botany* 1:104–116.
14. Cole, MM; Brown, R. (1976) The vegetation of the Ghanzi area of western Botswana. *Journal of Biogeography* 3:169–196.
15. O'Donnell, FC, et al. (2015) A quantitative description of the interspecies diversity of belowground structure in savanna woody plants. *Ecosphere* 6:1–15.
16. Zhou, Y, et al. (2020) Rooting depth as a key woody functional trait in savannas. *New Phytologist* 227:1350–1361.
17. Canadell, J, et al. (1996) Maximum rooting depth of vegetation types at the global scale. *Oecologia* 108:583–595.
18. Holdo, RM; Timberlake, J. (2008) Rooting depth and above-ground community composition in Kalahari sand woodlands in western Zimbabwe. *Journal of Tropical Ecology* 24:169–176.
19. Seymour, CL. (2008) Grass, rainfall and herbivores as determinants of *Acacia erioloba* (Meyer) recruitment in an African savanna. *Plant Ecology* 197:131–138.
20. Seghieri, J. (1995) The rooting patterns of woody and herbaceous plants in a savanna; are they complementary or in competition? *African Journal of Ecology* 33:358–365.
21. Case, MF, et al. (2020) Root-niche separation between savanna trees and grasses is greater on sandier soils. *Journal of Ecology* 108:2298–2308.
22. Smit, G; Rethman, N. (1998) Root biomass, depth distribution and relations with leaf biomass of *Colophospermum mopane*. *South African Journal of Botany* 64:38–43.
23. Verweij, RJT, et al. (2011) Water sourcing by trees in a mesic savanna: responses to severing deep and shallow roots. *Environmental and Experimental Botany* 74:229–236.
24. Shackleton, CM. (1997) The prediction of woody productivity in the savanna biome, South Africa. PhD thesis, University of the Witwatersrand, Johannesburg.

25. Dudley, JP. (2000) Seed dispersal by elephants in semiarid woodland habitats of Hwange National Park, Zimbabwe. *Biotropica* 32:556–561.
26. Bond, WJ; Keeley, JE. (2005) Fire as a global 'herbivore': the ecology and evolution of flammable ecosystems. *Trends in Ecology & Evolution* 20:387–394.
27. Higgins, SI, et al. (2012) Which traits determine shifts in the abundance of tree species in a fire-prone savanna? *Journal of Ecology* 100:1400–1410.
28. Bond, WJ, et al. (2017) Demographic bottlenecks and savanna tree abundance. In Cromsigt JPG, et al. (eds) *Conserving Africa's Mega-Diversity in the Anthropocene*. Cambridge University Press, Cambridge, pp. 161–188.
29. Owen-Smith, N. (1993) Woody plants, browsers and tannins in southern African savannas. *South African Journal of Science* 89:505–510.
30. Coley, PD, et al. (1985) Resource availability and plant antiherbivore defense. *Science* 230:895–899.
31. Wigley, BJ, et al. (2018) Defence strategies in African savanna trees. *Oecologia* 187:797–809.
32. Cooper, SM; Owen-Smith, N. (1986) Effects of plant spinescence on large mammalian herbivores. *Oecologia* 68:446–455.
33. Charles-Dominique, T, et al. (2017) The architectural design of trees protects them against large herbivores. *Functional Ecology* 31:1710–1717.
34. Wigley, BJ, et al. (2019) A thorny issue: woody plant defence and growth in an East African savanna. *Journal of Ecology* 107:1839–1851.
35. Skarpe, C, et al. (2000) Browsing in a heterogeneous savanna. *Ecography* 23:632–640.
36. Wigley, BJ, et al. (2015) Mammal browsers and rainfall affect acacia leaf nutrient content, defense, and growth in South African savannas. *Biotropica* 47:190–200.
37. Owen-Smith, N; Cooper, SM. (1987) Palatability of woody plants to browsing ruminants in a South African savanna. *Ecology* 68:319–331.
38. Lauenroth, WK; Gill, R. (2003). Turnover of root systems. In de Kroon, H; Visser, EJW (eds) *Root Ecology*. Springer, Berlin, pp. 61–89.
39. Mutanga, O, et al. (2004) Explaining grass-nutrient patterns in a savanna rangeland of southern Africa. *Journal of Biogeography* 31:819–829.
40. Rutherford, MC. (1980) Annual plant production–precipitation relations in arid and semi-arid regions. *South African Journal of Science* 76:53–57.
41. O'Connor, TG. (1994) Composition and population responses of an African savanna grassland to rainfall and grazing. *Journal of Applied Ecology* 31:155–171.
42. Osborne, CP; Freckleton, RP. (2009) Ecological selection pressures for C-4 photosynthesis in the grasses. *Proceedings of the Royal Society B – Biological Sciences* 276:1753–1760.
43. Ehleringer, JR; Monson, RK. (1993) Evolutionary and ecological aspects of photosynthetic pathway variation. *Annual Review of Ecology and Systematics* 24:411–439.
44. Hempson, GP, et al. (2015) Ecology of grazing lawns in Africa. *Biological Reviews* 90:979–994.
45. Olivier, RCD; Laurie, WA. (1974) Habitat utilization by hippopotamus in the Mara River. *African Journal of Ecology* 12:249–271.

46. McNaughton, SJ. (1984) Grazing lawns: animals in herds, plant form, and coevolution. *The American Naturalist* 124:863–886.
47. Waldram, MS, et al. (2008) Ecological engineering by a mega-grazer: white rhino impacts on a South African savanna. *Ecosystems* 11:101–112.
48. Van der Plas, F, et al. (2013) Functional traits of trees on and off termite mounds: understanding the origin of biotically-driven heterogeneity in savannas. *Journal of Vegetation Science* 24:227–238.
49. Hipondoka, MHT, et al. (2003) Vertical distribution of grass and tree roots in arid ecosystems of Southern Africa: niche differentiation or competition? *Journal of Arid Environments* 54:319–325.
50. Knoop, WT; Walker, BH. (1985) Interactions of woody and herbaceous vegetation in a southern African savanna. *The Journal of Ecology* 9:235–253.
51. Cramer, MD, et al. (2012) Belowground competitive suppression of seedling growth by grass in an African savanna. *Plant Ecology* 213:1655–1666.
52. Stevens, N, et al. (2018) Transplant experiments point to fire regime as limiting savanna tree distribution. *Frontiers in Ecology and Evolution* 6:137.
53. Morrison, TA, et al. (2019) Grass competition overwhelms effects of herbivores and precipitation on early tree establishment in Serengeti. *Journal of Ecology* 107:216–228.
54. Wakeling, JL, et al. (2015) Grass competition and the savanna-grassland 'treeline': a question of root gaps? *South African Journal of Botany* 101:91–97.
55. Tinley, K. (1982) The influence of soil moisture balance on ecosystem patterns in southern Africa. In Huntley, BJ; Walker, BH (eds) *Ecology of Tropical Savannas*. Springer, Berlin, pp. 175–192.
56. O'Connor, TG; Bredenkamp, GJ. (1997) Grassland. In Cowling, RM, et al. (eds) *Vegetation of Southern Africa*. Cambridge University Press, Cambridge, pp. 215–257.
57. Guan, K, et al. (2014) Terrestrial hydrological controls on land surface phenology of African savannas and woodlands. *Journal of Geophysical Research: Biogeosciences* 119:1652–1669.
58. February, EC, et al. (2019) Physiological traits of savanna woody species: adaptations to resource availability. In Scogings, PF; Sankaran, M (eds) *Savanna Woody Plants and Large Herbivores*. Wiley, Oxford, pp. 309–329.
59. Novellie, P. (1989) Tree size as a factor influencing leaf emergence and leaf fall in *Acacia nigrescens* and *Combretum apiculatum* in the Kruger National Park. *Koedoe* 32:95–99.
60. Shackleton, CM. (1999) Rainfall and topo-edaphic influences on woody community phenology in South African savannas. *Global Ecology and Biogeography* 8:125–136.
61. Schenk, HJ; Jackson, RB. (2002) Rooting depths, lateral root spreads and below-ground/above-ground allometries of plants in water-limited ecosystems. *Journal of Ecology* 90:480–494.
62. Holdo, RM, et al. (2018) Rooting depth varies differentially in trees and grasses as a function of mean annual rainfall in an African savanna. *Oecologia* 186:269–280.
63. Dohn, J, et al. (2013) Tree effects on grass growth in savannas: competition, facilitation and the stress-gradient hypothesis. *Journal of Ecology* 101:202–209.

64. Priyadarshini, KVR, et al. (2016) Seasonality of hydraulic redistribution by trees to grasses and changes in their water-source use that change tree–grass interactions. *Ecohydrology* 9:218–228.
65. Fisher, MJ, et al. (1996) Grasslands in the well-watered tropical lowlands. In Hodgson, J; Illius, AW (eds) *The Ecology and Management of Grazing Systems*. CAB International, Wallingford, pp. 393–425.
66. Simoes, M; Baruch, Z. (1991) Responses to simulated herbivory and water stress in two tropical C 4 grasses. *Oecologia* 88:173–180.

Chapter 8: Plant Demography and Dynamics: Fire Traps

Besides seasonal changes in biomass dynamics, plants also change in abundance through the establishment of new plants and eventually their death. These demographic processes are difficult to study because of the timescales involved: trees can live for centuries. Among grasses, populations can turn over between one year and the next. Moreover, new grass plants can arise either from seeds or through vegetative growth by established plants. Plants do not progress steadily through life-history stages like mammals and birds do. Seeds may remain dormant for many years until favourable conditions occur, and then all germinate. Most of the seedlings will fade and disappear, unless conditions remain favourable for long enough for roots to be established. Trees may be held in the juvenile stage for many years by recurrent fires. Grass tufts expand and contract through tiller production and death as well as through the establishment of new plants from seeds. Creeping grasses spread by sending out runners, which can root from nodes and break connections so as to appear as new plants. Notions of climax states and stable equilibria have little relevance for savanna vegetation dynamics. Much of plant competition operates like a lottery for access to gaps opened by the death of plants established earlier. Plants typically produce huge numbers of seeds because this increases their chances of finding a gap. Individual trees can vary enormously in size during their growth from newly germinated seedling to mature seed-producing adult, generating a huge shift in biomass without any numerical change.

Nevertheless, impending population changes can be inferred from changes in the stage structure of the plants constituting the population. Populations with lots of small plants are likely to be expanding, while those with mostly big trees and few potential recruits could be in imminent decline. But caution is needed, because size does not necessarily represent age.

Grasses may appear to be 'Lilliputs' in comparison with giant trees or 'Gullivers',[1] but this size contrast is misleading. Much of the mass of a tree is made up of accumulated dead wood, while grasses have a greater portion of their mass underground than appears above. A fairer comparison between the relative abundances of these plant types can be made in terms of leaf canopy cover, because energy captured by leaf surfaces drives the metabolic processes generating growth. Because the grass canopy is continuous while the tree canopy is intermittent, most of the annual production of plant matter in savannas takes place in the form of grasses.

This chapter outlines what is known about the life-history transitions and population dynamics of savanna trees, shrubs and grasses, fragmented in time and space though this knowledge is.

Woody Plant Populations

A healthily expanding tree population is expected to have a size class distribution resembling a reverse-J, i.e. lots of small plants in waiting to replace fewer big plants. Few savanna tree populations show this ideal pattern (Figure 8.1). Humped distributions with many intermediate-sized plants can be produced where seedling establishment takes place episodically, only in years when suitable conditions occur. This generates cohorts of plants that were established in the same year moving through the size classes (Figure 8.2). However, a similar size structure distribution could also arise if some factor blocks plants from growing beyond some size. A true-J distribution with mostly big trees and few recruits can be generated if intervals between successful establishment become sufficiently long. Trees need to produce only 1–2 replacement trees during a lifetime potentially spanning several centuries in order to maintain the population.

Seedlings or juvenile woody plants under 0.5 m in height tend to be undercounted because they are hidden among the grasses, unless the grass cover is removed by grazing, but they then become exposed to being browsed. However, caution is needed in interpreting whether plants are genuinely seedlings, i.e. under a year old. Some little trees could have enormous rootstocks built up over many years, awaiting an opportunity to shoot up beyond the height where they get burnt back. Enormous numbers of seedlings may germinate in favourably wet years, but most fall by the wayside before the end of their first dry season.

Marula trees can show particularly puzzling size structure profiles. A study contrasting the situation between nature reserves with elephants and communal rangelands lacking elephants found a dearth of small trees 2–4 m in height in both localities,[2] and this size class was also missing in nearby Kruger NP[3] (Figure 8.1). The missing size class was represented within Kruger NP inside fenced areas excluding elephants, but also other herbivores, as well as along road verges outside the park.[4] There was no apparent lack of juvenile marula plants <0.5 m in height, either in the nature reserves or in communal lands,[2] but inside Kruger NP some places had only big marula trees, whereas elsewhere

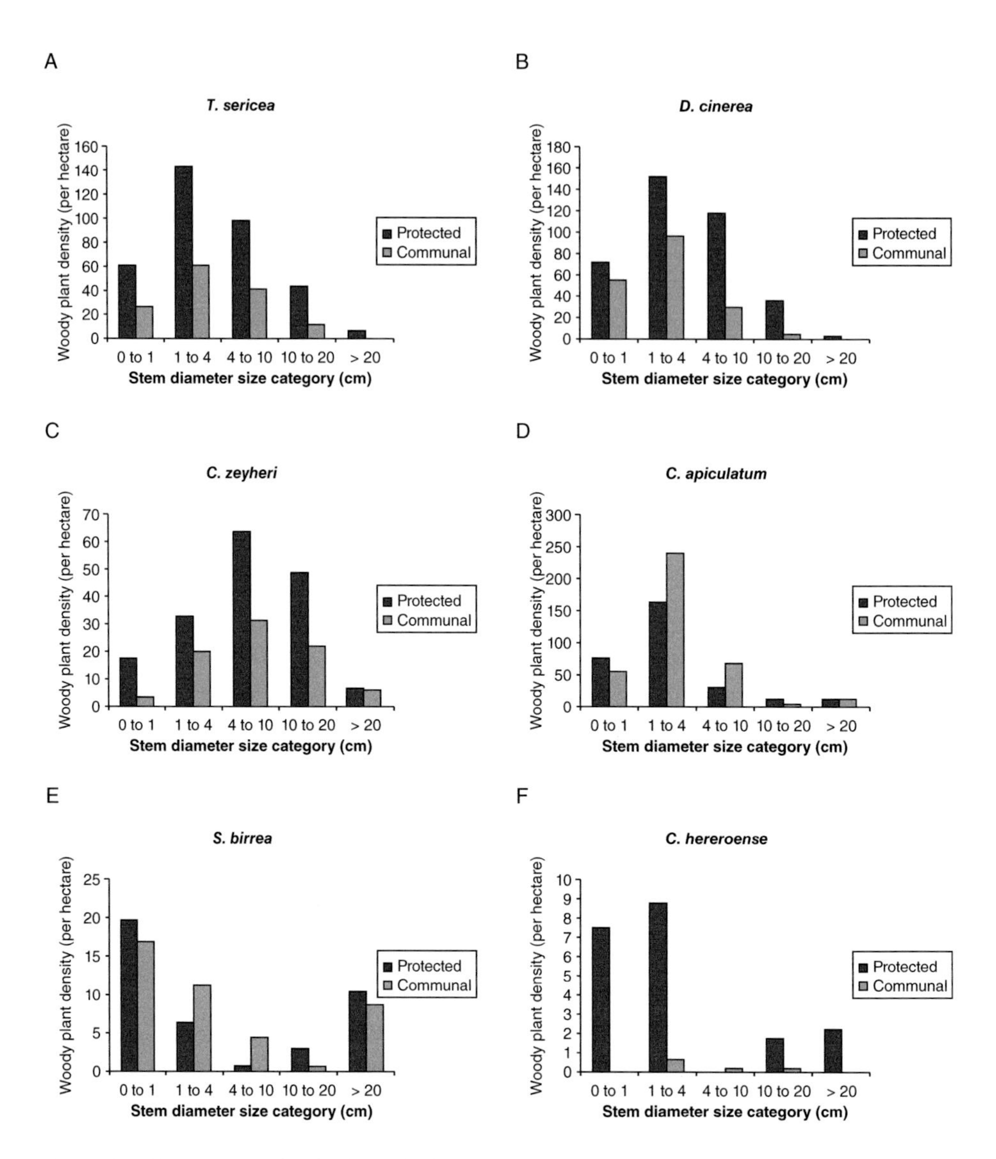

Figure 8.1 Size structure distributions of selected tree species comparing protected areas (black) with communal rangelands (grey) in the South African Lowveld region bordering Kruger NP. (A,B) Typical reverse-J distributions shown by silver clusterleaf and sickle bush; (C,D) humped distributions shown by large-leaf bushwillow (*Combretum zeyheri*) and red bushwillow; (E,F) gap in intermediate size classes shown by marula and russet bushwillow (from Kirsten Neke, PhD thesis, University of the Witwatersrand, 2005).

juvenile-sized plants were common.[3,5] How might these patterns be interpreted? (1) Do elephants selectively break plants in the missing size range? Or (2) do plants adaptively grow rapidly through the missing size range to reach the height at which they become less vulnerable to elephant damage? It is also possible that big marula trees are a legacy of a recruitment wave before

Figure 8.2 Even-aged stand of fever trees 50 years after a flood provided favourable conditions for their establishment in northern Kruger NP, still densely thronged although almost 20 m tall (photo by Johna Turner).

elephant numbers became substantial, but other tree species similarly vulnerable to elephant damage did not show this missing size class.

In order to reach seed-producing maturity, all savanna trees must escape the fire trap holding them below 2–3 m in height. Variable intervals between hot fires may eventually allow saplings to grow above 3 m.[6,7] Juvenile acacias may remain within the fire trap for decades.[8] Cooler conditions slow growth rates, and thus contribute to the absence of trees from high-altitude grasslands.[9]

Savanna trees respond adaptively to the fire trap by growing rapidly in height relative to stem diameter until they get above 3 m and thereafter shift their growth investment towards expanding in stem diameter (Figure 8.3).[2] Marula trees exhibited the lowest height relative to diameter during the sapling stage among the tree species measured, despite eventually growing into the biggest trees.

Baobab trees also show spatially variable size structure profiles. In Kruger NP, plants found on rocky hills showed a range in size, while few small trees were present on the plains below.[10] This could be explained either because fewer elephants ascended the hills, or because fires penetrated less frequently, due to lower grass cover among the rocks. Whatever the reason, this pattern suggests a source–sink relationship, with seeds produced on the hills getting dispersed to replace the diminishing population on the plains. Tree species sensitive to being set back by giraffe browsing can likewise become restricted to rocky hillsides.[11]

Within Serengeti NP, there have evidently been two recruitment episodes of acacia trees. The more recent one began during the 1970s when the extent of

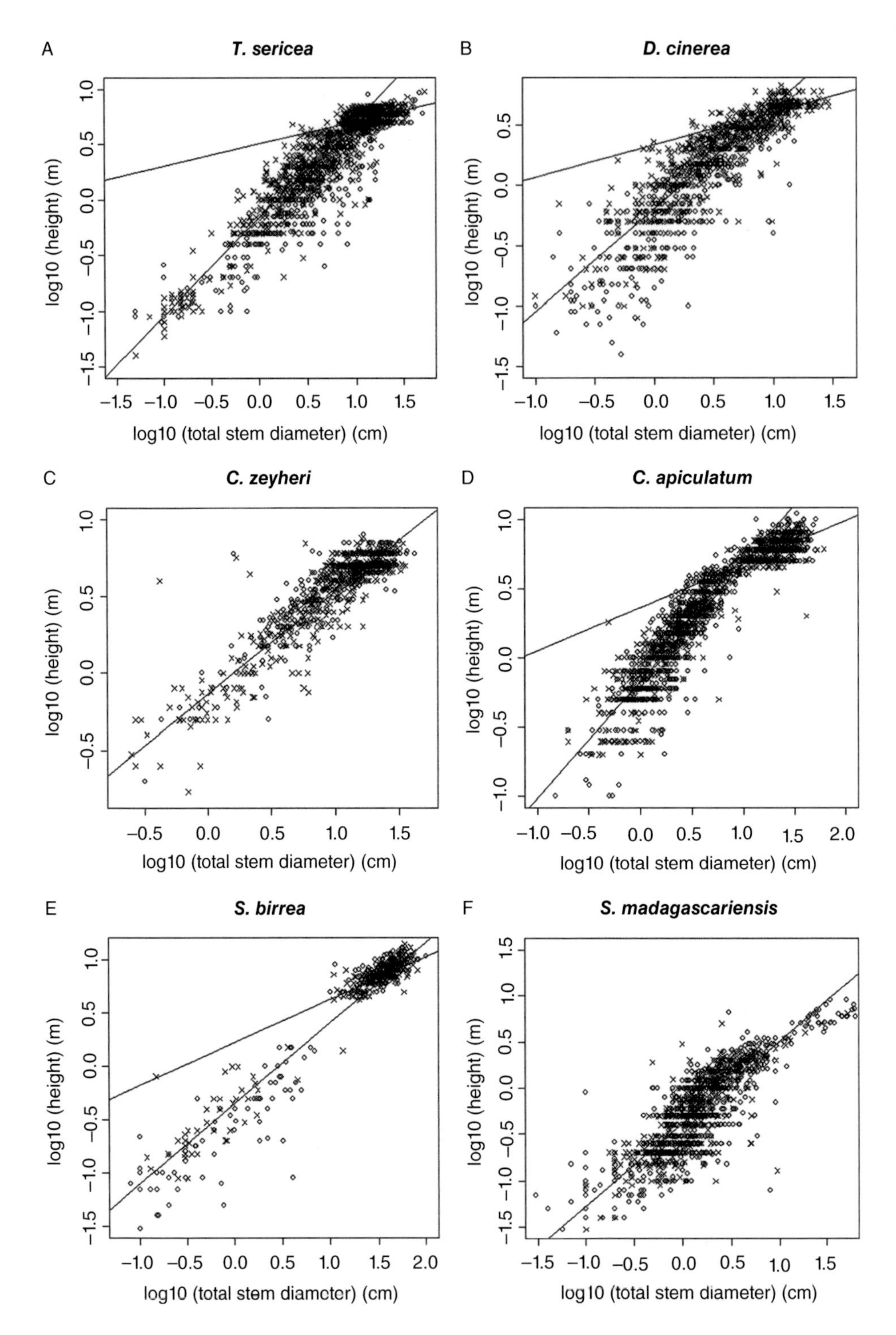

Figure 8.3 Allometry of growth in height versus stem diameter for selected tree species combining protected areas (circles) with communal rangelands (crosses) in the South African Lowveld region

the park area burnt annually became greatly reduced by grazing pressure from the resurgent wildebeest population.[12,13] An earlier wave of tree establishment apparently occurred during the 1890s, following the rinderpest panzootic and consequent loss of livestock by Maasai herders. Distinct cohorts of splendid acacias were generated during these two periods. Currently, few juvenile acacias occur in the understorey. Instead, shrubs of other species predominate, indicating an impending population turnover.[14]

Growth rings shown by acacia trees within the Hluhluwe-iMfolozi Park, calibrated for age using radiocarbon dating, revealed spatial variation in age structure profiles along with temporal variation in time of establishment.[15] Establishment periods were not synchronised spatially and differed among tree species. They could not be coupled with any particular climatic pattern, fire incidence or change in herbivore populations. In this park, a 2-km strip along the boundary along with areas of dense bush had been cleared of all woody plants during the 1940s to confine tsetse flies (the vector for cattle sleeping sickness) within the park, before aerial spraying with insecticides took place. The cleared areas all reverted to the vegetation formations that had been there previously – acacia trees where there had been acacia savanna, evergreen shrubs in the form of gwarrie bushes (*Euclea divinorum*) in the lowland pockets where shrub thickets had been present. Some tree species seem to be undergoing a widespread expansion wave presently. They include mopane, silver clusterleaf and tamboti in South Africa's savanna bushveld, while sweet thorn has been spreading into grassland through the Eastern Cape.[16] Ongoing turnover among acacia species has been observed over decades in the Hluhluwe-iMfolozi Park.[17]

Trees eventually die from competition for resources,[18] soil moisture deficits,[19] wind throw,[20,21] fires burning into their trunks, disease or from cavities in their trunks generated by soil water deficiencies.[22] Mortality rates typically decrease with increasing tree size, from 5 percent to 10 percent annually among small saplings towards 1–2 percent among large trees. Wind throw may topple quite large trees, especially those with trunk decay.[23] The potential lifespans of savanna trees or shrubs can vary hugely, from over a thousand years for leadwood trees attaining a trunk circumference up to 2.2 m to 30–40 years for sicklebush and sweet thorn.[15,24] No acacia tree sampled in Hluhluwe-iMfolozi was older than a century,[15] although giraffe thorn trees can live up to 240 years of age elsewhere.[25] The oldest reliable age for a baobab

Figure 8.3 (*cont.*) bordering Kruger NP. Species names as in Figure 8.1, plus monkey orange (*Strychnos madagascariensis*). Note how plants grow initially upwards then expand basal diameter (and canopies) once they emerge above the flame zone (from Kirsten Neke, PhD thesis, University of the Witwatersrand, 2005).

tree is ~2500 years.[26] During an exceptionally severe drought experienced in Kruger NP, tree mortality was locally high but spatially variable.[27] Shrubs and saplings incurred less mortality than taller trees.

Grass Populations

Grass populations turn over much faster than tree populations. During extreme droughts, the ground surface can become locally bared of plants by large grazers plus the activities of harvester termites (Figure 7.3) and yet be covered by green grass within a few weeks after rains resume.[28,29] Of course, the grass species may be different – composed mainly of annual grasses, or less nutritious needle grasses, which produce huge numbers of small seeds.[30] Various forbs may also contribute to the post-drought cover.[28,31] Nevertheless, I have observed three instances where the grass re-establishing was mainly Guinea grass, among the most palatable species for large herbivores (see Chapter 13). Even red grass, with comparatively large seeds, can rapidly fill gaps opened by drought-related mortality among tufts or tillers, provided rodents and seed-eating birds have not consumed all the seeds.[32,33] But where do these grass species find refuge during droughts so as to re-establish their seed banks? Low bushes may provide some protection from grazing and desiccation.

Population turnovers among savanna grasses can be surprisingly rapid.[34,35] Grass turnover is less in upland or montane grasslands where rainfall varies proportionately less and individual grasses fluctuate in abundance mainly through tuft expansion and contraction. Little is known about the potential lifespan of a grass tuft. Grazing pressure can make a big difference, as will be discussed in Chapter 13, after the herbivores have been introduced.

Overview

Savanna tree populations may expand or contract in response to shifting climatic conditions or other influences. Progression through life-history stages by tree saplings is governed largely by variable intervals between fires. Notions of compositionally stable climax states have little relevance. While some savanna tree species can live for several centuries, others have lifespans briefer than a century. While the grass biomass produced annually depends on rainfall, the species contributing can vary from one year to the next.

Studies elsewhere in the world have documented how pine tree woodlands in North America were established episodically at intervals of several hundred years, governed by variable precipitation and fire return intervals.[36] This pattern is likely to be widespread, although difficult to document.

In the following chapter, I will expand the time horizon to consider changes in the vegetation cover that have taken place since savanna formations appeared during the late Miocene.

Suggested Further Reading

O'Connor, TG; Everson, TM. (1998) Population dynamics of perennial grasses in African savanna and grassland. In Cheplick, P (ed.) *Population Biology of Grasses*. Cambridge University Press, Cambridge, pp. 333–365.

Staver, AC, et al. (2011) History matters: tree establishment variability and species turnover in an African savanna. *Ecosphere* 2(4):Art 49.

References

1. Bond, WJ; van Wilgen, BW. (1996) *Fire and Plants*. Springer, Berlin.
2. Neke, KS. (2005) The regeneration ecology of savanna woodlands in relation to human utilisation. PhD thesis, University of the Witwatersrand, Johannesburg.
3. Helm, CV; Witkowski, ETF. (2012) Characterising wide spatial variation in population size structure of a keystone African savanna tree. *Forest Ecology and Management* 263:175–188.
4. Helm, CV; Witkowski, ETF. (2013) Continuing decline of a keystone tree species in the Kruger National Park, South Africa. *African Journal of Ecology* 51:270–279.
5. Biggs, R; Jacobs, OS. (2002) The status and population structure of the marula in the Kruger National Park. *South African Journal of Wildlife Research* 32:1–12.
6. Higgins, SI, et al. (2000) Fire, resprouting and variability: a recipe for grass–tree coexistence in savanna. *Journal of Ecology* 88:213–229.
7. Wakeling, JL, et al. (2011) Simply the best: the transition of savanna saplings to trees. *Oikos* 120:1448–1451.
8. Bond, WJ. (2008) What limits trees in C-4 grasslands and savannas? *Annual Review of Ecology Evolution and Systematics* 39:641–659.
9. Wakeling, JL, et al. (2012) The savanna–grassland 'treeline': why don't savanna trees occur in upland grasslands? *Journal of Ecology* 100:381–391.
10. Edkins, MT, et al. (2008) Baobabs and elephants in Kruger National Park: nowhere to hide. *African Journal of Ecology* 46:119–125.
11. Bond, WJ; Loffell, D. (2001) Introduction of giraffe changes acacia distribution in a South African savanna. *African Journal of Ecology* 39:286–294.
12. Sinclair, ARE, et al. (2007) Long-term ecosystem dynamics in the Serengeti: lessons for conservation. *Conservation Biology* 21:580–590.
13. Sinclair, ARE, et al. (2008) Historical and future changes to the Serengeti ecosystem. In Sinclair, ARE, et al. (eds) *Serengeti III: Human Impacts on Ecosystem Dynamics*. University of Chicago Press, Chicago, pp. 7–46.

14. Anderson, TM, et al. (2015) Compositional decoupling of savanna canopy and understory tree communities in Serengeti. *Journal of Vegetation Science* 26:385–394.
15. Staver, AC, et al. (2011) History matters: tree establishment variability and species turnover in an African savanna. *Ecosphere* 2:1–12.
16. Bond, WJ; Midgley, GF. (2012) Carbon dioxide and the uneasy interactions of trees and savannah grasses. *Philosophical Transactions of the Royal Society B – Biological Sciences* 367:601–612.
17. Bond, WJ, et al. (2001) Acacia species turnover in space and time in an African savanna. *Journal of Biogeography* 28:117–128.
18. Sea, WB; Hanan, NP. (2012) Self-thinning and tree competition in savannas. *Biotropica* 44:189–196.
19. Chase, MF, et al. (2019) Severe drought limits trees in a semi-arid savanna. *Ecology* 100:e02842.
20. Spinage, CA; Guinness, FE. (1971) Tree survival in the absence of elephants in the Akagera National Park, Rwanda. *Journal of Applied Ecology* 8:723–728.
21. Williams, RJ; Douglas, M. (1995) Windthrow in a tropical savanna in Kakadu National Park, northern Australia. *Journal of Tropical Ecology* 11:547–558.
22. Shackleton, CM. (1997) The prediction of woody productivity in thie savanna biome, South Africa. PhD thesis, University of the Witwatersrand, Johannesburg.
23. Teren, G, et al. (2018) Elephant-mediated compositional changes in riparian canopy trees over more than two decades in northern Botswana. *Journal of Vegetation Science* 29:585–595.
24. Fuls, A; Vogel, JC. (2005) The life-span of leadwood trees. *South African Journal of Science* 101:98–100.
25. Seymour, CL. (2008) Grass, rainfall and herbivores as determinants of *Acacia erioloba* (Meyer) recruitment in an African savanna. *Plant Ecology* 197:131–138.
26. Patrut, A, et al. (2018) The demise of the largest and oldest African baobabs. *Nature Plants* 4:423–426.
27. Swemmer, AM. (2020) Locally high, but regionally low: the impact of the 2014–2016 drought on the trees of semi-arid savannas, South Africa. *African Journal of Range & Forage Science* 37:31–42.
28. Wilcox, KR. (2020) Rapid recovery of ecosystem function following extreme drought in a South African savanna grassland. *Ecology* 101:e02983.
29. Wigley-Coetsee, C; Staver, AC. (2020) Grass community responses to drought in an African savanna. *African Journal of Range & Forage Science* 37:43–52.
30. Abbas, HA, et al. (2019) The worst drought in 50 years in a South African savannah: limited impact on vegetation. *African Journal of Ecology* 57:490–499.
31. Novellie, PA; Bezuidenhout, H. (1994) The influence of rainfall and grazing on vegetation changes in the Mountain Zebra National Park. *South African Journal of Wildlife Research* 24:60–71.
32. O'Connor, TG. (1991) Local extinction in perennial grasslands: a life-history approach. *The American Naturalist* 137:753–773.

33. O'Connor, TG; Pickett, GA. (1992) The influence of grazing on seed production and seed banks of some African savanna grasslands. *Journal of Applied Ecology* 29:247–260.
34. O'Connor, TG. (1993) The influence of rainfall and grazing on the demography of some African savanna grasses: a matrix modelling approach. *Journal of Applied Ecology* 30:119–132.
35. Anderson, TM. (2008) Plant compositional change over time increases with rainfall in Serengeti grasslands. *Oikos* 117:675–682.
36. Brown, PM; Wu, R. (2005) Climate and disturbance forcing of episodic tree recruitment in a southwestern ponderosa pine landscape. *Ecology* 86:3030–3038.

Chapter 9: Paleo-savannas: Expanding Grasslands

Savannas did not feature during the dawn of the age of mammals following the demise of the dinosaurs 66 Ma. The first grasses had appeared by then, but remained a miniscule constituent of the vegetation. Through the Palaeocene and Eocene, forests and closed woodlands blanketed the Earth during the extremely hot and wet conditions that prevailed from the equator to the poles.[1] Most of the modern tree families were represented by then, including the legumes. Grasses became prominent first in South America during the Oligocene around 30 Ma.[2] By the mid-Miocene 12 Ma, grasslands had become sufficiently widespread in North America to support a herbivore fauna with dentition honed for grazing, dominated by horses. Grasses came to the fore in Africa during the late Miocene around 10 Ma, including species showing the C_4 photosynthetic pathway, as recorded by plant wax biomarkers in marine sediments.[3,4] After 8 Ma, C_4 grasslands had expanded throughout the tropics and subtropics, under climates that became cooler and drier, especially in Africa.[5] By the mid-Pliocene 3.5 Ma, savannas dominated by C_4 grasses were widely established through most of Africa.

However, the record of past vegetation in Africa is fragmentary in time and space, dependent on where sediments retain plant remains. Studies have been focused on rift valley settings in eastern Africa, where sediments can be dated quite precisely from volcanic layers. However, gaps exist in the sediment record. Marine sediments deposited offshore provide a broad regional record of past vegetation in catchments of rivers based on the pollen and other biomarkers they contain, although these can get shifted around by ocean currents. Plant remains from the limestone caves in interior southern Africa where fossils are preserved are more difficult to date. Sites documenting vegetation changes from pollen and other residues become more widely available from late in the Pleistocene.

Grasses can be identified as a group from their pollen, but are not readily distinguished at tribal or generic level. Silica bodies (or phytoliths) contained in grass leaves support only tribal distinctions. However, C_4 grasses leave a characteristic signature in the form of carbon isotope ratios, recorded in soil carbonates, organic matter and bones and teeth of the animals feeding on

Figure 9.1 Fossil wood in Luangwa Valley, Zambia.

them (see Chapter 7). Woody plant families or genera can be distinguished from their pollen. Trees that are wind-pollinated produce profuse pollen, but those pollinated by insects, like the acacias, are more sparing of pollen and thus under-represented. Further records of past vegetation are provided by fossilised leaves or trunks (Figure 9.1) as well as from biomarkers like leaf waxes.

Savanna Expansion

Carbon isotope ratios in soils document the expansion of C_4 grasses through the Pliocene into the early Pleistocene, from 5 to 2 Ma. This increase was probably partly at the expense of grasses with the C_3 pathway.[6] A further indicator of grassland expansion through this period comes from wear patterns on the teeth of herbivores supported by carbon isotope ratios in dental enamel. I will describe the evidence separately for eastern and southern regions of Africa because of the differences in the sources used.

Eastern Africa

During the late Oligocene around 28 Ma, broad-leaved deciduous woodland resembling miombo was present both in Ethiopia and Tanzania, and presumably in between.[7,8,9] Acacias were also present by that time in Tanzania.[10] Wet forest occurred near Lake Victoria ~18 Ma, but by 12 Ma had become replaced by a heterogeneous mix of forest, woodland and even savanna.[9,10] However, the grass pollen at that time came solely from C_3 grasses. By the late Pliocene ~3 Ma, C_4 grasses had achieved overwhelming dominance.[3,4,9,11] Tectonic movements related to rift valley formation contributed to shifts in local climates during the Pliocene, causing the eastern African interior to be drier than projected from global temperature conditions,[12] supported by increased evidence of fires.[13] This brought about a reduction in forest cover over eastern

Africa.[9] Spines and other structural defences against large herbivores became a feature of several plant lineages during this time.[14] This infers increased representation of woody plants typical of dry/eutrophic savannas with structural rather than chemical defences against herbivores. The projected woody plant cover, estimated from stable carbon isotope ratios in soils, had fallen below 40 percent at Awash in Ethiopia, and perhaps around half of this in the Turkana region of northern Kenya, by 6 Ma.[15] By the mid-Pliocene ~4 Ma, most of the modern plant families could be recognised from fossil wood at Laetoli in central Tanzania, including acacias.[16] Genera present today that seem missing locally at that time include *Balanites* (desert date), *Combretum* (bushwillows), *Croton* and *Euclea* (gwarri).

The onset of Pleistocene glaciations after 2.7 Ma was associated with a sharp drop in tree pollen and corresponding rise in C_4 grass pollen in marine deposits off eastern Africa.[9] Chloridoid grasses indicative of dry conditions became prominent between 2.7 and 2.5 Ma. Acacias and wild myrrhs, typical of dry savanna, were notable in the woody vegetation around Olduvai, along with pollen from forest trees such as yellowwoods and junipers, probably present in nearby highlands.[12,17] Within the Omo valley, grass pollen increased substantially in its representation between 2.8 and 2.35 Ma.[18] Tree species representative of dry savanna woodland had become established there by 2 Ma. Vegetation to the east of Lake Turkana retained a local predominance of riparian woodland species ~1.5 Ma interspersed mosaically with more grassy formations.[19] Underground trees or geoxyles, adapted to withstand frequent fires, became prominent constituents of the vegetation during the course of the early Pleistocene.[20]

Southern Africa

During the late Miocene and Pliocene, pollen from marine deposits off the coasts of Angola and Namibia show that both broad-leaved and fine-leaved savanna trees were present in the adjacent interior,[21] but the white rhinos found in the early Pliocene fauna at Langebaanweg in the south-western Cape around 5 Ma still grazed only C_3 grasses.[22] The savanna bushveld around Makapansgat in South Africa's northern Limpopo Province also contained only C_3 plants at that time.[23] The first indication of C_4 grasses at Makapansgat comes from stable isotopes in small mammal fossils dated around 3.4 Ma, although ostrich eggshells indicate that C_4 grasses had been present in Namibia a little earlier.[24] C_4 grasses expanded greatly in their representation at Makapansgat only after 2 Ma, somewhat later than in eastern Africa.[25] In the vicinity of the dolomite caves in the South African Highveld, an open savanna of protea bushes along with riverine forest patches existed in

Figure 9.2 Protea shrub savanna, a relict highland grassland formation resistant to fire. (A) Drakensberg foothills, South Africa; (B) Nyika Plateau, Malawi, post-fire.

the late Pliocene/early Pleistocene, similar to the modern-day grassland–forest mosaic prevailing in the Drakensberg foothills (Figure 9.2).[21,26] Montane grasslands with protea shrubs and heathlands on rocky slopes are distributed further north on high-lying remnants of the African Erosion Surface into eastern Africa.[27]

Pleistocene Oscillations

The onset of glacial advances and retreats initiating the Pleistocene after 2.6 Ma was associated with oscillations between extremes of aridity and wetness over much of Africa. Initially, these were driven by the 41 kyr periodicity in the Earth's angle of rotation, but the 21 kyr period generated by precession was particularly influential for rainfall in the tropics.

Eastern Africa

Over eastern Africa, cooler glacial periods were associated with expansions of grassy savanna, while tree cover increased during warmer interludes.[28] In the Serengeti region, rainfall projected from plant wax biomarkers apparently fluctuated between extremes of 400–900 mm, and thus from semi-arid to mesic, during the 200 kyr interval between 1.9 and 1.7 Ma when climatic variation became accentuated.[29] Despite rain-shadow effects intensified by the rise of Ngorongoro and other large volcanoes in the east, a lake became established at Olduvai due to blocked drainage.[12] C_4 grasses or sedges were prominent in the vegetation there in lowlands, while Afromontane forest was present on nearby highlands. The mixture of dry savanna, bushland and wooded grassland around Olduvai after 1.7 Ma closely resembled the heterogeneous mosaic shown by the modern dry savanna.[12,30] Lake Olduvai dried up following rift movements around 1.3 Ma but became re-established intermittently later.

Further north in Kenya at Turkana West, the C_4 grass contribution apparently fluctuated between extremes as wide as 5 percent to 100 percent, as inferred from wax biomarkers in soils, between 2.3 and 1.7 Ma.[31] In East Turkana the woody vegetation around 1.95 Ma, as revealed by silicified wood, comprised species typical of riverine woodland.[32] Thereafter, grassy savanna expanded generally.[9,15] Conditions became temporarily moister in eastern Africa between 1.1 and 0.9 Ma, but the consequences for vegetation changes have not been elucidated. Savanna grasslands expanded at the expense of rainforests through western and central Africa, even into the Congo basin, during and following the coldest period of the last glaciation around 20 ka.[33]

Southern and South-Central Africa

Sediments deposited in the Indian Ocean by the Zambezi, Limpopo and other rivers draining southern parts of Africa record the vegetation in their catchments during the Middle and Late Pleistocene, based on fossil pollen and plant leaf waxes.[34,35,36] In the Limpopo catchment, montane forest including yellowwood trees gave way to open heathland resembling Cape fynbos during extreme glacial conditions. Savanna woodland expanded during the brief interglacial interludes. Pollen from C_4 grasses and sedges was more constantly present. Sediments deposited by the Zambezi River show a closely similar pattern, including the presence of ericaceous heathland as well as montane forest during glacial periods, but with greater representation of mopane during transitional periods. Miombo woodland does not feature, perhaps simply because its trees are poor pollen producers (this applies also to acacias). On the western side of the subcontinent, sediments off the Angolan coast indicate the presence of open grassy or shrubby vegetation during glacial periods and an expansion of miombo woodland during interglacials.[21]

Sediments contained within Lake Malawi reflect abrupt vegetation changes over quite short periods after 135 ka, when low lake levels reflected mega-drought conditions.[37] Yellowwood pollen gave way to grass pollen after 105 ka, while pollen from miombo woodland increased in prominence between 75 and 65 ka. This was followed by an expansion by yellowwood forest after 60 ka when global temperatures were descending towards the LGM. Following the LGM, miombo woodland increased at the expense of montane forest.[38]

In the Cradle region of South Africa's Highveld, shrubby grassland with riverine woodland flanking rivers persisted through the early Pleistocene from 1.8 to 1.2 Ma.[39,40] In the vicinity of Wonderwerk Cave located in the Northern Cape, close to the current transitions between the arid savanna, grassland and

Nama-Karoo biomes, a mixture of C_4 and C_3 grasses persisted from 1.95 until 0.99 Ma, as shown by pollen trapped in stalagmites and hyrax dung deposits.[41,42] Thereafter, C_4 grasses increased in prominence as aridity intensified.

Within interior South Africa, the vegetation near Florisbad after 300 ka, shortly preceding the earliest modern humans, fluctuated between being grassy with karoo shrubs during warm periods and fynbos shrubbery when cooler conditions ensued.[43] Two sites located in the northern bushveld region of South Africa, Wonderkrater spring and Tswaing impact crater, provide an exceptionally detailed record of vegetation shifts in their vicinity after 200 ka, derived from fossil pollen.[44,45,46] At Tswaing, the local vegetation shifted from dry savanna woodland with bushwillow prominent to yellowwood forest or fynbos as glacial conditions set in after 190 Ma. Vegetation near Wonderkrater was evidently a grassy savanna prior to 45 ka, approaching the LGM, with both acacias and bushwillows prominent along with camphor bush (*Tarchonanthus camphoratus*).[47] By 35 ka, with conditions becoming cooler but wetter, C_3 grasslands and ericaceous shrubs took over and C_4 grasses were absent. Pollen from yellowwood (*Podocarpus* spp.), wild olive (*Olea africana*) and boer-bean (*Schotia brachypetala*) trees indicated the nearby presence of evergreen forest. There was a hiatus in pollen through the LGM around 20 ka when temperatures were perhaps 5°C cooler than at present and conditions became very dry. After 14 ka, savanna woodland with acacia trees became established once more as conditions warmed into the current interglacial period. On the high Lesotho plateau, low shrubland including plant genera currently found much further south prevailed during and following the LGM 23–17 ka, indicating a northward shift by winter rainfall conditions. In the environs of Sibudu Cave, located close to the KwaZulu-Natal coast, yellowwood forest was more prominent prior to 60 ka than it is today, indicating cooler and wetter conditions.

Overview

The record of past vegetation in Africa, records three noteworthy features: (1) a progressive trend towards greater grass cover through the Pliocene and Pleistocene; (2) wide oscillations in the local presence of different vegetation formations, especially during the late Pleistocene; and (3) greater representation of grassland–forest interspersion along with ericaceous shrubs during the cooler glacial periods that predominated during the greater portion of the Pleistocene.

Savanna formations first appeared within Africa during the Miocene, before C_4 grasses became overwhelmingly dominant. Spinescent acacias, characteristic of dry/eutrophic savannas, were prominent in eastern Africa by the mid-Pliocene. Nevertheless, riparian woodlands prevailed in the Omo and Turkana basins throughout the early Pleistocene, adjoining the lakes that formed and rivers that flowed. C_4 grasses became dominant somewhat later in subtropical southern Africa than in tropical eastern regions. During the Middle and Late Pleistocene, ericaceous heathland (fynbos) became prominent in fossil pollen as far north as the Zambezi River catchment during glacial extremes, along with widened representation of evergreen forests. Nevertheless, grass pollen remained abundant, while pollen from miombo woodland trees was surprisingly uncommon. Montane grassland–forest mosaics were more widely prevalent during the cooler glacial conditions that predominated during the Pleistocene than they are at present over southern and south-central Africa, replacing broad-leaved savanna woodlands. Fine-leaved, acacia-dominated savanna had become established in parts of eastern Africa by this time and was well represented in the south-west within present-day Namibia.

Wide fluctuations in the local vegetation within quite short periods are strikingly evident in the finer temporal resolution provided by sedimentary accumulations in South Africa and Malawi through the late Pleistocene. They are probably responses to the 21 kyr oscillation in the Earth's rotational wobble. Whatever the composition, grassy vegetation remained predominant throughout eastern and southern Africa from the late Pliocene onward. An additional factor beside climate contributing to a more open tree cover through most of the Pleistocene was the prevalently low atmospheric concentration of carbon dioxide.[48] This restricts the growth rate of woody plants and hence their chances of growing beyond the fire trap.

Suggested Further Reading

Bamford, MK, et al. (2016) Pollen, charcoal and plant macrofossil evidence of Neogene and Quaternary environments in southern Africa. In Knight, J; Grab, SW (eds) *Quaternary Environmental Change in Southern Africa*. Cambridge University Press, Cambridge, pp. 306–323.

Barboni, D. (2014) Vegetation of northern Tanzania during the Plio-Pleistocene: a synthesis of paleobotanical evidences from Laetoli, Olduvai and Peninj hominin sites. *Quaternary International* 322–323:264–276.

Bonneville, R. (2010) Cenozoic vegetation, climate changes and hominid evolution in tropical Africa. *Global Planetary Change* 72:390–412.

Jacobs, BF, et al. (2010) A review of the Cenozoic vegetation history of Africa. In Werderlin, L; Sanders, WJ (eds) *Cenozoic Mammals of Africa*. University of California Press, Berkeley, pp. 57–72.

References

1. Jacobs, BF, et al. (1999) The origin of grass-dominated ecosystems. *Annals of the Missouri Botanical Garden* 86:590–643.
2. Strömberg, CAE. (2011) Evolution of grasses and grassland ecosystems. *Annual Review of Earth and Planetary Sciences* 39:517–544.
3. Uno, KT, et al. (2016) Neogene biomarker record of vegetation change in eastern Africa. *Proceedings of the National Academy of Sciences of the United States of America* 113:6355–6363.
4. Polissar, PJ, et al. (2019) Synchronous rise of African C4 ecosystems 10 million years ago in the absence of aridification. *Nature Geoscience* 12:657–660.
5. Cerling, TE, et al. (1997) Global vegetation change through the Miocene/Pliocene boundary. *Nature* 389:153–158.
6. Feakins, SJ, et al. (2013) Northeast African vegetation change over 12 my. *Geology* 41:295–298.
7. Jacobs, BF. (2004) Palaeobotanical studies from tropical Africa: relevance to the evolution of forest, woodland and savannah biomes. *Philosophical Transactions of the Royal Society of London Series B: Biological Sciences* 359:1573–1583.
8. Jacobs, BF, et al. (2010) A review of the Cenozoic vegetation history of Africa. In Werdelin, L; Sanders, WJ (eds) *Cenozoic Mammals of Africa*. University of California Press, Berkeley, pp. 57–72.
9. Bonnefille, R. (2010) Cenozoic vegetation, climate changes and hominid evolution in tropical Africa. *Global and Planetary Change* 72:390–411.
10. Bobe, R. (2006) The evolution of arid ecosystems in eastern Africa. *Journal of Arid Environments* 66:564–584.
11. Edwards, EJ, et al. (2010) The origins of C4 grasslands: integrating evolutionary and ecosystem science. *Science* 328:587–591.
12. Barboni, D. (2014) Vegetation of Northern Tanzania during the Plio–Pleistocene: a synthesis of the paleobotanical evidences from Laetoli, Olduvai, and Peninj hominin sites. *Quaternary International* 322:264–276.
13. Keeley, JE; Rundel, PW. (2005) Fire and the Miocene expansion of C4 grasslands. *Ecology Letters* 8:683–690.
14. Charles-Dominique, T, et al. (2016) Spiny plants, mammal browsers, and the origin of African savannas. *Proceedings of the National Academy of Sciences of the United States of America* 113: E5572–E5579.
15. Cerling, TE, et al. (2011) Woody cover and hominin environments in the past 6 million years. *Nature* 476:51–56.
16. Bamford, MK. (2011) Fossil woods. In Harrison, T (ed.) *Paleontology and Geology of Laetoli: Human Evolution in Context*. Springer, Dordrecht, pp. 217–233.
17. Andrews, P; Bamford, M. (2008) Past and present vegetation ecology of Laetoli, Tanzania. *Journal of Human Evolution* 54:78–98.
18. Bobe, R; Eck, GG. (2001) Responses of African bovids to Pliocene climatic change. *Paleobiology* 27:1–47.

19. Bamford, MK. (2017) Pleistocene fossil woods from the Okote Member, site FwJj 14 in the Ileret region, Koobi Fora Formation, northern Kenya. *Journal of Human Evolution* 112:134–147.
20. Maurin, O, et al. (2014) Savanna fire and the origins of the 'underground forests' of Africa. *New Phytologist* 204:201–214.
21. Neumann, FH; Bamford, MK. (2015) Shaping of modern southern African biomes: Neogene vegetation and climate changes. *Transactions of the Royal Society of South Africa* 70:195–212.
22. Franz-Odendaal, TA, et al. (2002) New evidence for the lack of C4 grassland expansions during the early Pliocene at Langebaanweg, South Africa. *Paleobiology* 28:378–388.
23. Hopley, PJ, et al. (2007) Orbital forcing and the spread of C4 grasses in the late Neogene: stable isotope evidence from South African speleothems. *Journal of Human Evolution* 53:620–634.
24. Ségalen, L, et al. (2007) Timing of C4 grass expansion across sub-Saharan Africa. *Journal of Human Evolution* 53:549–559.
25. Hopley, PJ, et al. (2007) High- and low-latitude orbital forcing of early hominin habitats in South Africa. *Earth and Planetary Science Letters* 256:419–432.
26. Luyt, CJ; Lee-Thorp, JA. (2003) Carbon isotope ratios of Sterkfontein fossils indicate a marked shift to open environments c. 1.7 Myr ago. *South African Journal of Science* 99:271–273.
27. Meadows, ME; Linder, HP. (1993) Special paper: a palaeoecological perspective on the origin of afromontane grasslands. *Journal of Biogeography* 20:345–355.
28. Dupont, LM, et al. (2011) Glacial–interglacial vegetation dynamics in South Eastern Africa coupled to sea surface temperature variations in the Western Indian Ocean. *Climate of the Past* 7:1209.
29. Magill, CR, et al. (2013) Ecosystem variability and early human habitats in eastern Africa. *Proceedings of the National Academy of Sciences of the United States of America* 110:1167–1174.
30. Albert, RM, et al. (2015) Vegetation landscape at DK locality, Olduvai Gorge, Tanzania. *Palaeogeography, Palaeoclimatology, Palaeoecology* 426:34–45.
31. Uno, KT, et al. (2016) A Pleistocene palaeovegetation record from plant wax biomarkers from the Nachukui Formation, West Turkana, Kenya. *Philosophical Transactions of the Royal Society B: Biological Sciences* 371:20150235.
32. Bamford, MK. (2011) Late Pliocene woody vegetation of Area 41, Koobi Fora, East Turkana Basin, Kenya. *Review of Palaeobotany and Palynology* 164:191–210.
33. Malhi, Y, et al. (2013) African rainforests: past, present and future. *Philosophical Transactions of the Royal Society B: Biological Sciences* 368:20120312.
34. Dupont, L. (2011) Orbital scale vegetation change in Africa. *Quaternary Science Reviews* 30:3589–3602.
35. Castañeda, IS, et al. (2016) Middle to Late Pleistocene vegetation and climate change in subtropical southern East Africa. *Earth and Planetary Science Letters* 450:306–316.
36. Dupont, LM; Kuhlmann, H. (2017) Glacial–interglacial vegetation change in the Zambezi catchment. *Quaternary Science Reviews* 155:127–135.
37. Beuning, KRM, et al. (2011) Vegetation response to glacial–interglacial climate variability near Lake Malawi in the southern African tropics. *Palaeogeography, Palaeoclimatology, Palaeoecology* 303:81–92.

38. Ivory, SJ, et al. (2012) Effect of aridity and rainfall seasonality on vegetation in the southern tropics of East Africa during the Pleistocene/Holocene transition. *Quaternary Research* 77:77–86.
39. Reed, KE. (1997) Early hominid evolution and ecological change through the African Plio–Pleistocene. *Journal of Human Evolution* 32:289–322.
40. Bamford, MK, et al. (2016) Pollen, charcoal and plant macrofossil evidence of Neogene and Quaternary environments in southern Africa. In Knight, J; Grab, SW (eds) *Quaternary Environmental Change in Southern Africa*. Cambridge University Press, Cambridge, pp. 306–323.
41. Ecker, M, et al. (2018) The palaeoecological context of the Oldowan–Acheulean in southern Africa. *Nature Ecology & Evolution* 2:1080–1086.
42. Brook, GA, et al. (2010) A 35 ka pollen and isotope record of environmental change along the southern margin of the Kalahari from a stalagmite and animal dung deposits in Wonderwerk Cave, South Africa. *Journal of Arid Environments* 74:870–884.
43. Scott, L; Neumann, FH. (2018) Pollen-interpreted palaeoenvironments associated with the Middle and Late Pleistocene peopling of Southern Africa. *Quaternary International* 495:169–184.
44. Scott, L. (1999) Vegetation history and climate in the Savanna biome South Africa since 190,000 ka: a comparison of pollen data from the Tswaing Crater (the Pretoria Saltpan) and Wonderkrater. *Quaternary International* 57:215–223.
45. Scott, L, et al. (2003) Age interpretation of the Wonderkrater spring sediments and vegetation change in the Savanna Biome, Limpopo province, South Africa. *South African Journal of Science* 99:484–488.
46. Scott, L. (2016) Fluctuations of vegetation and climate over the last 75 000 years in the Savanna Biome, South Africa: Tswaing Crater and Wonderkrater pollen sequences reviewed. *Quaternary Science Reviews* 145:117–133.
47. Backwell, LR, et al. (2014) Multiproxy record of late Quaternary climate change and Middle Stone Age human occupation at Wonderkrater, South Africa. *Quaternary Science Reviews* 99:42–59.
48. Bond, WJ, et al. (2003) The importance of low atmospheric CO_2 and fire in promoting the spread of grasslands and savannas. *Global Change Biology* 9:973–982.

Part II: Synthesis: Savanna Structure and Dynamics

The ecology of savanna vegetation in Africa is governed largely by the ecology of grasses. Savanna grasses with the C_4 photosynthesis pathway are superbly adapted to cope with erratic rainfall. They can grow rapidly during times when soil moisture is adequately available, exploiting pulses of nutrient release while overcoming the limitation on growth by low atmospheric concentrations of carbon dioxide. Hence they can out-compete tree seedlings in extracting soil water from the topsoil layer where most nutrients occur. For juvenile trees to establish amid the dense mat of grass roots, they must extend their roots to greater depths in the soil and expand their growth to times earlier and later in the wet season than grasses. Tree saplings are repeatedly set back by having the accumulated carbon burnt back. To escape the fire trap, tree saplings must build up sufficient resources below ground to enable them to elevate their foliage above the flame zone. Grasses lose only dead top-hamper to fires and suffer from its accumulation in the absence of fires or grazing. Grasses grow both beneath and between tree canopies. Dense evergreen foliage would be required to shade out grasses, but such trees are restricted to locations that seldom burn.

Not all woody plants have deep roots and grasses vary in how tall they grow and hence how much fuel they generate to support hot fires. Too little research has been done on individual species ecologies, especially in their root deployment below-ground. Soil fertility along with rainfall underlie a functional subdivision between dry/eutrophic savanna typified by thorny acacias with finely subdivided leaves and broad-leaved savanna woodlands prevalent where soils are sandy and rather infertile. This savanna subdivision is not replicated outside of Africa.

Within African savannas the tree cover can vary locally from open grasslands to embedded patches of forest and thicket. This is governed largely by the redistribution of rainwater within local landscapes and its effects on plant growth and fire incidence. The annual production of grass biomass above ground is almost linearly dependent on the annual rainfall total. If rainfall is halved during droughts, there is an equivalent reduction in the amount of food produced to support grass-dependent herbivores. Deciduous trees produce their new leaf crop each year ahead of the rains and thereby remain less influenced by the current season's precipitation. Tree populations turn over slowly, in response to long-term climatic trends, although shrubs are shorter-lived. The grass cover responds more dynamically to between-year changes in rainfall. Little attention has been paid to the growth patterns of the plants labelled 'forbs', which encompass a range in growth forms from annual flowers to underground trees and much in between.

Trees get excluded from upland regions on ancient land surfaces where weathering has produced hardpan layers and climatic conditions are cooler, although not too dry to support woody plants. Elsewhere, water logging prevents trees from growing. The relative roles of temperature, precipitation, soil structure and fire patterns in generating extreme grassy states still need to be disentangled. Pleistocene oscillations in tree pollen indicate that soils do not form a rigid restriction. Variable levels of atmospheric carbon dioxide potentially make an additional contribution.

Our perception of Africa's savanna vegetation is biased by the fact that we live within an interglacial interlude, a brief respite from the cooler and drier conditions that prevailed through most of the Pleistocene. Not only that, but industrialisation and fossil fuel burning have pushed atmospheric carbon dioxide to levels last reached during the Pliocene 5 million years ago. Our human ancestors evolved under conditions that were somewhat different from those we modern humans have experienced during the Holocene.

These are the take-home messages from the chapters forming this section of the book:

1. The prevalence of savanna vegetation formations is climatically controlled by seasonal restrictions in rainfall.

2. Spatial heterogeneity in the woody vegetation cover is an inherent functional feature.

3. The dry/eutrophic savanna subdivision governed by bedrock geology is exclusive to Africa.

4. The expansion of C_4 grasses was primarily an adaptive response to intermittent soil moisture, generating pulsed releases of soil nutrients.

5. The C_4 pathway enables savanna grasses to dominate tree seedlings in competition for soil moisture.

6. Recurrent fires contribute additionally to suppressing tree growth towards maturity.

7. Grassland–forest–heathland mosaics were more widespread during glacial advances than during the current interglacial interlude.

8. African savannas are fundamentally disequilibrium ecosystems that persist in shifting patch mosaics at various scales.

Figure II.4 Luxuriant grassland of red grass, central Serengeti NP, Tanzania, pending grazing by large herbivores.

Part III: The Big Mammal Menagerie: Herbivores, Carnivores and Their Ecosystem Impacts

Africa's large herbivore assemblage is the product of the environments that also nurtured human origins and these animals contributed to evolutionary transitions in our hominin lineage. Around 90 species of large herbivore can be tallied on the continent, around half of them associated with the savanna biome (Figure III.1; the rest live in forests or deserts). All of them are ungulates, representing the orders Artiodactyla (with even-toes) and Perissodactyla (with odd-toes), except for the African elephant (Proboscidea). They span a range in body mass from ~5000-kg elephants down to 5-kg dikdiks. Over half of the

Figure III.1 Diversity of extant large herbivores in Africa in various tribes represented by grazers, browsers and mixed feeders. The number of species within each family or tribe in each feeding category, across all habitats, is in parentheses.

savanna inhabitants obtain their diet primarily from grasses, while the remainder feed mostly on the leaves of trees, shrubs and other plant types. In doing so they must also adjust to the seasonal variation in wetness versus dryness that is the defining feature of savanna environments, along with midday heat loads and seasonally restricted sources of drinking water. At the same time their survival is threatened by numerous large carnivores. Grazing and browsing have ramifying effects on the vegetation cover, modify the spread of fires, and accelerate the recycling of the mineral nutrients back to the soil. However, the diversity of large herbivores that we see today was exceeded back in time during the period when humans evolved.

Several of these animal species have been the focus of my research. During my doctoral study of white rhinos in the Hluhluwe-iMfolozi Park, I wandered along the pathways that these mega-grazers followed, inspected what grasses they ate, unravelled their social relationships, tracked their journeys to water and back, and observed how they interacted with other animals, during daylight and darkness (Figure III.2A). My observations led me to recognise the distinctive features that rhinos shared with other 'megaherbivores' (animals weighing more than 1000 kg once adult), such as elephants and hippos.[1] Their very large body size helps them resist environmental variation, but makes them vulnerable to human overkill. The conservation problem underlying my study was the transforming influence of white rhinos on the vegetation cover, affecting habitat conditions for other species.

The limited (3.5-year) duration of this study left unanswered the fundamental question of how large herbivores might achieve some form of balance with the plant cover. An opportunity to address it was provided by my post-doctoral study on the population dynamics of kudus in Kruger National Park (NP), undertaken ensconced mostly in a 4×4 vehicle (Figure III.2B). Each kudu could be recognised from variation in stripe patterns, photographically recorded, allowing me to document population changes by registering individual births and deaths. This study revealed how sensitively recruitment and mortality responded to annual variation in rainfall.[2]

To explore in more detail what governed the plant species that kudus and other browsers chose to eat, I followed habituated young animals in the Nylsvley Nature Reserve, recording every plant they consumed from sunrise to sunset each day (Figure III.2C,D). This revealed how plant secondary metabolites, particularly condensed tannins, influenced diet selection and hence the potential impact of browsing in suppressing bush encroachment.[3,4]

Subsequent studies by my students explored aspects of the ecology of black as well as white rhinos, sable and roan antelope (*Hippotragus niger* and *H. equinus*, respectively), wildebeest and zebra, buffalo, eland, gemsbok (*Oryx gazella*) and elephant, the latter mainly through what they do to trees.

Figure III.2 (A) Observing white rhinos on foot in Mfolozi Game Reserve; (B) watching kudus in Kruger National Park from a Land Rover; (C) recording what an impala in Nylsvley Nature Reserve ate, using a keyboard coupled to a tape recorder; (D) following after a foraging kudu at Nylsvley.

Field observations became incorporated into computer models linking diet selection to population dynamics.[5,6]

This Part III of the book will cover features of the ecology of Africa's large herbivores in the context of savanna vegetation and underlying physical features typical of Africa, as documented in Parts I and II. This provides the foundation for identifying how these animals interact with ecosystem processes such as fire spread and nutrient cycling and contribute to spatiotemporal heterogeneity. The time horizon is then extended back into the past to review how large herbivore assemblages and associated carnivores have changed since the Miocene epoch when Africa's modern fauna along with savanna-inhabiting hominins originated.

References

1. Owen-Smith, RN. (1988) *Megaherbivores: The Influence of Very Large Body Size on Ecology*. Cambridge University Press, Cambridge.
2. Owen-Smith, N. (1990) Demography of a large herbivore, the greater kudu *Tragelaphus strepsiceros*, in relation to rainfall. *The Journal of Animal Ecology* 59:893–913.
3. Owen-Smith, N; Cooper, SM. (1987) Palatability of woody plants to browsing ruminants in a South African savanna. *Ecology* 68:319–331.
4. Owen-Smith, N. (1994) Foraging responses of kudus to seasonal changes in food resources: elasticity in constraints. *Ecology* 75:1050–1062.
5. Owen-Smith, RN. (2002) *Adaptive Herbivore Ecology: From Resources to Populations in Variable Environments*. Cambridge University Press, Cambridge.
6. Owen-Smith, N. (2009) *Dynamics of Large Herbivore Populations in Changing Environments: Towards Appropriate Models*. John Wiley & Sons, Chichester.

Chapter 10: Niche Distinctions: Resources Versus Risks

The prevalence of grass in the savanna vegetation cover is associated with the diversity and abundance of especially grazers in Africa's large herbivore fauna. But how is it that all of these species came to coexist, dependent on the same basic food resource? Ecological theory maintains that species cannot coexist in the same places unless they differ sufficiently in (1) the food types that they consume, (2) how they evade falling prey to predators, or (3) how they cope with extremes in physical conditions. Fundamental differences in digestive anatomy and physiology distinguish grazers from browsers, but what further distinctions exist among grazers and browsers to enable numerous species to coexist, including several rather similar in body size? How do they cope with seasonal variation in the nutritional value of the remnants of the plant cover that remain into the dry season? How do they gain sufficient security from predation despite being conspicuously exposed in mostly open grasslands? How do seasonally diminishing sources of surface water restrict where they can be during the dry season? This chapter addresses how these requirements contribute to niche partitioning among Africa's large herbivores, particularly the numerous grazers.

Grazers Versus Browsers: Seasonal Diets

During wet seasons, vastly more potential food confronts large herbivores than they could potentially consume. Then rains cease, most trees shed their leaves and grasses desiccate and turn brown. Browsers seeking tree leaves encounter an absolute shortage of food because so few leaves remain within reach on trees and shrubs and most are chemically defended. Grazers face especially a reduction in forage quality, because the grass parts that remain above ground are dry, fibrous and hence difficult to digest. Thus dietary distinctions become heightened during the dry season.

Under wet season conditions of abundant food, the grass-dependent species form a dietary cluster in the lower right quadrant of Figure 10.1A. They include antelope representing three tribes (Alcelaphini, Reduncini and Hippotragini), along with African buffalo (*Syncerus caffer*), zebras (*Equus* spp.), warthogs

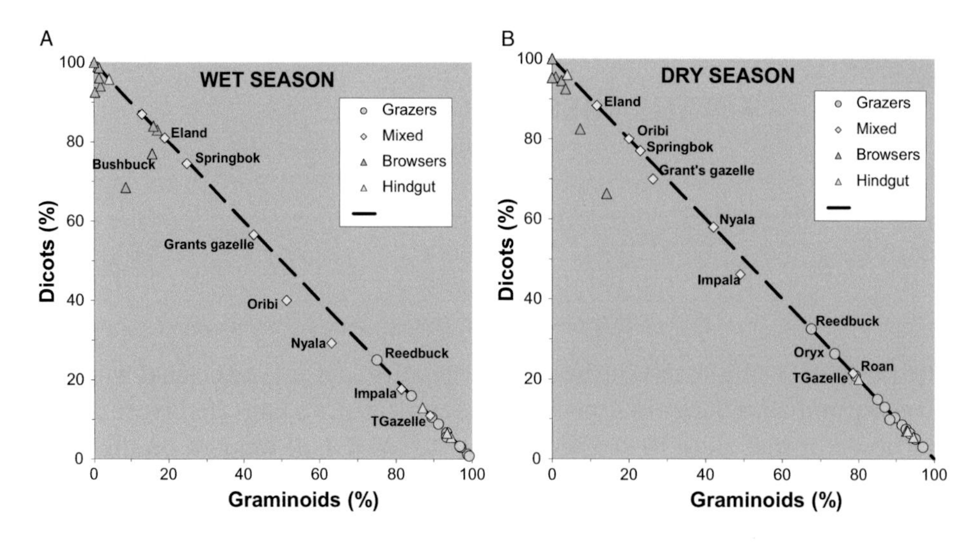

Figure 10.1 Seasonal diet composition of African large herbivores, comparing contributions from dicotyledonous trees, shrubs and forbs ('dicots') against that from graminoids (grasses and sedges). Deviations below the dashed line indicate contributions from fruits, pods and other plant parts. Note the distinct clusters of grazers and browsers with relatively few species falling within the 'mixed feeder' divide (modified and updated from Owen-Smith (1997) *Zeitschrift fur Saugetierkunde* 62, Suppl.II: 176–191).

(*Phacochoerus africana*), hippo (*Hippopotamus amphibius*) and white rhino (*Ceratotherium simum*; Figure 10.2). Browsers augment woody plant leaves with varying contributions from forbs, flowers and fruits, but consume only small amounts of grass. They include most of the tragelaphine (spiral-horned) antelope, various dwarf antelope, gerenuk (*Litocranius walleri*), giraffe (*Giraffa camelopardalus)* and black rhino (*Diceros bicornis*; Figure 10.3). Then there are the mixed feeders, encompassing various gazelles, along with impala (*Aepyceros melampus*), nyala (*Tragelaphus angasi*) and oribi (*Ourebia ourebi*), which consume a variable mix of tree, shrub and grass leaves. Nevertheless, they can be subdivided between those dependent mainly on grass and those consuming mainly browse. Note that forbs, encompassing non-grassy herbs, form part of the browse component. The African elephant (*Loxodonta africana*) is not represented in the figure because it is in a league on its own, consuming not only both grass and browse but also fibrous bark and roots not eaten by other large herbivores.[1]

During the dry season, the browsers concentrate more narrowly on browse while the mixed feeders shift their diets towards a greater contribution from woody plant foliage (Figure 10.1B). Africa's savanna browsers seldom consume leafless twigs, unlike moose in northern latitudes. Some grazers consume a little more browse in the dry season, notably the hippotragine antelope locally,

Figure 10.2 Grazers grazing grass mostly of similarly low height. I could have chosen an image of a waterbuck also grazing short grass, but the picture showing it grazing reeds indicates its ability to feed on quite tall graminoids when little else remains. (A) Blue wildebeest; (B) red hartebeest (*Alcelaphus buselaphus*); (C) topi (*Damaliscus lunatus*); (D) sable antelope; (E) roan antelope; (F) gemsbok (*Oryx gazella*); (G) waterbuck; (H) Uganda kob (*Kobus kob*); (I) common reedbuck; (J) plains zebra (*Equus quagga*); (K) African buffalo; (L) white rhino.

but the combined dicot contribution (trees plus forbs) to the diet of these grazers rarely exceeds 10–15 percent. Warthogs retain a primarily graminoid (grasses plus sedges) diet through the dry season by using their snouts to dig up corms and rhizomes. Porcupines (*Hystrix africaeaustralis*) also seek roots and bulbs, but these large rodents do not fall size-wise within the large herbivore category. Elephants fall back on woody plant parts like bark and roots in the

Figure 10.3 Various browsers and mixed feeders. (A) Giraffe; (B) black rhino; (C) eland; (D) greater kudu; (E) nyala; (F) bushbuck; (G) impala; (H) Grant's gazelle (*Gazella granti*); (I) Thomson's gazelle (*Eudorcas thomsonii*); (J) springbok (*Antidorcas marsupialis*); (K) steenbok; (L) dikdik.

dry season, along with grass roots.[1,2] The group of specialist grazers retain their primary dependency on grass, however dry, through the dry season.

Within the dietary component categorised broadly as grass or browse, particular vegetation constituents can play a key role in supporting animals through crucial periods of the year.[3] During my white rhino study, I noted how these mega-grazers shifted their grazing among grassland types during the course of the year (Figure 10.4). Rhinos concentrated on lawn grasslands as long as the predominant short grasses retained adequate foliage.[4] After the intensifying dry season caused plant regrowth to cease, rhinos turned their

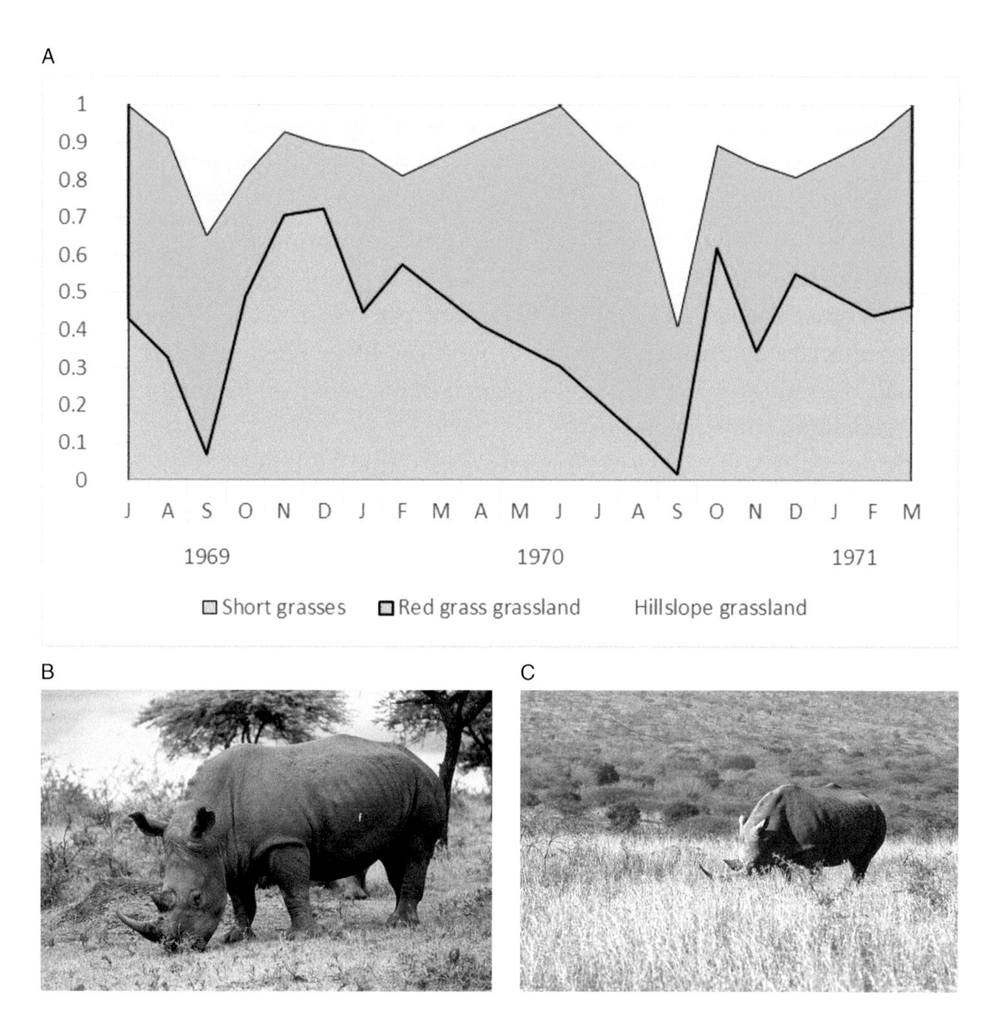

Figure 10.4 Shifting grazing among grassland types by white rhinos over the seasonal cycle in western Mfolozi GR. (A) Changing proportion of white rhinos grazing in different grassland types, shifting from short grass lawns during the wet season towards taller red grass grassland reserves during the dry season; (B) white rhino grazing short grass lawn; (C) white rhino grazing in hillslope grassland.

attention to the grasses growing under tree canopies, which retained green leaves longest. After these grasses became flattened, rhinos shifted their grazing to the extensive red grass grasslands. After these taller grasslands had mostly become grazed down, the rhinos moved up onto hillslopes where tall grass remained abundant. Wildebeest (*Connochaetes taurinus*) show similar shifts over the seasonal cycle, concentrating in grazing lawns as long as these retain adequate foliage.[5] Other grazers, like sable antelope (*Hippotragus niger*), buffalo and zebra, favour somewhat taller grass during the wet season and seek localities where grasses retain some green leaves in the dry season.[6] Eventually, all of the grass consumed may be in the form of dry, brown 'necromass', however inadequate nutritionally.

For both grazers and browsers, lowlands where gravitated soil moisture supports some green foliage retention can serve as key resource areas in the dry season.[3] Grazers also benefit from places where fires have removed the accumulated dry material (or 'top-hamper'), promoting some green grass regrowth before the wet season commences. In very dry years, grazers may fall back on grass species they normally do not eat.[7] Such forage reserves help buffer animals against starvation. When too little grass remains locally, animals may wander further afield in search of places where local rain showers have generated some green flush.[8]

The browsing kudus (*Tragelaphus strepsiceros*) that I followed in the Nylsvley Nature Reserve showed a more intricate sequence of dietary shifts over the seasonal cycle (Figure 10.5).[9,10,11] These habituated young animals ranged freely in a fairly large (213 ha) enclosure and selected their diets from the savanna plants available there. The vegetation on offer included 60 tree and shrub species plus over 100 forb and grass species. These could be grouped into relative palatability classes based on seasonal variation in how they were utilised.[9] While food remained plentiful through the wet season into the early dry season, the staple food component consumed by the kudus was constituted by the foliage of various broad-leaved deciduous trees and shrubs, including several species of bushwillow plus wild raisins. Fine-leaved thorn trees made only a small contribution because they were scarce in the enclosure; elsewhere, acacia species feature prominently in the wet season diet of kudus. However, what these kudus especially sought were various forbs and creepers lacking the woody twigs of trees and shrubs. Fruits and flowers were eaten when available, especially the large monkey oranges that ripened late in the dry season, as well as the pods of the acacias. After forbs had withered and deciduous trees had mostly shed their leaves, the kudus turned to the foliage remaining on evergreen or semi-evergreen trees and shrubs. Near the end of the dry season when the relatively palatable evergreens had developed browse lines with no leaves left within reach, the kudus began consuming the leaves of several deciduous species that they had previously ignored, which flushed new foliage ahead of the rains. Chemical analyses revealed that these apparently unpalatable species had high contents of condensed tannins in their leaves. Thus, each of these plant groups played a distinct key role in supporting the kudus at different stages of the seasonal cycle.[10]

These observations led me to generalise the following categories of functionally complementary food types over the seasonal cycle: (1) *sought out foods* of high nutritional value, represented for kudus by forbs, flowers and fruits; (2) *staple foods*, providing the bulk of the wet season diet, constituted by

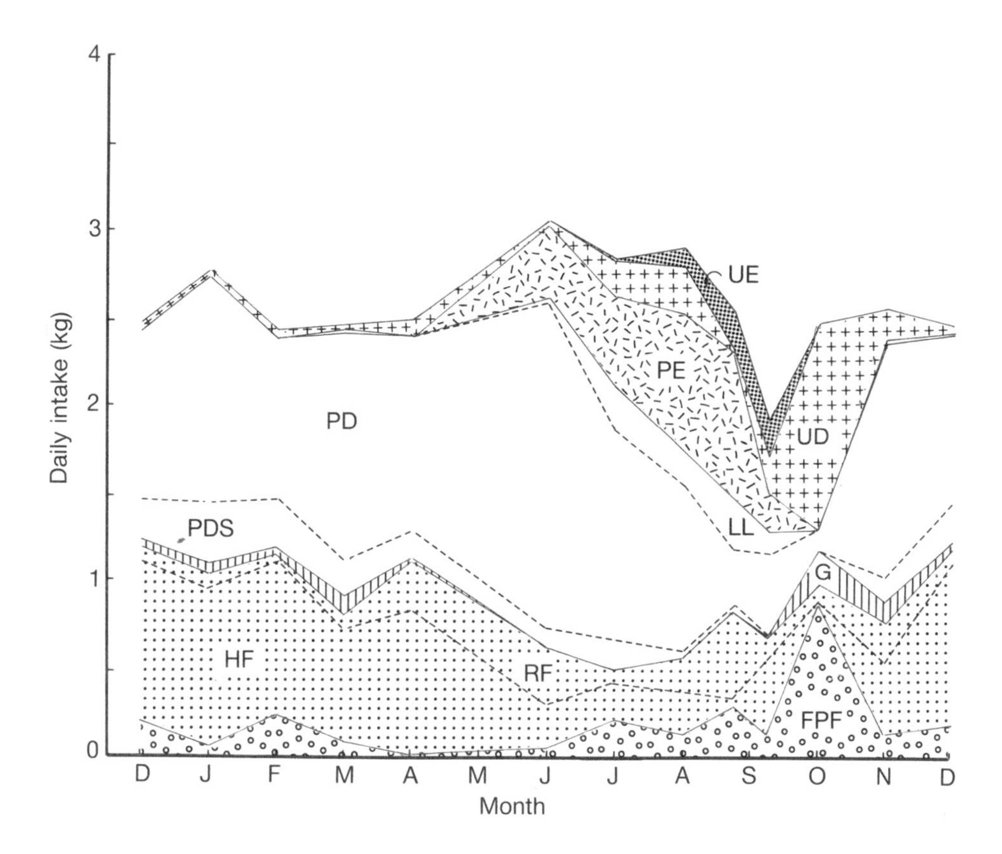

Figure 10.5 Monthly shifting diet of kudus in broad-leaved savanna woodland in Nylsvley Nature Reserve by plant types from the wet season through the dry season. FPF, fruits, pods and flowers; HF, herbaceous forbs; RF, robust forbs; G, grass; PDS, palatable deciduous spinescent trees; PD, palatable deciduous trees; LL, fallen leaf litter; PE, palatable evergreen trees; UD, unpalatable deciduous trees; UE, unpalatable evergreen trees (from Owen-Smith & Cooper (1989) *Journal of Zoology* 219:29–43).

palatable deciduous trees and shrubs; (3) *reserve foods*, added to the diet after the deciduous species had started shedding their leaves, constituted by relatively palatable evergreens; (4) *bridging foods*, consumed during transitional periods when little other edible food remained, formed by the early flushing deciduous species; and (5) *buffer foods*, eaten only when little other food remained late in the dry season, constituted by unpalatable evergreen shrubs.[12] Functionally, the high-quality and staple food types govern reproductive performance and offspring growth. Reserve food sources supply marginally adequate nutrition for survival through the dry season. Bridging and buffer resources slow rates of starvation during particularly adverse periods. Accordingly, each of these vegetation components serves as a 'key resource' under different conditions. A missing category could render a local habitat uninhabitable. This helps explain the absence of kudus from Serengeti NP where deciduous acacias are overwhelmingly dominant and evergreen browse sparse.

Reserve resources can be equated with the 'fall-back' foods recognised by primatologists. However, the importance of buffer resources, eaten only in years when extremes of food depletion are experienced, is widely overlooked. If these are missing, animals soon starve and die unless they have adequate fat reserves.

Besides protein and digestible energy, animals require mineral nutrients, most especially sodium and phosphorus. In Serengeti, 'hot spots' where herbivores concentrate are associated with locally elevated levels of sodium along with other mineral nutrients in grasses.[13,14] Some of the grasses prevalent in grazing lawns accumulate much higher levels of sodium in their leaves than is typical of most plants.[15] In Kruger NP, faecal phosphorus levels of grazers occupying basaltic soils were substantially higher than those of animals of the same species found on granitic soils.[16] In contrast, faecal nitrogen levels tended to be higher in regions with granitic geology, perhaps because grasses retain more green foliage into the dry season where soils are sandy and the tree cover is greater.[16,17]

Digesting Grass Versus Browse

Dietary distinctions among large herbivores are supported by physiological and anatomic features. Plant leaves are difficult to digest because the cell contents containing nutrients are enclosed within cell walls composed of cellulose and strengthened by lignin (the substance of wood). In order to extract sufficient nutrition from a leafy diet, large herbivores depend on microbial fermentation to break down the cellulose into volatile fatty acids, which can be metabolised to provide energy. To increase the surface area for bacterial action, herbivores must macerate plant tissues into fine enough particles. The leaves of trees and shrubs present somewhat different properties from the leaves and stems of grasses. Leaves of woody plants digest more rapidly, but less completely than grass leaves because much of the fibre they present is strengthened by lignin. The more fibrous leaves of grasses take longer to digest, but are ultimately more digestible, unless very stemmy. The digestibility of C_4 tropical grasses is restricted by the thickened walls surrounding the bundle sheath cells (see Chapter 6). This feature, along with the dry season desiccation, helps explain why the dietary distinction between grazers and browsers is especially clear-cut in Africa.

The incisor teeth play a frontline role in plucking bite-sized clusters of leaves. Ruminants lack incisors in their upper jaws and grasp plant parts between the lower incisors and a hard pad on the upper palate. The lower incisors of grazers tend to protrude forwards, allowing tougher stems to slide out while leaves are retained. Grazers have relatively wider incisor arcades and correspondingly broader muzzles than browsers,[18,19] facilitating cropping grass tillers. The narrower muzzles and more upright incisors of browsers enable them to pluck individual leaves or clusters of leaves from woody branches. Mixed feeders resemble browsers in their dental features. The reduncine antelope associated with wetland grasslands have somewhat narrower muzzles than wildebeest and other alcelaphine grazers. Non-ruminants like zebras clip plant parts between their upper and lower incisors, and thus ingest a higher proportion of grass stems than ruminants. The very largest herbivores crop grasses using their widened lips, like rhinos and hippos, or use extended noses to gather sufficient material, as elephants do. The projecting incisor teeth of elephants, forming tusks, aid in removing bark and in digging up roots of trees in addition to serving as weapons during fights among males. Warthogs dig with their fibrous snouts, not their tusks.

The molar teeth comminute the plant material ingested into finer particles. Browsers have elevated cusps on their molar surfaces, which help in puncture-crushing tree leaves to release the nutrients encapsulated. Grazers have complex patterns of enamel ridges that are more effective for shredding grass leaves. The height of the molar crown above the gum is taller (more 'hypsodont') among grazers than in browsers, so as to cope with the abrasive silica bodies contained within grass leaves and the grit that coats low-lying tufts.[20,21] Gazelles, which also feed at low levels but on forbs and shrublets, likewise have relatively high-crowned molars. There are also distinctions among grazers in degree of hypsodonty and relative size of molars versus premolars.[22] The alcelaphines, especially wildebeest, have higher molar crowns and reduced premolars compared with reduncines and hippotragines. Roan antelope (*Hippotragus equinus*) converge on zebra in some of their dental features, in line with their dependency on somewhat tall grass. Wear patterns on tooth surfaces provide further indications of the kind of material chewed: pits indicate browsing and scratches grazing.[23,24]

All large herbivores, apart from elephants, rely upon digestive fermentation to gain sufficient energy from plant parts consumed. They differ in where the fermentation chamber is located in the digestive tract and in features of the anatomy of this chamber related to the diet composition. Ruminants have enlarged compartments situated between the oesophagus and the true stomach or abomasum (Figure 10.6).[25] The largest of these chambers, called the rumen, forms the main fermentation chamber. The preceding one (the reticulum) receives the ingested food, and the next one (the omasum) absorbs excess water from the mix passed on to the true stomach (abomasum), where protein

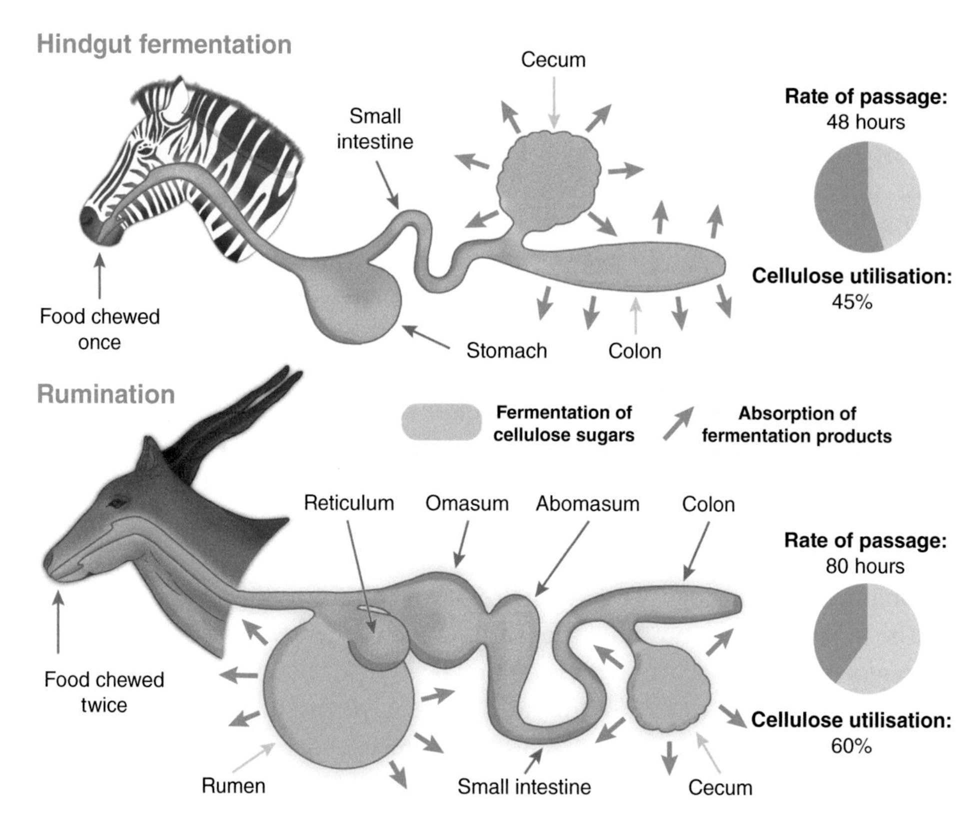

Figure 10.6 Comparative digestive anatomy and physiology of foregut and hindgut fermenters (artwork by Nuria Morales Garcia guided by Christine Janis).

gets digested under acid conditions. An advantage of the foregut fermentation is that food can be regurgitated and chewed further (ruminated) to reduce particle sizes for efficient digestion. The contents of the rumen must be kept neutral or only slightly acid, despite the release of fatty acids, otherwise the bacteria get digested. The pH is maintained by copious secretion of slightly alkaline saliva. Food residues are held in the rumen until particles become fine enough to be passed on. Grass residues tend to form a floating raft until they are reduced in size sufficiently to sink.[26] Hence grazers have relatively larger rumens than browsers, enabling them to retain food residues for longer. Browsers ingesting woody stems along with leaves must allow larger particles to pass from the rumen than grazers do. They are better adapted than grazers to cope with bloat caused by gases released by rapidly fermenting starches and sugars. Consequently, grazers seldom consume succulent fruits. Mixed feeding impala, which switch their diet from mostly grass to mainly browse seasonally, increase the surface area formed by papillae lining the interior of the rumen to cope with rapidly fermenting browse.[27] Although the eland is commonly classified as a mixed feeder, because in places it consumes substantial amounts of grass, its digestive anatomy is typical of that of a browser. Much of their low-level feeding is actually on various forbs and dwarf shrubs amid the grasses. The coupled features of digestive anatomy and physiology constrain the relative portions of grass and browse that large herbivores can consume.[28] Because of their foregut fermentation, ruminants release gases generated during fermentation by burping.

Non-ruminants, or hindgut fermenters, have an expanded and sacculated large intestine along with an enlarged side compartment called the cecum, which is situated at the junction between the small and large intestines (homologous with our appendix).[29] An advantage of hindgut fermentation is that plant proteins as well as soluble carbohydrates get digested in the true stomach and small intestine before fermentation occurs, meaning that their products are absorbed more completely. However, because chewing post-ingestion is precluded, hindgut fermenters digest cell wall constituents less completely than do ruminants. In compensation, they pass food material through the gut faster and are less adversely affected by indigestible fibre.[30,31] Differences in gut anatomy among grazing and browsing non-ruminants, such as the two African rhinos, have yet to be documented. Hindgut fermenters release gases by farting.

Hippos have a capacious foregut that is partly compartmentalised, but do not ruminate. Instead they rely on prolonged retention of the forage they consume to extract sufficient nutrition for their comparatively low metabolic requirements.[32] White rhinos are able to digest grass fibre nearly as effectively as several medium-sized ruminants, because larger size prolongs digestive retention.[4] Elephants pass food especially rapidly through their digestive tracts, absorbing carbohydrates and other nutrients from highly fibrous plant parts like tree bark and roots, without digesting much fibre.[32] Their digestive physiology is basically similar to that of much smaller herbivores like large rodents and hares.

Because of their dental and anatomic adaptations, browsers digest grass poorly and thus gain less nutritional value when they consume grass than do grazers. When the kudus that I followed did graze, they chose the leafiest species or post-burn regrowth. Browsing ruminants would starve with rumens clogged by floating mats if they ate much grass. Grazers are limited in the amounts of rapidly fermenting forbs and leaves they can consume because they are less able to release the bloating gases produced. Furthermore, they have relatively smaller livers than browsers[27] and thus less capacity to deal with the potentially toxic chemicals commonly present in tree, shrub and forb leaves, but rare among grasses. Tree leaves commonly contain tannins,[33,34] which bind with proteins (both plant proteins and digestive enzymes) to further inhibit digestive degradation. Some browsers have relatively large salivary glands, secreting proteins that bind effectively to tannins, thereby reducing their effects. Browsers also seem to differ in their tolerance for particular kinds of chemicals. Kudus readily feed on bushwillows, which test positive for polyphenols, but elephants frequently discard leaves of plants in this genus.[1] Black rhinos commonly browse succulent euphorbias despite the toxic milky sap they contain.

Body Size

Body size influences diet composition because of how metabolic requirements scale with body mass. The resting metabolic rate is a function of body mass raised to the power 0.75, rather than being directly proportional to mass (i.e. exponent of 1).[35] This means that, per unit of body mass, larger animals require less energy than smaller animals ($M^{0.75}/M^1 = M^{-0.25}$). Accordingly, larger ungulates can subsist on poorer-quality diets higher in fibre and lower in digestible energy and protein than smaller ones.[36,37] Nevertheless, larger ungulates could satisfy their lower metabolic requirements by consuming relatively less food per day, rather than food of lower quality. While a buffalo or a wildebeest needs to eat about 2–3 percent of its body mass per day (as plant dry mass over animal live mass), elephants and rhinos typically consume only 1–1.5 percent of their body mass daily. As a result, very large herbivores like elephants, rhinos and giraffes may show diets similar in protein content to those eaten by smaller ungulates at times when food is plentiful. Their metabolic tolerance for poor-quality food comes into play during the dry season when the green leaf supply fades. Larger herbivores can subsist on poorer-quality food, but smaller herbivores can get by on less food.

Among browsers, a mass ratio by a factor of ~2.5 is apparent in the series duiker→bushbuck (*Tragelaphus scriptus*)→nyala or lesser kudu (*T. imberbe*)→ greater kudu (*T. strepsiceros*)→eland (*T. oryx*)→giraffe. Besides metabolism, body size influences the height reach while bite size requirements of browsers are restricted by leaf size. A 180-kg kudu nibbling the tiny (1.5 g) leaves of umbrella thorn trees all day would starve, whereas a 45-kg impala would be adequately nourished. Leaves that had been out of reach up in tree canopies become available to browsers when shed during the dry season, until the wind disperses them. Kudus gain less from fallen leaves than impalas because the rate of food intake they obtain by plucking single leaves is too slow for their greater quantitative requirements.[38] This allows impalas to be abundant in places where kudus are absent.

Mouth size and hence bite dimensions also influence how browsers cope with the physical defences presented by plants to restrict losses to herbivores.[39] Small antelope can nibble the leaves between thorns, but potentially

Figure 10.7 Browsing actions. (A) Giraffe stripping leaves among thorns with its lower incisors; (B) kudus plucking thorny branch tips; (C) kudu pruning forb branchlets; (D) impala nibbling leaves among thorns.

get entangled by hooked thorns (see Chapter 7). Larger browsers requiring bigger bites can manage hooked thorns by swallowing the twigs the right way. They are more greatly restricted by long straight thorns or spines. Giraffes use their tongues and lower incisors to strip multiple leaves from branch tips of fine-leaved acacia trees, thereby largely overcoming the restriction posed by thorns (Figure 10.7).

For grazers, the grass height preferentially grazed has been related to body size.[36] However, there is no consistent relationship. The two smallest grazers – mountain reedbuck (*Redunca fulvorufula*) and oribi – both occupy quite tall grasslands.[40] Impalas graze quite a wide range in grass height.[41] White rhinos and hippos actually graze the shortest grass, despite their large size. Among the set of grazers weighing between 100 and 250 kg, some favour short grass (notably wildebeest) and others quite tall grass (like sable and roan antelope). The ability of roan antelope to handle tall and hence fibrous grass is facilitated by digestive passage rates more similar to those of browsers than other grazers,[42] although the supporting anatomy has yet to be investigated. Among the reduncine grazers, an intriguing sequence in relative body mass is apparent: from mountain reedbuck to reedbuck (*Redunca* spp.), various wetland kobs (*Kobus* spp.) and waterbuck (*K. ellipsiprimnus*). The various species of wetland kobs replace one another geographically, except in northern Botswana where a remnant number of puku (*K. vardoni*) exists close alongside much larger numbers of red lechwe (*K. leche*).[43]

Coping With Heat and Aridity

While seeking food, savanna herbivores must cope with high heat loads towards midday and, in higher southern latitudes, temperatures that may drop towards freezing overnight during the austral winter. Ungulates usually seek shade through midday (Figure 10.8A,B),[44] but where shady trees are sparse they must rely on the reflectance of their hair coat or pelage to reduce the radiant heat load (Figure 10.8C).[45] Thermal loading restricts the amount of time that animals have available to feed, especially during the midday period when conditions are hottest. The activity of kudus was curtailed on days when the maximum daily temperature exceeded 36°C in the wet season and 30°C in the dry season.[46] The seasonal difference was due to a change from a shaggier hair coat for the winter cold to a sleeker one to cope with summer heat. Kudus became compromised when cold spells followed hot days during the transition between the dry season and the wet season, when food is particularly sparse, and some may die of hypothermia.[47] Lethal conditions depended not on how low temperatures dropped overnight, but on the persistence of cool, wet and windy conditions through the day, inhibiting feeding. Grazers seem less vulnerable to cold-related mortality than browsers, perhaps because they benefit from heat generated by the fermentation of dry grass.

The very largest herbivores, such as elephants, rhinos and buffalos, are sparsely haired, because their problem is getting rid of excess heat generated by activity. As body size increases, the surface area available to dissipate heat decreases relative to the body mass producing the heat. Hence these big animals seek cooling by wallowing in pools or mud, when opportunities are presented (Figure 10.8E,F). Warthogs, although much smaller, are also rather hairless, because they seek security underground through the night and must be active through midday to meet their food needs.

White rhinos have exceptionally large sweat glands, which can drench the body surface when they overheat. Elephants lack sweat glands, but do lose water through their skin, which is corrugated to increase surface area.[48] Their large ears contribute to body cooling via blood circulated through them, with ear-flapping promoting convectional heat losses. While zebras and most antelope possess sweat glands, they restrict sweating because of the water loss incurred. Smaller antelope conserve water using nasal panting rather than sweating so that condensed fluid in the nasal passages can be recycled. Heat

Figure 10.8 Coping with midday heat and sunshine. (A) Wildebeest occupying shade in the Kalahari; (B) gemsbok herd clustered in shade in the Kalahari; (C) red hartebeest displaying its shiny pelage reflecting radiant heat in Kalahari; (D) zebra enduring sunshine laterally with stripes supposedly ameliorating the heat load, Kruger NP; (E) buffalo immersed in mud wallow, Manyaleti; (F) white rhino plastered with mud after wallow, Mfolozi GR.

generated by muscular activity coupled with exposure to radiant energy can restrict how long animals continue foraging into the heat of midday.

However, large herbivores gain a substantial amount of water from their food. In addition to water absorbed from green leaves, further water is released as a product of carbohydrate fermentation. Ruminants reabsorb water from digested residues passing through the large intestine, thereby producing dry dung pellets.[49] Grazing ruminants inhabiting wet grasslands, including the reduncine antelope as well as buffalo, have less-lengthy large intestines than dry savanna grazers like the alcelaphines and thus produce moister dung pats. The dung pats of hindgut fermenters are also moist. The kidneys assist further by concentrating urine. Ungulate species differ in the effectiveness of their adaptations,[50] and water needs also depend on diet. Grazers consuming dry grass need to drink more frequently than browsers able to find some green leaves even during the dry season. Hindgut fermenters have a lesser capacity to reabsorb moisture than ruminants, so that rhinos, zebras and elephants have a greater dependence on surface water than most ruminants.

Ultimately, water lost from sweat glands or via panting must be replaced, which requires travel to and from places retaining surface water. Non-ruminant grazers like zebras generally need to visit surface water sources daily or every other day during the dry season (Figure 10.9), while ruminants producing dry dung pellets may drink only at intervals of 3–4 days.[51]

For much of the year, grazers can access water from ephemeral pans that retain pools for varying periods into the dry season.[52] Browsers with access to

Figure 10.9 Drinking at waterholes in the form of pans. (A) Buffalo at dam, Kruger NP; (B) zebras at pan, Tarangire NP, Tanzania; (C) elephants at pan, Hwange NP; (D) kudus at pan, Hwange NP.

green leaves may not need to drink at all, unless conditions become very dry. Supremely independent of surface water are the various dwarf antelope, like steenbok (*Raphicerus campestris*) and dikdik (*Madoqua* spp.), which remain in small home ranges far from surface water year-round.[53] Among larger grazers, the oryxes (*Oryx* spp.) can also remain independent of surface water except during extremely dry conditions.[54] South African oryxes (*O. gazella*) can overcome the lack of surface water in the semi-desert areas they inhabit by consuming various moisture-containing melons.

Travel to and from water restricts time available for feeding and habitat occupation during the dry season, especially among grazers.[51] Most grazers must remain within 4 km of rivers or pools retaining water.[55,56] The salinity of the water remaining can also affect its use for drinking and contributes to the seasonal movement of migratory grazers from the Serengeti plains.[57] Concentrations of water-dependent species near drinking places draw predators, increasing the exposure of these ungulates to being killed. With food also running short locally near water, the stress levels experienced by grazers are greatly elevated during the dry season. As will be discussed in Chapter 12, this can lead to population crashes.

Body Fat Reserves

The fat stores of African ungulates seldom exceed 5–10 percent of body mass,[58,59] much less than the 10–20 percent typical of deer inhabiting high northern latitudes. This limitation may seem surprising, considering the lean conditions experienced by savanna dwellers through the dry season. For grazers, the problem is not the amount of potential food remaining, but rather how to digest the dry grass. The gut microbes remain active only if supplied with adequate protein, which can be obtained via breakdown of the muscular tissues of the host. This helps explain why savanna ungulates depend more on body protein rather than fat in the dry season and recycle urea via the hindgut to maintain the activities of the gut bacteria in digesting cellulose.[60] A further consideration is that fat stored subcutaneously interferes with the ability of animals to get rid of excess body heat. The fat stores of African ungulates are located mainly in the mesenteries around the gut and near the kidneys, where it is less insulating.

Furthermore, body fat is needed not only to enable survival through the dry season, but also to subsidise particular activities. For male ungulates, the peak expenditure occurs around the time when competition for mating opportunities occurs, typically early in the dry season. Male impalas deplete most of their fat reserves during this period and enter the remainder of the dry season with little kidney fat, although much fat is still retained in bone marrow.[61]

Female impala retain body fat through the dry season and drawn on it to support the peak energy demands of late gestation and early lactation around the beginning of the wet season. Once animals have used up their mesenteric fat and start calling on bone marrow reserves, they are at risk of starvation.[62]

White rhinos were renowned among early hunters for the amount of subcutaneous fat they yielded, while hippos can also carry substantial fat stores.[63] Eland carry more body fat than other ruminants.

Evading Predation

Exposure to predation contributes to niche partitioning by restricting habitat occupation. The first line of defence by potential prey entails detecting an approaching predator before it comes close enough to launch a successful attack. Herd formation increases security by providing more eyes for surveillance. Grazing ungulates are particularly vulnerable while cropping grass head-down. They must lift their heads and look around periodically in order to detect approaching carnivores. However, this vigilance is at the expense of feeding time.

Speedy running helps evade an attack. The fastest antelope are topi (*Damaliscus lunatus cokei*) and hartebeest (*Alcelaphus buselaphus*), as discovered by big game hunters pursuing them on horseback. However, a fast gallop is less effective when there are trees in the way. Hence these alcelaphine antelope typically occupy open savanna grasslands. Wildebeest are almost as speedy, but their special feature is endurance locomotion, facilitating migration.[64] They strongly avoid entering thicker patches of woodland.[65] Larger ungulates run less fast and must rely on overt defence to ward off attacks by lions. Zebras attempt to deflect attacks from behind with flailing hooves. Buffalos resist predation by closing ranks in large herds. Mega-herbivores weighing over 1000 kg are invulnerable to being killed by lions once they reach maturity, but must still protect their offspring. Sable antelope and oryx deploy their horns, possessed by females as well as males, to ward off hyenas or wild dogs trying to take their young, but this defence is ineffective against lions. Consequently, these hippotragines are found in habitats where their risk of predation is reduced because there are fewer lions – sable antelope in savanna woodlands and oryx in arid regions. Roan antelope occupy moist savannas where low overall herbivore densities support few lions. Hartebeest also seem quite vulnerable to predation[66,67] and are commonly present in regions with fewer other grazers. Kob and their congeners inhabit wetland grasslands where waterbodies hamper chases by lions and other carnivores. Waterbuck, generally found on dry land near rivers, have a novel deterrent – a musky skin secretion, which seems to make them less palatable to carnivores as well as humans.[68]

Browsers living in wooded habitats evade predation by jumping or dodging and are elusive because of their lower densities than grazers. Impalas are outstanding long-jumpers, while eland and kudu are renowned high-jumpers. Gazelles and steenboks are particularly adept at dodging round bushes when chased by cheetahs (*Acinonyx jubatus*).[69] Healthy antelope may advertise this by pronking or stotting up and down, signalling to the predator that chasing after them would be wasted effort.[70] These differences in predator evasion tactics are supported by physical differences. The alcelaphine grazers have high shoulders forming a pivot, while the tragelaphine browsers have high rumps, powering jumping.

The risk of being killed by a predator increases at night, when darkness obscures stalking lions and leopards and hyenas are also most active. Most of the larger ungulates forage for less time and move less at night than during the day.[4,71] At night, wildebeest keep away from the edges of the open glades with short grass that they inhabit, where vegetation cover increases.[72] Ungulates typically schedule their journeys to water during mid-morning while visibility is great and rising temperatures inhibit carnivore activity. Only the largest ungulates, from buffalo to rhinos and elephants, drink frequently around or after sun-down when temperatures are cooler. They are also equally active day and night. Small antelope the size of a steenbok or dikdik have to contend with diurnal predation by large eagles as well as cheetahs and wild dogs, and so may be more active nocturnally when they are better concealed.

Trade-offs must be made when food runs short. Animals cannot survive very long without food and so must take a chance on meeting up with a predator in more risky habitats in these circumstances.[62] Hence, the risk of being killed increases during the dry season for many species.[73] The conundrum is, should this mortality be ascribed 'top-down' to predation, or 'bottom-up' to resource limitations? These two influences on niche occupation are inextricably entangled.[74]

Overview

Large herbivores are adapted to their specific plant-based diets through a combination of dentition, digestive anatomy, toxin physiology and body size. These features come into play especially during the dry season when little food remains and most of it is low in nutritional value and chemically or structurally defended. However, there is much overlap in the plant species favoured among both grazers[41,75,76] and browsers.[9] We need to look beyond dietary adaptions to consider other features of species niches.

Dietary partitioning alone is inadequate to explain the regional coexistence among medium–large ruminants of rather similar size. The role of predation in restricting habitat occupation, particularly in relation to the woody plant cover, becomes of overriding importance. Habitat features become less important for herbivores from zebra and buffalo size upwards. They occupy wooded as well as open savanna habitats and rely on their capability to ward off lion attacks. Surface water requirements form an additional constraint on where grazers can be during the dry season. The large grazers capable of defending themselves against lions happen to be the most water-dependent. They attract predation when they concentrate near water in the late dry season and they become weakened from food deficiencies.

Africa's current diversity of medium–large grazers is unmatched on other continents. However, for fair comparisons we must look prior to the end-Pleistocene extinctions. Earlier, South America contained more species of large herbivore than Africa does today. However, those that grazed were almost only either equids or mega-sized, adapted via hindgut fermentation to handle tall fibrous grasses.[77] The only wild grazer still found there is the vicuna (*Vicugna vicugna*), a camelid occupying fertile meadows in high Andean plateaus. The pampas deer (*Ozotoceros bezoarticus*) is a mixed feeder inhabiting the low-lying pampas grasslands of Argentina and Pantanal wetlands rather than the cerrado savanna. North America retained a large grazing bovine in the form of the American bison (*Bison bison*), once hugely abundant through its grasslands; but the deer found there are either browsers or mixed feeders. No deer anywhere in the world is as specialised anatomically for a purely grass diet as the grazing bovids. Wild equids disappeared from North America soon after humans arrived. Various mammoths had been the major grazers, as they were over most of Eurasia outside the tropics, coupled there with woolly rhinos. Tropical Asia retains an intact large mammal fauna, with grazers represented by various forms of wild cattle, but none of these feeds exclusively on grass, and neither does the European bison. None of Australia's large kangaroos feeds narrowly on grass – they all consume a somewhat mixed diet. The kangaroos that went extinct following the arrival of modern humans were mostly large browsers.

African savannas uniquely feature the ecology of large grazers, vastly more diverse than found in other continents even before the late Pleistocene extinctions. The next chapter will familiarise you with the ecology of Africa's large carnivores, the predators that threaten these herbivores and also, of course, the larger primates.

Suggested Further Reading

Du Toit, JT; Cumming, DHM. (1999) Functional significance of ungulate diversity in African savannas and the ecological implications of the spread of pastoralism. *Biodiversity and Conservation* 8:1643–1661.

Kingdon, J; Hoffmann, M (eds) (2013) *The Mammals of Africa*. Vols I–VI. A & C Black, London.

Owen-Smith, N. (2002) *Adaptive Herbivore Ecology. From Resources to Populations in Variable Environments*. Cambridge University Press, Cambridge.

References

1. Owen-Smith, N; Chafota, J. (2012) Selective feeding by a megaherbivore, the African elephant (*Loxodonta africana*). *Journal of Mammalogy* 93:698–705.
2. Barnes, RFW. (1982) Elephant feeding behaviour in Ruaha National Park, Tanzania. *African Journal of Ecology* 20:123–136.
3. Illius, AW; O'Connor, TG. (2000) Resource heterogeneity and ungulate population dynamics. *Oikos* 89:283–294.
4. Owen-Smith, RN. (1988) *Megaherbivores: The Influence of Very Large Body Size on Ecology*. Cambridge University Press, Cambridge.
5. Yoganand, K; Owen-Smith, N. (2014) Restricted habitat use by an African savanna herbivore through the seasonal cycle: key resources concept expanded. *Ecography* 37:969–982.
6. Macandza, VA, et al. (2012) Habitat and resource partitioning between abundant and relatively rare grazing ungulates. *Journal of Zoology* 287:175–185.
7. Macandza, VA. (2003) Forage selection of African buffalo through the late dry season in the Satara region of the Kruger National Park. *South African Journal of Wildlife Research* 34: 113–121.
8. Staver, AC, et al. (2019) Grazer movements exacerbate grass declines during drought in an African savanna. *Journal of Ecology* 107:1482–1491.
9. Owen-Smith, N; Cooper, SM. (1987) Palatability of woody plants to browsing ruminants in a South African savanna. *Ecology* 68:319–331.
10. Owen-Smith, N; Cooper, SM. (1989) Nutritional ecology of a browsing ruminant, the kudu (*Tragelaphus strepsiceros*), through the seasonal cycle. *Journal of Zoology* 219:29–43.
11. Owen-Smith, N. (1994) Foraging responses of kudus to seasonal changes in food resources: elasticity in constraints. *Ecology* 75:1050–1062.
12. Owen-Smith, RN. (2002) *Adaptive Herbivore Ecology: From Resources to Populations in Variable Environments*. Cambridge University Press, Cambridge.
13. McNaughton, SJ. (1988) Mineral nutrition and spatial concentrations of African ungulates. *Nature* 334:343–345.
14. McNaughton, SJ, et al. (1997) Promotion of the cycling of diet-enhancing nutrients by African grazers. *Science* 278:1798–1800.
15. Stock, WD, et al. (2010) Herbivore and nutrient control of lawn and bunch grass distributions in a southern African savanna. *Plant Ecology* 206:15–27.
16. Grant, CC, et al. (2000) Nitrogen and phosphorus concentration in faeces: an indicator of range quality as a practical adjunct to existing range evaluation methods. *African Journal of Range and Forage Science* 17:81–92.

17. Grant, CC, et al. (1996) The usefulness of faecal phosphorus and nitrogen in interpreting differences in live-mass gain and the response to P supplementation in grazing cattle in arid regions. *Onderstepoort Journal of Veterinary Research* 63:121–126.
18. Gordon, IJ; Illius, AW. (1988) Incisor arcade structure and diet selection in ruminants. *Functional Ecology* 2:15–22.
19. Janis, CM; Ehrhardt, D. (1988) Correlation of relative muzzle width and relative incisor width with dietary preference in ungulates. *Zoological Journal of the Linnean Society* 92:267–284.
20. Mendoza, M; Palmqvist, P. (2008) Hypsodonty in ungulates: an adaptation for grass consumption or for foraging in open habitat? *Journal of Zoology* 274:134–142.
21. Damuth, J; Janis, CM. (2011) On the relationship between hypsodonty and feeding ecology in ungulate mammals, and its utility in palaeoecology. *Biological Reviews* 86:733–758.
22. Codron, D, et al. (2008) Functional differentiation of African grazing ruminants: an example of specialized adaptations to very small changes in diet. *Biological Journal of the Linnean Society* 94:755–764.
23. Schubert, BW, et al. (2006) Microwear evidence for Plio-Pleistocene bovid diets from Makapansgat Limeworks Cave, South Africa. *Palaeogeography, Palaeoclimatology, Palaeoecology* 241:301–319.
24. Ungar, PS. (2017) *Evolution's Bite: A Story of Teeth, Diet, and Human Origins*. Princeton University Press, Princeton, NJ.
25. Hofmann, RR; Stewart, DRM. (1972) Grazer or browser: a classification based on the stomach-structure and feeding habits of East African ruminants. *Mammalia* 36:226–240.
26. Clauss, M, et al. (2008) The morphophysiological adaptations of browsing and grazing mammals. In Gordon, IJ; Prins, HHT (eds) *The Ecology of Browsing and Grazing*. Springer, Berlin, pp. 47–88.
27. Hofmann, RR. (1989) Evolutionary steps of ecophysiological adaptation and diversification of ruminants: a comparative view of their digestive system. *Oecologia* 78:443–457.
28. Codron, D; Clauss, M. (2010) Rumen physiology constrains diet niche: linking digestive physiology and food selection across wild ruminant species. *Canadian Journal of Zoology* 88:1129–1138.
29. Langer, P. (1988) *The Mammalian Herbivore Stomach: Comparative Anatomy, Function and Evolution*. Gustav Fischer, Jena.
30. Duncan, P, et al. (1990) Comparative nutrient extraction from forages by grazing bovids and equids: a test of the nutritional model of equid/bovid competition and coexistence. *Oecologia* 84:411–418.
31. Meyer, K. (2010) The relationship between forage cell wall content and voluntary food intake in mammalian herbivores. *Mammal Review* 40:221–245.
32. Clauss, M, et al. (2007) The relationship of food intake and ingesta passage predicts feeding ecology in two different megaherbivore groups. *Oikos* 116:209–216.
33. Cooper, SM; Owen-Smith, N. (1985) Condensed tannins deter feeding by browsing ruminants in a South African savanna. *Oecologia* 67:142–146.

34. Owen-Smith, N. (1993) Woody plants, browsers and tannins in southern African savannas. *South African Journal of Science* 89:505–510.
35. Farrell-Gray, CC; Gotelli, NJ. (2005) Allometric exponents support a 3/4-power scaling law. *Ecology* 86:2083–2087.
36. Bell, RHV. (1971) A grazing ecosystem in the Serengeti. *Scientific American* 225:86–93.
37. Jarman, P. (1974) The social organisation of antelope in relation to their ecology. *Behaviour* 48:215–267.
38. Owen-Smith, N; Cooper, SM. (1985) Comparative consumption of vegetation components by kudus, impalas and goats in relation to their commercial potential as browsers in savanna regions. *South African Journal of Science* 81:72–76.
39. Cooper, SM; Owen-Smith, N. (1986) Effects of plant spinescence on large mammalian herbivores. *Oecologia* 68:446–455.
40. Owen-Smith, N. (1985) Niche separation among African ungulates. In Vrba, ES (ed.) *Species and Speciation*. Transvaal Museum Monograph 4. Transvaal Museum, Pretoria, pp. 167–171.
41. Arsenault, R; Owen-Smith, N. (2008) Resource partitioning by grass height among grazing ungulates does not follow body size relation. *Oikos* 117:1711–1717.
42. Heitkonig, IMA. (1993) Feeding strategy of roan antelope in a low nutrient savanna. PhD thesis, University of the Witwatersrand, Johannesburg.
43. O'Shaughnessy, R, et al. (2014) Comparative diet and habitat selection of puku and lechwe in northern Botswana. *Journal of Mammalogy* 95:933–942.
44. Fuller, A, et al. (2016) Towards a mechanistic understanding of the responses of large terrestrial mammals to heat and aridity associated with climate change. *Climate Change Responses* 3:1–19.
45. Finch, VA. (1972) Thermoregulation and heat balance of the East African eland and hartebeest. *American Journal of Physiology* 222:1374–1379.
46. Owen-Smith, N. (1998) How high ambient temperature affects the daily activity and foraging time of a subtropical ungulate, the greater kudu (*Tragelaphus strepsiceros*). *Journal of Zoology* 246:183–192.
47. Owen-Smith, N. (2000) Modeling the population dynamics of a subtropical ungulate in a variable environment: rain, cold and predators. *Natural Resource Modeling* 13:57–87.
48. Dunkin, RC, et al. (2013) Climate influences thermal balance and water use in African and Asian elephants: physiology can predict drivers of elephant distribution. *Journal of Experimental Biology* 216:2939–2952.
49. Woodall, PF; Skinner, JD. (1993) Dimensions of the intestine, diet and faecal water loss in some African antelope. *Journal of Zoology* 229:457–471.
50. Kihwele, ES, et al. (2020) Quantifying water requirements of African ungulates through a combination of functional traits. *Ecological Monographs* 90:e01404.
51. Cain III, JW, et al. (2012) The costs of drinking: comparative water dependency of sable antelope and zebra. *Journal of Zoology* 286:58–67.
52. Naidoo, R, et al. (2020) Mapping and assessing the impact of small-scale ephemeral water sources on wildlife in an African seasonal savannah. *Ecological Applications* 30:e02203.

53. Manser, MB; Brotherton, PNM. (1995) Environmental constraints on the foraging behaviour of a dwarf antelope (*Madoqua kirkii*). *Oecologia* 102:404–412.
54. Knight, M. (2013) *Oryx gazella*. In Kingdon, J; Hoffman, AM (eds) *The Mammals of Africa VI Pigs, Hippopotamuses, Chevrotain, Giraffes, Deer and Bovids*. A & C Black, London, pp. 572–576.
55. Western, D. (1975) Water availability and its influence on the structure and dynamics of a savannah large mammal community. *African Journal of Ecology* 13:265–286.
56. Smit, IPJ, et al. (2007) Do artificial waterholes influence the way herbivores use the landscape? Herbivore distribution patterns around rivers and artificial surface water sources in a large African savanna park. *Biological Conservation* 136:85–99.
57. Gereta, E; Wolanski, E. (1998) Wildlife–water quality interactions in the Serengeti National Park, Tanzania. *African Journal of Ecology* 36:1–14.
58. Ledger, HP. (1968) Body composition as a basis for a comparative study of some East African mammals. *Symposia of the Zoological Society of London* 21:289–310.
59. Smith, NS. (1970) Appraisal of condition estimation methods for East African ungulates. *East African Wildlife Journal* 8:123–129.
60. Sinclair, ARE; Duncan, P. (1972) Indices of condition in tropical ruminants. *African Journal of Ecology* 10:143–149.
61. Dunham, KM; Murray, MG. (1982) The fat reserves of impala, *Aepycevos melampus*. *African Journal of Ecology* 20:81–87.
62. Sinclair, ARE; Arcese, P. (1995) Population consequences of predation-sensitive foraging: the Serengeti wildebeest. *Ecology* 76:882–891.
63. Selous, FC. (1881) *A Hunter's Wanderings in Africa*. R. Bentley & Son, London.
64. Curtin, NA, et al. (2018) Remarkable muscles, remarkable locomotion in desert-dwelling wildebeest. *Nature* 563:393–396.
65. Stabach, JA, et al. (2016) Variation in habitat selection by white-bearded wildebeest across different degrees of human disturbance. *Ecosphere* 7:e01428.
66. Georgiadis, NJ, et al. (2007) Savanna herbivore dynamics in a livestock-dominated landscape. II: Ecological, conservation, and management implications of predator restoration. *Biological Conservation* 137:473–483.
67. Ng'weno, CC, et al. (2017) Lions influence the decline and habitat shift of hartebeest in a semiarid savanna. *Journal of Mammalogy* 98:1078–1087.
68. Spinage, CA. (2013) *Kobus ellipsiprymnus*. In Kingdon, J; Hoffmann, M (eds) *Mammals of Africa VI Pigs, Hippopotamuses, Chevrotain, Giraffes, Deer and Bovids*. A & C Black, London, pp. 461–468.
69. Mills, MGL. (2017) *Kalahari Cheetahs: Adaptations to an Arid Region*. Oxford University Press, Oxford.
70. Caro, TM. (1986) The functions of stotting: a review of the hypotheses. *Animal Behaviour* 34:649–662.
71. Owen-Smith, N; Goodall, V. (2014) Coping with savanna seasonality: comparative daily activity patterns of African ungulates as revealed by GPS telemetry. *Journal of Zoology* 293:181–191.

72. Owen-Smith, N; Traill, LW. (2017) Space use patterns of a large mammalian herbivore distinguished by activity state: fear versus food? *Journal of Zoology* 303:281–290.
73. Owen-Smith, N. (2008) Changing vulnerability to predation related to season and sex in an African ungulate assemblage. *Oikos* 117:602–610.
74. Owen-Smith, N. (2015) How diverse large herbivores coexist with multiple large carnivores in African savanna ecosystems: demographic, temporal and spatial influences on prey vulnerability. *Oikos* 124:1417–1426.
75. Kleynhans, EJ, et al. (2011) Resource partitioning along multiple niche dimensions in differently sized African savanna grazers. *Oikos* 120:591–600.
76. Pansu, J, et al. (2019) Trophic ecology of large herbivores in a reassembling African ecosystem. *Journal of Ecology* 107:1355–1376.
77. Owen-Smith, N. (2013) Contrasts in the large herbivore faunas of the southern continents in the late Pleistocene and the ecological implications for human origins. *Journal of Biogeography* 40:1215–1224.

Chapter 11: Big Fierce Carnivores: Hunting Versus Scavenging

Making a living as a large carnivore poses quite a challenge. A sufficiently big herbivore must be found and killed every few days, year-round, despite the defences of these animals and competition from other carnivores for the most vulnerable prey. But carnivory need not be so demanding. All herbivores eventually die, and their carcasses become available to whichever meat-eater arrives first. Few carnivores pass by opportunities to scavenge on dead animals that they encounter. However, in order to secure sufficient food year-round, it may be necessary to kill animals a little sooner, before they die of advancing age or malnutrition. Hence mammalian carnivores regarded as scavengers also kill at times, generally selecting animals that are weakened and about to die soon anyway. The only obligate scavengers are birds like vultures able to scan vast areas on the wing to find the remains of animals wherever these might be located, including carcasses left by predators. Carnivores also take advantage of opportunities to steal carcass remains from those that did the killing, labelled 'kleptoparasitism'.[1] Thus, among large mammals, there is not such a clear division between predators that kill and scavengers that seek already dead animals.

Among primarily hunters, a distinction can be made between those that lie in ambush and those that chase down prey. The big cats like lions (*Panthera leo*) and leopards (*P. pardus*) typically hide in ambush, or stalk stealthily to within pouncing distance. Cheetahs rely on brief but fast chases to capture their prey. African wild dogs (*Lycaon pictus*) engage in long chases, as is typical of canids. Spotted hyenas (*Crocuta crocuta*) also undertake prolonged chases, but quite selectively, and shift flexibly between mainly hunting or largely scavenging in different regions. The next largest carnivore, the brown hyena (*Hyaena brunnea*), is primarily a scavenger, covering distances up to 40–50 km nightly to find carcasses.[2]

The prey species preferred by each of these predators is governed by relative size and whether the predator hunts in groups, or solitarily. Lions and wild dogs, and sometimes also spotted hyenas and cheetahs, hunt in groups, while leopards are strictly solitary hunters. Differences in hunting tactics affect how strongly kills are biased towards prey weakened by advancing age, illness or injury. Felids stalking stealthily kill a higher proportion of healthy animals than the cursorial chasers. In this chapter, I elaborate on the comparative hunting tactics and prey selection by these large predators, as well as their impacts on prey abundance.

Large Carnivore Profiles

The African lion (adult female weight 125 kg, males around 200 kg; Figure 11.1A,B) is the dominant predator over most of savanna Africa where large ungulates occur. Lions live socially in prides generally including 3–8 adult females and several cubs, with one or more adult males attached. Group coordination enables lions to hunt successfully in open savannas providing little concealment. Their tawny brown colour is clearly adapted to blend with dry grass, and even with green grass for colour-blind ungulates (Figure 11.2A). Lions preferentially kill prey in the size range 190–550 kg, particularly wildebeest, zebra, buffalo and, in more arid regions, gemsbok.[3,4] However, smaller antelope like impala and springbok are under-recorded in kills because their carcasses get consumed completely. Male lions hunting separately from prides concentrate proportionately more on buffalo than pride females do.[5] Adult buffalo were hunted only by prides containing five or more lionesses in Serengeti.[6] Large prides can overcome even half-grown elephants weighing around 1500 kg,[7] and in some places young hippos and adult giraffes are frequently killed. Lions focus their kills on adult ungulates, because calves provide too little food for more than the male lions in the pride. Young ungulates feature increasingly prominently among their kills of prey from zebra size upwards.[8,9] In Ngorongoro Crater in Tanzania, lions depend to a greater extent on carcasses stolen from hyenas than is observed elsewhere.[10] Baboons and monkeys rarely feature among lion kills.

Lions hunt mainly at night, but may lurk in ambush near waterholes where ungulates congregate in the dry season during daylight.[11,12] Their relative prey choice can shift in response to the changing vulnerability of particular ungulate species, dependent on rainfall conditions as well as habitat features.[4] Buffalo are more prominent in kills made during the dry season and in dry years, when these formidable animals are weakened by food shortfalls. Wildebeest and zebra are less easily captured in dry years when the grass cover is reduced, because they can detect stalking lions at greater distances. Female ungulates feature more strongly in kills when handicapped by near-term foetuses or while giving birth, while male ungulates become more vulnerable to being killed while distracted by contests for mating opportunities.[13] Shifting prey availability and vulnerability can generate oscillations in relative predator and prey abundance.[14]

Figure 11.1 The large carnivores. (A,B) African lion; (C) leopard; (D) cheetah; (E) African wild dog; (F) spotted hyena.

The leopard (males 65 kg, females 40 kg; Figure 11.1C) is a typical ambush hunter relying on vegetation cover for concealment before pouncing on prey (Figure 11.2C). Their spotted pelage provides camouflage. They hunt mainly, but not exclusively, under cover of darkness. Leopards are distributed widely throughout Africa and beyond through tropical Asia, in both savannas and forests. Their preferred prey size range is 10–45 kg, i.e. medium-small or small antelope plus calves of medium-large ungulates.[15] Primates feature prominently among their kills in rainforests, but baboons and monkeys contribute a

Figure 11.2 Carnivore hunting approaches. (A) Lion camouflaged in long grass, Maasai Mara NR; (B) lioness with giraffe brought down on a road, Kruger NP; (C) leopard melting into bush with impala it has ambushed, Kruger NP; (D) cheetah scanning open plains for prey to chase, Serengeti NP.

much lesser portion than is commonly assumed in savanna habitats. Smaller carnivores, like jackals and domestic dogs, may also be killed. Leopards stash the carcasses of animals they have killed high up in trees to place them beyond the reach of lions and hyenas.

The cheetah (males 55 kg, females 45 kg; Figure 11.1D) is unusual among felids in being adapted for high-speed chases, including adept turns, to capture prey.[16] They hunt typically, but not exclusively, in open grassland habitats (Figure 11.2D). Hunts take place mainly diurnally in the early morning or late afternoon, but occasionally also at night.[17] Cheetahs frequently approach in full view, freezing when the prey look up, until they get within a range of 60–70 m before launching into a sprint. They typically reach speeds of 85 km/h, with a maximum of 105 km/h recorded. Cheetahs kill mostly prey in the size range 23–56 kg,[18] thereby overlapping with leopards. Male cheetahs frequently hunt in coalitions of 2–3 individuals and can kill prey as large as adult oryx or kudu. Primates are absent from cheetah kills. Cheetahs typically feed only partially on carcasses before abandoning them. They seldom scavenge and are easily displaced from their kills by spotted hyenas or lions. Their ranges cover large areas in search of vulnerable prey.

The African wild dog (weight 25 kg; Figure 11.1E) fills the cursorial canid niche and hunts in both open and wooded savannas in large packs. Wild dog packs undertake hunts mostly diurnally in the early morning or late afternoon, thereby restricting losses of kills to hyenas and lions. Their prey size range extends from dikdik to female kudu (5–160 kg), influenced by pack size.[19] Gazelles, impala and young kudu feature most prominently. Their chases after prey can be as short as 50 m, but may extend over 4 km. Animals in poor body condition feature more prominently in their kills than in the general populations of these species.[20] Wild dogs rarely scavenge in the wild, but readily accept carcasses offered in captivity.

The spotted hyena (weight 70 kg; Figure 11.1F) operates either mainly as a hunter or more as a scavenger, depending on circumstances. Their dentition and cranial musculature are adapted for crushing bones left in carcasses abandoned by other carnivores. Spotted hyenas generally hunt nocturnally, probably due to thermal constraints, because they frequently dip in pools of water. They selectively probe for animals that are debilitated in some way - young, old or injured.[10] Nevertheless, chases can extend over several kilometres, with a maximum of 24 km recorded. Hyena kills generally span the size range 50–180 kg, concentrated on the most abundant ungulate species available.[21] In Ngorongoro Crater, packs of spotted hyenas actively hunt even adult wildebeest and zebra,[10,22] perhaps enabled by the cool temperatures in this high-altitude caldera. In areas where lions are more effective hunters, hyenas obtain around half of their food by scavenging on lion kills.[23,24] Spotted hyenas are less common in wooded savannas, perhaps because trees hamper their prolonged pursuits. In Serengeti, hyenas commute over distances of 30–80 km between their dens and the places where wildebeest congregate.

For all of these carnivores, food is most readily obtained during the dry season when herbivores lose body condition and take greater risks seeking whatever food and water remains. Carnivores experience lean times during the wet season when ungulates are in prime health and hence less easily captured, particularly by pursuit predators. Newly born calves can serve as a bridging resource, but only for a short period before these calves can also run speedily. All of the carnivores can obtain sufficient water from the carcasses of the animals they consume, and are thus not dependent on surface water, although they drink opportunistically.

In addition, there are the two small hyenas, functioning mostly as scavengers. The brown hyena (*Hyaena brunnea*) is restricted to southern Africa but has a wide range from the Highveld extending to the Kalahari and Namib coast. It covers distances averaging 30 km nightly in search of small carcasses, including remains of seals along coastlines, but only occasionally kills small animals.[2] Brown hyenas supplement their largely meat diet with desert cucumbers, probably more for moisture than for food value. The striped hyena (*H. hyaena*) is more broadly omnivorous, scavenging for fruits and vegetable matter as well as various animal remains, plus hunting opportunistically.[25] Its wide range extends from eastern Africa into parts of the Middle East. Its nightly forays covered mean distances of around 20 km in Serengeti. Both of these small hyenas have the ability to crush and digest smaller bones. Among the jackals, the black-backed jackal (*Canis mesomelas*) supplements hunting for mice, birds, baby gazelles and invertebrates with scavenging on remains of large carnivore kills, while the side-striped jackal (*C. adjustus*) is more omnivorous, consuming a mix of fruits, small vertebrates, invertebrates and carrion.

Killing Rates

Predator abundance along with rates of killing per predator determine the mortality rates imposed on prey populations. Allowance needs to be made for meat obtained from animals that had died of other causes. In Serengeti NP, half of the food eaten by lions was scavenged from carcasses of ungulates that had died or been killed by hyenas,[26] but elsewhere lions generally kill the greater proportion. Moreover, not all hunting attempts are successful. The success rate of hunts by lions typically ranges between 15 and 30 percent.[27] Prides typically numbering around eight individuals generally kill two or more wildebeest-sized ungulates per week, thus accounting for over 100 animals during the course of a year. This equates to 12 or more medium–large ungulates killed per lion (including cubs) per year. The ungulate population needed to sustain this offtake is 650 or more wildebeest-sized ungulates plus their offspring. If the exclusive core territory of the pride covers around 50 km^2, this requires a year-round ungulate density of 15 animals per km^2, across all prey species larger than impala. If the overall ungulate density is less than this, a larger core territory would be required, raising maintenance costs for patrolling. Hence lion prides establish territorial residence only in places where sufficient prey remain resident year-round. Lions may be encountered more widely, but these animals will be ‘nomads’, still seeking an opportunity to claim a territory.

Leopards attain a success rate per hunt ranging between 15 and 35 percent, almost identical to lions.[28] Leopards need to kill an impala-sized animal almost weekly, and share kills only with offspring. Their projected kill rate is hence around 18 animals per leopard (including cubs) per year. The population of impala-sized prey needed to support each leopard is around 100 animals.

Cheetahs preferentially seek young ungulates less speedy than adults. In the Kalahari, fewer than half of the chases launched by cheetahs after adult steenbok or springbok were successful, but success rates approached 100 percent when juveniles of these small antelope were targeted.[16] Because they do not consume carcasses completely, cheetahs need to kill every 2–3 days, projecting a kill rate per individual exceeding 50 animals per year. This requires a prey availability of around 300 ungulates per cheetah. Each cheetah thus needs to move over a much larger home range than each leopard. While the impact of cheetahs per capita on prey populations can be huge, it is alleviated by the low densities that cheetahs attain. Moreover, the flesh left on the carcasses they abandon helps feed other carnivores.

Wild dogs achieve a high prey capture success (around 45 percent on average)[29] but remain surprisingly rare. They hunt almost daily. Allowing for pack size (typically about 10 adults plus pups), the annual kill rate per individual amounts to about 28 animals of impala size per year. However, their prey base is more narrowly restricted to the old and young segments of prey populations than is the case for the felids.[30] Nevertheless, by hunting in a group, wild dogs can kill larger ungulates than female cheetahs or leopards do. Mortality inflicted by lions and leopards on ungulates of prime age reduces the proportion of herbivore populations surviving to old age, and hence the prey base for wild dogs. This explains why packs of wild dogs wander over home ranges encompassing 150–2000 km^2 seeking vulnerable animals. However, wild dogs become restricted in the area they can cover while pups remain behind in dens.

The kill rate per spotted hyena, allowing for opportunistic switching between hunting and scavenging, may amount to around 9 medium-sized ungulates per year. Because of their highly selective hunting, testing for animals that are debilitated in some way, most of the mortality they inflict would be on old animals near the end of their lifespans. Consequently, the impact on prey populations would be lowered. However, newly born ungulates are especially vulnerable to predation by hyenas before they can run fast enough to escape, so in this aspect hyenas could have a much greater additive impact.

Overview

Only Africa still retains a full suite of stalking, coursing and scavenging carnivores. Lions are adapted through their social groups to hunt open savanna grazers and generally dominate the carnivore guild in prey biomass consumed (Figure 11.3). They are almost the sole predators on ungulates weighing over 150 kg in most places. Spotted hyenas shift flexibly between hunting and

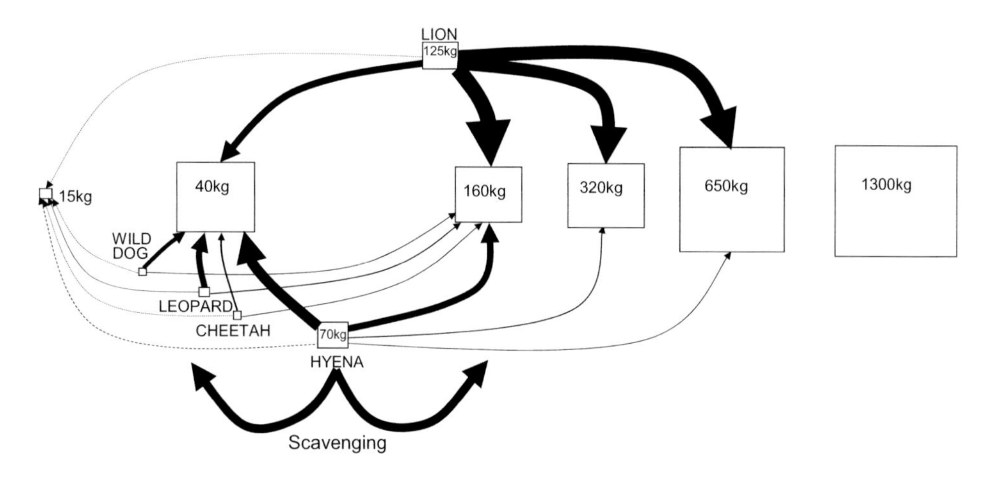

Figure 11.3 Size-structured predator–prey web in Kruger NP. Herbivore prey are grouped in body mass categories (shown in boxes) while the width of the arrows indicates how much prey biomass in these categories is consumed by each carnivore.

scavenging, using their bone-crunching jaws to subsist on carcass remains where lions dominate. Leopards, cheetahs and wild dogs hunt most effectively on ungulates smaller than those generally killed by lions, with their aggregate impact restricted by their low numbers. Kills by the cursorial hunters tend to be concentrated on the young and the old. Competition among these carnivores operates not simply through overlap in prey species killed; hunting by lions also restricts the proportion of prey populations surviving into the old age bracket when they become vulnerable to the cursorial hunters. Cheetahs and wild dogs range widely to locate the more vulnerable prey segments, constraining their abundance.

In South America, the carnivores hunting large mammals are restricted today to two solitarily hunting felids: jaguars (*Panthera onca*) in forests and wetlands, and pumas (*Puma concolor*) in woodlands and mountain slopes. Maned wolves (*Chrysocyon brachyurus*) obtain half of their diet from fruits and kill mainly rodents and rabbits. Spectacled bears (*Tremarctos ornatus*) found in forests are largely herbivorous, but supplement their vegetarian intake with some carrion. Australia once contained a carnivore called the thylacine, or Tasmanian wolf (*Thylacinus cynocephalus*), but it got exterminated following the arrival of the feral dogs known as dingos. During the Pleistocene, lions were present through most of Eurasia and even in the extreme north of North America, but today they persist beyond Africa only as a relict in dry bush in India. Cheetahs were formerly widespread beyond Africa through the Middle East, and cheetah-equivalents occurred in Europe and North America during the Pleistocene. North America retains cursorial canids in the form of the grey wolf (*Canis lupus*) and coyote (*C. latrans*) plus a single stalking felid in the form of the puma or mountain lion. Grizzly bears (*Ursus arctos*) do hunt baby ungulates before they can run fast, but otherwise operate as omnivores with opportunistic scavenging, like black bears (*Ursus americanus*). Tigers (*Panthera tigris*) and leopards are widely spread through Asia, while dholes (*Cuon alpinus*) remain locally in India as a wild dog equivalent. Largely scavenging carnivores, like hyenas, are missing from tropical Asia, perhaps because the wide travel necessary for this lifestyle is inhibited by the prevalent woodland. Wolverines (*Gulo gulo*) fill this role in the far north of Europe, as coyotes do to some extent in North America. Missing from Africa today is a tiger-equivalent, i.e. a solitary stalking felid matching the African lion in size.

Suggested Further Reading

Mills, MGL; Biggs, H. (1993) Prey apportionment and related ecological relationships between large carnivores in the Kruger National Park. *Symposium of the Zoological Society of London* 65:253–268.

Radloff, FGT; du Toit, JT. (2004) Large predators and their prey in a southern African savanna: a predator's size determines its prey size range. *Journal of Animal Ecology* 73:410–423.

Schaller, GB. (1972) *The Serengeti Lion. A Study of Predator–Prey Relations*. University of Chicago Press, Chicago.

References

1. DeVault, TL, et al. (2003) Scavenging by vertebrates: behavioral, ecological, and evolutionary perspectives on an important energy transfer pathway in terrestrial ecosystems. *Oikos* 102:225–234.
2. Mills, MGL. (1990) *Kalahari Hyaenas*. Blackburn Press, Caldwell, NJ.
3. Hayward, MW; Kerley, GIH. (2005) Prey preferences of the lion (*Panthera leo*). *Journal of Zoology* 267:309–322.
4. Owen-Smith, N; Mills, MGL. (2008) Shifting prey selection generates contrasting herbivore dynamics within a large-mammal predator–prey web. *Ecology* 89:1120–1133.
5. Funston, PJ, et al. (1998) Hunting by male lions: ecological influences and socioecological implications. *Animal Behaviour* 56:1333–1345.
6. Scheel, D. (1993) Profitability, encounter rates, and prey choice of African lions. *Behavioral Ecology* 4:90–97.
7. Joubert, D. (2006) Hunting behaviour of lions (*Panthera leo*) on elephants (*Loxodonta africana*) in the Chobe National Park, Botswana. *African Journal of Ecology* 44:279–281.
8. Grange, S, et al. (2004) What limits the Serengeti zebra population? *Oecologia* 140:523–532.
9. Owen-Smith, N; Mills, MGL. (2006) Manifold interactive influences on the population dynamics of a multispecies ungulate assemblage. *Ecological Monographs* 76:73–92.

10. Kruuk, H. (1972) *The Spotted Hyaena: A Study of Predation and Social Behavior*. Chicago University Press, Chicago, IL.
11. De Boer, WF, et al. (2010) Spatial distribution of lion kills determined by the water dependency of prey species. *Journal of Mammalogy* 91:1280–1286.
12. Davidson, Z, et al. (2012) Environmental determinants of habitat and kill site selection in a large carnivore: scale matters. *Journal of Mammalogy* 93:677–685.
13. Owen-Smith, N. (2008) Changing vulnerability to predation related to season and sex in an African ungulate assemblage. *Oikos* 117:602–610.
14. Ogutu, JO; Owen-Smith, N. (2005) Oscillations in large mammal populations: are they related to predation or rainfall? *African Journal of Ecology* 43:332–339.
15. Hayward, MW, et al. (2006) Prey preferences of the leopard (*Panthera pardus*). *Journal of Zoology* 270:298–313.
16. Mills, MGL. (2017) *Kalahari Cheetahs: Adaptations To An Arid Region*. Oxford University Press, Oxford.
17. Cozzi, G, et al. (2012) Fear of the dark or dinner by moonlight? Reduced temporal partitioning among Africa's large carnivores. *Ecology* 93:2590–2599.
18. Hayward, MW, et al. (2006) Prey preferences of the cheetah (*Acinonyx jubatus*) (Felidae: Carnivora): morphological limitations or the need to capture rapidly consumable prey before kleptoparasites arrive? *Journal of Zoology* 270:615–627.
19. Hayward, MW, et al. (2006) Prey preferences of the African wild dog *Lycaon pictus* (Canidae: Carnivora): ecological requirements for conservation. *Journal of Mammalogy* 87:1122–1131.
20. Pole, A, et al. (2003) African wild dogs test the 'survival of the fittest' paradigm. *Proceedings of the Royal Society of London Series B: Biological Sciences* 270:S57.
21. Hayward, MW. (2006) Prey preferences of the spotted hyaena (*Crocuta crocuta*) and degree of dietary overlap with the lion (*Panthera leo*). *Journal of Zoology* 270:606–614.
22. Cooper, SM, et al. (1999) A seasonal feast: long-term analysis of feeding behaviour in the spotted hyaena (*Crocuta crocuta*). *African Journal of Ecology* 37:149–160.
23. Henschel, JR; Skinner, JD. (1990) The diet of the spotted hyaenas *Crocuta crocuta* in Kruger National Park. *African Journal of Ecology* 28:69–82.
24. Périquet, S, et al. (2015) Spotted hyaenas switch their foraging strategy as a response to changes in intraguild interactions with lions. *Journal of Zoology* 297:245–254.
25. Kruuk, H. (1976) Feeding and social behaviour of the striped hyaena (*Hyaena vulgaris* Desmarest). *African Journal of Ecology* 14:91–111.
26. Scheel, D; Packer, C. (1995) Variation in predation by lions: tracking a movable feast. In Sinclair, ARE, et al. (eds) *Serengeti II: Dynamics, Management, and Conservation of an Ecosystem*. University of Chicago Press, Chicago, IL, p. 299.
27. West, PM; Packer, C. (2013) *Panthera leo*. In Kingdon, J; Hoffmann, M (eds) *Mammals of Africa V*. Bloomsbury, London, pp. 149–159.
28. Hunter, L, et al. (2013) *Panthera pardus*. In Kingdon, J; Hoffmann, M (eds) *Mammals of Africa V*. Bloomsbury, London, pp. 159–167.

29. McNutt, JW; Woodroffe, R. (2013) *Lycaon pictus*. In Kingdon, J; Hoffmann, M (eds) *Mammals of Africa V*. Bloomsbury, London, pp. 51–58.
30. Fitzgibbon, CD; Fanshawe, JH. (1989) The condition and age of Thomson's gazelles killed by cheetahs and wild dogs. *Journal of Zoology* 218:99–107.

Chapter 12: Herbivore Abundance: Bottom-up and Top-down Influences

Food resources constitute an ultimate limit on the population size attained, but how this operates is not so simple and direct. For large herbivores, food production depends on rainfall, which varies seasonally and between years. During the wet season, there is more green vegetation than herbivores could possibly eat. Towards the end of the dry season, animals may be starving because the little plant material that remains is mostly dead or dormant, largely indigestible and perhaps toxic (Figure 12.1). They may be forced to crowd near remaining water sources, increasing their exposure to predation. They may have to take greater risks in seeking what food remains. Many large herbivores migrate seasonally, while others remain resident year-round. Demographic segments make distinct contributions to population changes generated by the balance between births and deaths. Over multi-year periods, populations expand and contract in their distribution ranges, governed by various influences. In such spatially and temporally variable contexts, how can any stable abundance level, or 'carrying capacity', be maintained? These are the topics to be addressed in this chapter.

Seasonal Food Limitations

How herbivore populations respond to changing environmental conditions may not follow the expected patterns. The large ungulate that is most responsive to annual rainfall variation, in Kruger NP[1,2] and elsewhere,[3] is the African buffalo, despite its large size and digestive efficiency. This is because, as bulk grazers, they are most directly dependent on the amount of grass produced annually. Moreover, being especially water-dependent, they are confined to the vicinity of long-lasting water bodies through the dry season, where grass gets locally depleted. Buffalo herds must then undertake regular journeys between places that retain grass and where water still remains. During the severe drought of 1991/2, when the annual rainfall total was only half of the mean, Kruger Park's buffalo population fell by 45 percent.[2] In the Mara region of northern Serengeti, buffalo numbers dropped by almost 50 percent during the 1984/5 drought, and by over 75 percent during 1993/4 when rainfall

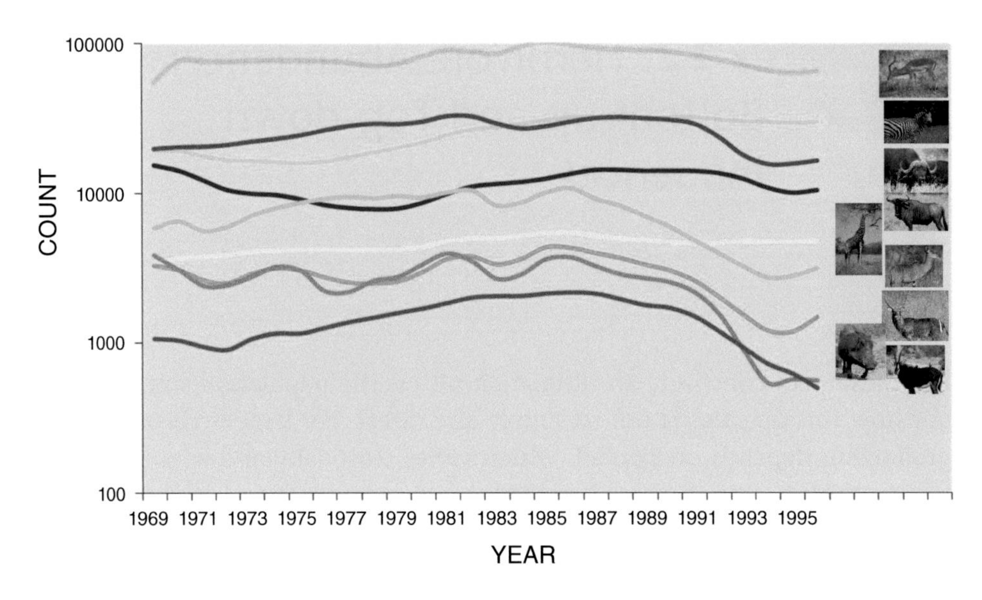

Figure 12.1 Trends in large herbivore populations in Kruger NP, from annual aerial counts over 25 years. The five most common species maintained fairly stable populations fluctuating twofold in abundance, while four less common species showed substantial downturns in abundance during the early 1990s. Note the log scale on the *y*-axis enabling populations to be compared on the same proportional scale.

deficiencies were worsened by an influx of cattle into the protected area.[3,4] In Ngorongoro Crater, drought-related mortality among buffalo in 2000 accounted for 30 percent of the population and was associated with complete loss of offspring in the current year plus a lack of calves in the following year.[5] However, buffalo can be quite mobile and, during the exceptionally severe drought that prevailed through 2014–16 in Kruger NP, herds moved to places where local rain showers had generated more grass, alleviating mortality.[6]

Plains zebras, able to exploit stemmy grass remnants through their non-ruminant digestion, appeared least sensitive to rainfall variation among Kruger Park's grazers.[7,8] Their numbers have also varied little in Serengeti.[9] However, their population dynamics were more responsive to annual rainfall variation in private conservancies in central Kenya.[10] In Serengeti, the survival of zebra foals actually improved during years with lower rainfall.[9]

After buffalo, the ungulate next most responsive to annual rainfall variation in Kruger NP is the greater kudu, a browser.[1,11] However, the mechanism involved is not simply the lack of food affecting survival through the dry season, but rather the effect of wet season rainfall on calf survival, presumably acting through maternal influences on late foetal growth and milk supply after birth. This is probably due to the influence of rainfall on forb production in the herbaceous layer.

The population dynamics of wildebeest responded only to the dry season component of the annual rainfall, both in Kruger NP[7,8] and in Serengeti.[12] Dry season rainfall prolongs the period over which green growth persists among short grasses. Higher wet season rainfall producing taller grass may also result in greater cover for stalking lions. In Kruger NP, other grazing ruminants exploiting somewhat taller grass than wildebeest responded positively to both the wet and dry season components of the annual rainfall total.[8]

The megaherbivores, including giraffe along with elephant and white rhino, responded little to annual rainfall variation in their population trends in Kruger NP[7] and elsewhere,[13,14] despite their bulk food requirements. Although the elephant population in Tsavo East NP crashed by 20–30 percent during very severe drought conditions,[15] this mortality was exacerbated because elephants had been compressed within the park by settlements outside its bounds. Drought-related mortality of this magnitude has not been recorded among elephants elsewhere. In Amboseli NP in Kenya, severe drought conditions caused mortality especially, although not exclusively, among calves.[14] Hippos are more prone to drought-related die-offs than other megaherbivores[16] because their grazing is restricted to river or lake margins, where forage gets depleted by concentrations of water-dependent ruminants, as well as by what hippos eat. The very largest herbivores have greater metabolic inertia to buffer their population trends against annual variation in food availability, which only temporarily affects fertility and calf survival.

Nevertheless, there are circumstances in which less-large herbivores can incur quite severe population crashes, amounting to 50 percent or more.[17] In the Central Kalahari region of Botswana, 90 percent of the wildebeest died when fences blocked access to surface water during the extreme 1982/3 drought, while hartebeest, eland and kudu numbers fell by 50 percent or more.[18] In the Klaserie Private Nature Reserve near Kruger NP, >90 percent of wildebeest and warthog, >80 percent of buffalo and zebra, and 70 percent of impala died during this same drought.[19] The exacerbating feature there was excessive provision of surface water in the form of the dams constructed on private properties, elevating densities of the water-dependent grazers before the drought. With water no longer limiting spatially, the grass cover became denuded everywhere, removing the resource buffer that might have persisted remote from water. The browsers, specifically kudu and giraffe, incurred less mortality than the grazers because woody plants produced new foliage at the end of the dry season despite the lack of rainfall, alleviating starvation.

In summary, most herbivore populations, apart from the very largest species, fluctuate in abundance in response to annual rainfall variation, indicating the controlling influence of changing food production. However, the range in variation seems typically no more than about twofold, achieving such approximate constancy in the numbers of animals supported from year to year. Greater seasonal food depletion during severe droughts, at least locally near surface water, can generate extreme population crashes, especially where animals cannot move freely to exploit spatial variability in rainfall. Demographic inertia in the maximum population growth rate (see below) limits the extent to which large herbivore populations can grow beyond what the prevailing food resources can support in successive years. However, via their feeding and other impacts these herbivores, especially the largest, can alter the vegetation cover, for better or for worse (see Chapter 13). Predation can potentially restrict how closely herbivore populations approach the food ceiling, at least locally.

What Difference Does Predation Make?

All animals ultimately die, but their deaths may occur somewhat sooner through the agency of a carnivore and reduce their reproductive contribution to maintaining a population. However, with fewer mouths left more food is available to feed the animals that remain and promote their reproductive success. This mechanism compensates to some extent for the losses of animals killed by predators. Hence the additive effect of predation on prey abundance is somewhat less than commonly surmised. The indirect effects of the risk of predation on where herbivores can safely feed and rest can be much greater.[20]

Moreover, predation usually interacts with resource limitations.[21] While food remains plentiful, herbivores can restrict their movements to places where they are most secure from predation. When food runs short, in the dry seasons of dry years, they are forced to take greater risks while seeking food, increasing their exposure to predation. Hence although almost all herbivores die through the agency of a predator, the mortality incurred by the population depends on the prevailing rainfall conditions.[7] The main effect of predation is to restrict habitat occupation to more secure places, most of the time. For example, in Kruger NP wildebeest herds remain settled within open glades with short grass providing little cover for predators for most of the year, day and night, until little grass remains, before venturing out into more densely wooded areas.[22,23]

The reason why wildebeest are orders of magnitude more numerous in Serengeti than in Kruger NP is due not simply to better-quality grazing, but also to the vast extent of open grassland where they are relatively secure from being ambushed by lions. The huge concentrations of migratory ungulates overwhelm the capacity of lions to have much additive impact. About 15 percent of wildebeest deaths and 30 percent of zebra deaths result directly from starvation, and many of the wildebeest eaten by lions are scavenged from hyena kills.[24] Nevertheless, the densities of resident ungulates, including topi, impala and warthog, increased in northern Serengeti after lion numbers there were reduced as a consequence of illegal hunting concentrated on buffalo, despite the fact that the species responding positively were not the primary prey species of lions.[25]

However, there are circumstances in which predation can have a substantial impact on prey abundance, via a nexus of interactions among alternative prey species. In Kruger NP, several of the less-common herbivore species declined to 20 percent or less of their former abundances, shortly following a severe drought (Figure 12.2). Their downward trends were associated with a two- to threefold increase in adult mortality, implicating increased predation specifically by lions.[26,27] Lions had increased following a doubling in zebra numbers. Zebras increased because park managers had provided additional waterholes, aimed at alleviating the effects of droughts. Lions turned to hunting the less-common ungulate species, which became more vulnerable to predation than their preferred prey, i.e. wildebeest, zebra and buffalo, as a consequence of the drought for various reasons. Notably, it was the alternative prey species, rather than the primary prey species, that were most susceptible to elevated predation. Furthermore, none of the ungulate species adversely impacted has yet gone locally extinct in the park, although the status of roan antelope is precarious.

This scenario revealed some of the mechanisms dampening fluctuations in prey populations.[21] Buffalo, the largest ungulate in the primary prey base supporting lions, gained demographic buffering because prime-aged animals are not easily killed, meaning that lion prides focus on young buffalo in the herds. Medium-sized ungulates gain security by occupying habitats where they are less vulnerable and by showing contrasting responses to rainfall variation. Relative population stability was conferred by the habitat template. When conditions are benign, animals can mostly occupy secure sites. During adverse conditions, they are forced to spend more time outside these refuges seeking whatever food and water remain. Predation and diminishing food availability thus interact to counteract population fluctuations.

Migration: Seasonal Relocation of Populations

All of Africa's migrants are grazers, because it is the grasses that respond most directly to gradients in rainfall via forage production and quality. Migratory populations attain vastly greater abundance levels than shown by those remaining resident year-round, for various reasons.[28,29] Seasonal concentrations swamp the capacity of resident predators, so that many carcasses are left uneaten.[30] Predator numbers are restricted by the dearth of prey left after the

Figure 12.2 Food depletion by the late dry season. (A) Grass cover grazed down to bare soil by wildebeest, basaltic region of Kruger NP; (B) zebra grazing remnants of taller grass in same region of Kruger NP; (C) buffalo grazing remnants of grass in Kruger NP; (D) buffalo trekking to water over area barren of grass, Kruger NP; (E) kudu nibbling flowers appearing on trees before leaves in Kruger NP; (F) skinny impala nibbling remaining leaves on a shrub during drought.

migrants have departed. Regular seasonal migrations over long distances are, or were, a feature of numerous wildebeest and many zebra populations across Africa.[31,32] Mass migrations are still undertaken by white-eared kob (*Kobus kob leucotis*) and tiang (*Damaliscus lunatus tiang*) east of the Nile River in southern Sudan.[33] Other grazers shift between seasonally separated ranges without moving as far, being confined near surface water in the dry season and dispersing further afield in the wet season.[34] Some ungulates move essentially nomadically, i.e. widely but less regularly; for example, eland and various gazelles.[35,36] Many ungulate species remain within the same home ranges year-round; for instance, sable and roan antelope as well as various reduncine grazers, plus most browsers. Some populations are partitioned between a portion that migrates and a remnant that stays behind, a phenomenon labelled partial migration.[37] Even in Serengeti, wildebeest populations remain resident year-round in the western corridor and in Ngorongoro Crater.

Several benefits accrue to migrants. They (1) gain access to high-quality but ephemeral resources (Figure 12.3), promoting reproductive success,[38,39] (2) locally dilute predation,[28] and (3) get dispersed more broadly than residents. Lack of year-round surface water can contribute to the abandonment of the wet season range. When migratory movements get blocked, herbivore populations collapse to much lower numbers.[40]

Genuine or 'mass' migrations represent one extreme of a spectrum characterised by regular seasonal shifts in home range occupation. Importantly, seasonal concentrations of migrants greatly exceed the local densities that could be supported locally year-round. The numerical density of wildebeest in Serengeti gets amplified almost 10-fold to nearly 500 animals per km^2 while they concentrate in the 2500 km^2 extent of the short grass plains during the time when calves are produced. Similar concentrations develop within the Maasai Mara Reserve in the north where the migrant wildebeest congregate during the dry season, benefiting from the green grass retained there due to locally high rainfall, along with water readily available from the Mara River. In southern Sudan, densities attained by kob and tiang on the Nile River floodplain can exceed 1000 animals per km^2 locally during the dry season.[29] Historically, huge numbers of ungulates of various species concentrated seasonally in Kenya's rift valley near Lake Nakuru, especially zebra and Thomson's gazelle, before hunting eliminated them.[41] Migrations were probably more widely prevalent in the past, before modern humans intruded.

Life History: Births, Deaths and Population Growth

Life-history features determine (1) how fast a population can potentially grow, and (2) the recruitment rate needed to counterbalance adult mortality through predation or other causes. Formal population models must accommodate (1) age at first reproduction, (2) number of offspring produced in each litter, (3) inter-birth intervals, (4) mortality incurred at different ages, and (5) the potential lifespan. Population models focus on the female segment, because the number of males makes little difference to the dynamics, unless there are very few or if the birth sex ratio deviates much from 50:50. The potential population growth rate depends on body size, because bigger

Figure 12.3 Migratory wildebeest and zebra in Serengeti. (A) Wildebeest dispersed over the eastern short grass plains early in the wet season; (B) zebra spread where the grass is taller further west; (C) zebra flanking a wildebeest concentration on the short grass plain; (D) wildebeest aggregated in the central woodlands during migration; (E) wildebeest spread in northern Serengeti towards the end of the dry season after most grass has been grazed short; (F) aggregation crossing the Mara River back south near the end of the dry season.

animals take longer to reach sexual maturity and have longer inter-birth intervals associated with longer gestation periods.[42] This is largely a consequence of how the metabolic rate scales with body size, as discussed in connection with food requirements in Chapter 10. Longevity is ultimately curtailed by the wearing down of tooth surfaces and consequent impairment of food processing.[43]

Ruminants from medium-sized impala (45 kg) to large eland (450 kg) all have gestation periods of 7–9 months, enabling them to synchronise annual births with the wet season.[44] Females typically first give birth at three years of age and potentially produce a single offspring annually during a lifespan of ~15 years. The maximum rate of population growth that can be sustained if all females survive with zero mortality from birth to the maximum lifespan is almost 30 percent per year. This rate could be exceeded temporarily if offspring mortality, perhaps in a drought year, leaves a population comprising mostly prime-aged females.

Smaller antelope like springbok and steenbok can attain reproductive maturity by one year of age, have gestation periods of 4–5 months, and may reproduce twice during some years, but seldom live longer than 10 years.[45,46] They can potentially attain a population growth rate as high as 40 percent per year. Buffalo (gestation period 11 months) and zebra (gestation period 12 months) cannot maintain annual synchrony of births with the wet season and periodically skip a year.[47,48] Furthermore, they both produce their first offspring only at 4–5 years of age, with longevity lengthened to 20–25 years to compensate for the slower recruitment. Their maximum rate of population growth is around 15 percent per year. Giraffe, with a gestation period of 15 months and birth intervals generally around 20 months, no longer maintain birth synchrony with the wet season and live up to 26 years of age. Rhinos have gestation periods of 15–16 months, give birth first around 7 years of age and have a mean birth interval of 2.5 years until they reach a maximum of about 40 years of age. Their maximum population growth rate is lowered to 9 percent per year. African elephants have a gestation period as long as 22 months, first give birth between 10 and 15 years of age and produce offspring with around a 4-year birth interval until 55–60 years of age. Their maximum population growth rate is reduced to ~6 percent per year. Hippos have a gestation period as short as 8 months, but still maintain birth intervals of around 2 years and typically do not reproduce until as late as 10 years of age in the wild. Hence their potential population growth rate would not much exceed that of rhinos. Intriguingly, it seems that each female can potentially produce a maximum of 12 offspring, should she live through her potential reproductive lifespan, whatever the life-history schedule.

The lower growth rates of megaherbivores mean that their populations take longer to recover from setbacks like drought-induced mortality than those of smaller species. However, on the other hand, they are less susceptible to elevated mortality during droughts because of their fibre-tolerant food requirements and greater metabolic inertia. Having mostly escaped predation by large carnivores once they reach maturity, megaherbivores have little capacity to sustain predation imposed by human hunters.

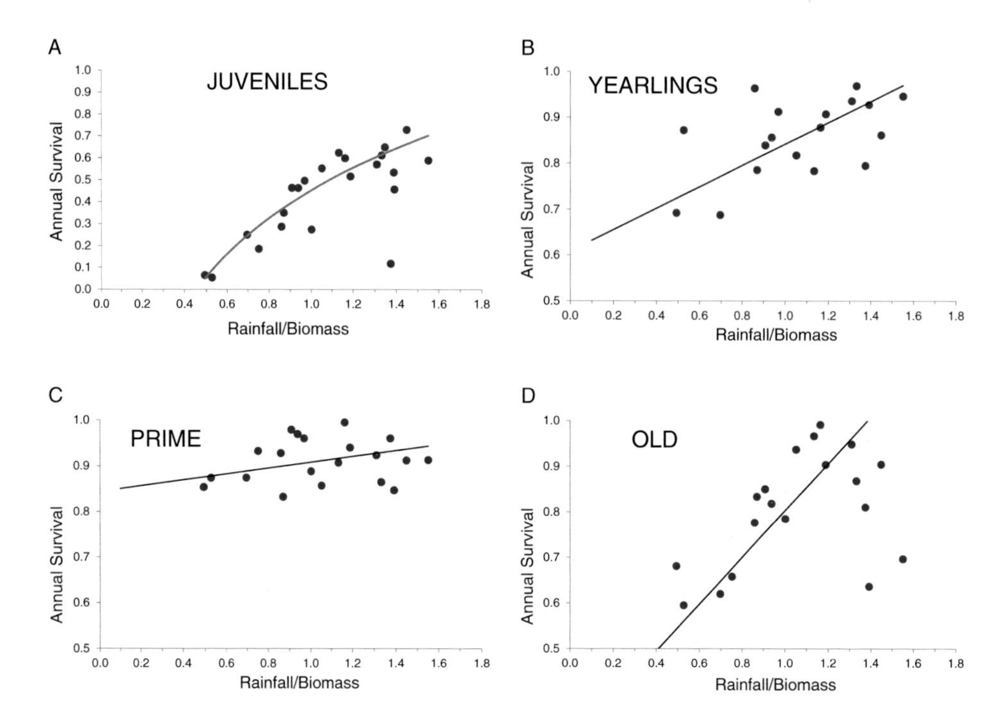

Figure 12.4 Annual survival rates of juvenile, yearling, prime and old kudus governed by annual rainfall relative to population biomass density in Kruger NP. Outlying points represent 1981 when a cold spell caused mortality despite high prior rainfall (from Owen-Smith (2002) *Adaptive Herbivore Ecology*).

Populations ultimately stop growing once recruitment into the adult segment is barely sufficient to counterbalance mortality incurred during adulthood. The survival of juveniles and old females is most sensitive to limitations in food availability related to rainfall as well as to the population density (Figure 12.4). Various combinations of adult mortality and offspring recruitment can generate zero population growth. If, hypothetically, no adult dies before reaching the maximum longevity, at least 15 percent of the offspring born must attain reproductive maturity in order for the population to persist. If half of the adults die or are killed annually, every single offspring born would need to survive to replace them. These scenarios represent the extreme bounds for population viability. A more realistic demographic combination producing zero population growth for a large ungulate would be 50 percent survival of offspring to maturity coupled with 25 percent annual mortality in the adult segment, including those dying of old age.

Observed combinations of adult mortality and juvenile recruitment generating population stasis in Kruger NP are not obviously influenced by body size, except towards the extremes (Figure 12.5). The highest rates of juvenile

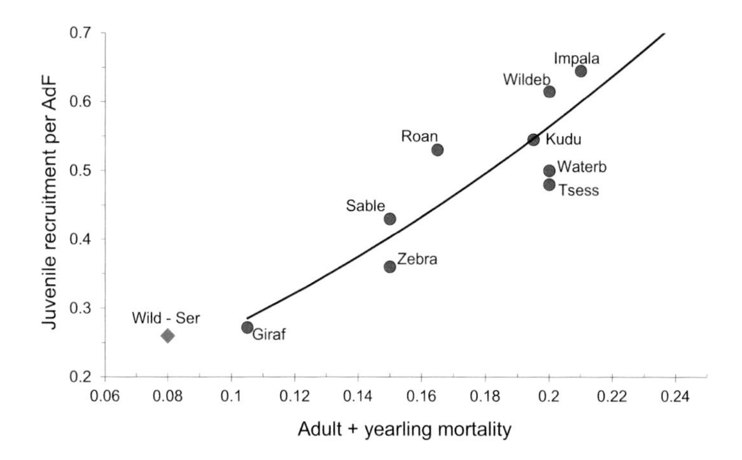

Figure 12.5 Combinations of adult mortality and juvenile recruitment generating zero population growth observed in Kruger NP plus one point (green) representing wildebeest in Serengeti.

recruitment were exhibited by the two ungulates that form the major prey species of their main predators, i.e. wildebeest for lions and impala for less-large carnivores. For both populations, females generally gave birth as early as two years of age.[49] In Serengeti, wildebeest exhibited a different combination, coupling low adult mortality with low juvenile recruitment.[12] This indicates that these wildebeest are food-limited, without much additive contribution from predation. Adult mortality averaged to 9 percent annually, meaning that few died much before 15 years of age. However, juvenile mortality exceeded 70 percent during the first year. For kudus, the annual mortality rate among the adult female segment was almost 20 percent, around twice the annual mortality incurred by the prime-aged segment alone.

Ungulates typically use changing daylength as a cue to schedule when mating and hence subsequent births occur. However, this cue does not exist near the equator because daylength remains a constant 12 h there. Consequently, although impala and hartebeest exhibit narrow birth peaks in southern Africa, they reproduce year-round in Serengeti close to the equator.[50,51] On the other hand, wildebeest and topi retain narrow birth peaks near the equator despite the lack of the daylength cue. Narrowly restricted births swamp predators so that more offspring survive.[52]

Intriguingly, in southern Africa different ungulate species time their birth peaks in different stages of the wet season. Tsessebe (*Damaliscus lunatus lunatus*) and hartebeest have birth peaks during October–November early in the wet season. Impala drop their lambs from late November into December. Wildebeest calve during December. Kudu produce offspring in January–February. Sable antelope give birth as late as February or even March.[51] Zebra and buffalo, with inter-birth intervals longer than a year, show rather weak pulses in births late in the wet season. Rhinos exhibit a diffuse birth peak early in the dry season, while elephants have births concentrated early in the wet season. Grazers favouring wetlands, where green grass remains available in the dry season, including waterbuck as well as their kob congeners, have births spread more widely through the year.[44]

For white rhinos, oestrous cycling is suppressed while food quality is poor, leading to a pulse in matings when nutrition improves after the start of the rains.[42] The result is a surge in births 16 months later, early in the dry season. A similar mechanism could be operating for wildebeest and topi in Serengeti. Female ungulates generally do not resume cycling until after they have regained body condition following weaning of the preceding calf.

Are Bigger or Smaller Herbivores More Abundant?

Whether smaller or bigger animals are more abundant depends on how abundance is measured. Smaller animals tend to be more numerous, because individually each requires less food. However, larger animals need less nutrition per unit of their biomass and so can attain greater biomass densities from the same amount of food. From metabolic scaling, numerical population density (N) should decrease with body mass (M) raised to the power three-quarters: $N = M^{-0.75}$, while biomass density should increase as a function of $M^{0.25}$ ($M^{1.0}/M^{-0.75}$). Expressed in words, abundance assessed in terms of biomass density, i.e. weighted by the body mass, should increase with body mass raised to the power one-quarter. However, there is much variation in the maximum population density levels recorded for individual herbivore species, related to their specific food requirements and other factors (Figure 12.6). Simply putting a regression line through the scatter of points can be misleading.

The herbivores most closely approaching the metabolic ceiling in local abundance set by food requirements are certain large grazers, notably wildebeest, buffalo and white rhino, along with mixed feeding elephant. Wetland grazers can also attain high densities locally, including kob in Uganda, and lechwe (*Kobus leche*) in Zambia. Maximum abundance levels attained by browsers are only about 20 percent of those shown by grazers of similar body mass. However, some grazers are no more numerous relative to their size than browsers, notably non-ruminants like zebra and warthog. Contrary to common assumptions, the numerical densities that elephants and white rhinos can attain, amounting to around 5 animals per km^2, are not lower than those maintained by many much-smaller ungulates. Below a body mass of around 50 kg, numerical densities decline with decreasing body mass, apart

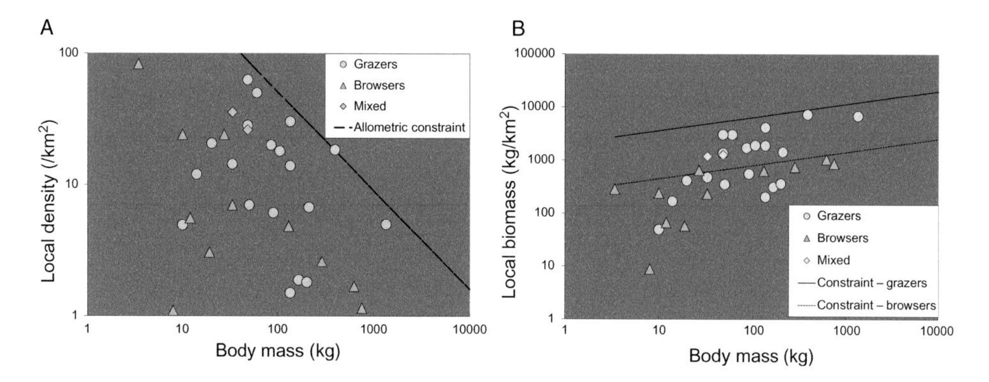

Figure 12.6 Allometric relationships between population density and body mass for various African ungulate populations, distinguishing grazers from browsers. (A) Scatter plot compiled from all available records of local densities within a home range or study area, for various African ungulates; the line represents the allometric constraint generated by a power coefficient of –0.75; (B) numerical densities translated into biomass; constraint lines have slope of 0.25.

from the outlying point representing the huge densities attained locally by little dikdiks in Kenya.[53]

The values plotted in Figure 12.6 represent the maximum abundance levels recorded for each species locally within a home range. If abundance levels are assessed over a broader region, like a national park or other large protected area, the relationship between population density and body mass falls away. This is because smaller herbivores are more narrowly restricted in their habitat occupation by their food or security requirements. Among the largest herbivores, regional and local densities converge.

Although the biggest herbivores attain the greatest standing biomass, the turnover of this biomass engendered by mortality is much slower than for smaller herbivores: ~5 percent versus 15–25 percent annually. Hence herbivores weighing between 100 and 500 kg actually produce most potential food for large carnivores, particularly the grazers among them. Moreover, the carcasses of big megaherbivores are utilised less effectively, leaving more flesh behind for scavengers and decomposers.

Distribution Patterns

Populations can grow and shrink additionally via expansions and contractions in their distribution ranges. Dispersal movements played a major role in re-establishing wildlife populations in South Africa's protected areas after the decimation by hunting plus the rinderpest epizootic. Elephants and various antelope moved into Kruger NP from secluded regions of Mozambique once protection was afforded. Large herbivores and carnivores filtered back into the

Umfolozi Game Reserve, where everything except rhinos had been shot in an attempt to eliminate tsetse flies, coming from the nearby Hluhluwe Game Reserve.[54] White rhinos reintroduced from there to Kruger NP had spread through most of the park before the recent poaching wave hit. Changes in the geographic distributions of species are engendered by shifts in the home ranges occupied by the individual animals or groups representing these species.

The drastic decline of the sable antelope population in Kruger NP was brought about by the disappearance of herds from previously occupied localities as well as shrinkage in numbers of the remaining herds.[55] Roan antelope had been even more restricted in their presence within the park, which was at the southern limit of their historical distribution range. They had occurred in two clusters of distinct herds in the far north plus three isolated herds further south, comprising 250–400 animals in total.[55] When their population crashed, one complete cluster plus all of the isolated herds disappeared, leaving a local remnant of just three herds, totalling fewer than 40 animals.[26] Both of these antelope species were threatened by the prospect of a downward ratchet toward local extinction.[7] While herd sizes remain small, it is more difficult for mothers to defend their calves jointly against attacks by hyenas and wild dogs.[55]

Overview

Savanna herbivore populations typically vary over perhaps a twofold range in abundance in response to annual variation in the effective carrying capacities, governed by the rainfall dependence of plant growth and hence food resources. Periodic droughts amplify food depletion in the dry season and can generate substantial population crashes, especially if movements are restricted. Direct effects of predation on prey populations seem quite minor, because of compensatory mechanisms. Indirect responses to the risk of predation can be huge, in contrast, by restricting habitat occupation and hence regional population densities. Impacts of resources and predation on herbivore populations are inextricably entangled, because when food runs out animals take greater risks. Megaherbivores, migratory antelope and wetland dwellers most closely approach the population ceilings set by metabolic requirements for food. Migration elevates populations largely by restricting exposure to resident predator populations. Although population biomass densities rise with body size, turnover of this biomass diminishes correspondingly. Accordingly, herbivores in the size range 100–500 kg are most productive in biomass generated annually. Browsers attain local density levels amounting to only 20 percent of those exhibited by grazers of similar size.

South America currently lacks concentrations of large herbivores. Australia has enormous numbers of kangaroos and wallabies, but populations fluctuate hugely due to die-offs during periodic droughts. Predation by dingos seems inconsequential.[56] Spectacular migratory concentrations of saiga (*Saiga tartarica*) and Mongolian gazelles (*Procapra gutturosa*) develop in the steppe grasslands of central Asia and among caribou/reindeer (*Rangifer tarandus*) herds in the arctic north. Bison (*Bison bison*) formerly achieved similarly enormous abundance in North American grasslands. Intrinsically generated population irruptions like those manifested by deer in North America have not been observed in Africa. Seasonal concentrations of grazers near water are a feature especially of African savannas.

Suggested Further Reading

Owen-Smith, N. (1988) *Megaherbivores. The Influence of Very Large Body Size on Ecology*. Cambridge University Press, Cambridge.

Owen-Smith, N. (2010) *Dynamics of Large Herbivore Populations in Changing Environments*. Wiley-Blackwell, Oxford.

References

1. Mills, MGL, et al. (1995) The relationship bewteen rainfall, lion predation and population trends in African herbivores. *Wildlife Research* 22:75–87.
2. Marshal, JP, et al. (2011) The role of El Niño–Southern Oscillation in the dynamics of a savanna large herbivore population. *Oikos* 120:1175–1182.
3. Dublin, HT; Ogutu, JO. (2015) Population regulation of African buffalo in the Mara–Serengeti ecosystem. *Wildlife Research* 42:382–393.
4. Ottichilo, WK, et al. (2000) Population trends of large non-migratory wild herbivores and livestock in the Masai Mara ecosystem, Kenya, between 1977 and 1997. *African Journal of Ecology* 38:202–216.
5. Estes, RD, et al. (2006) Downward trends in Ngorongoro Crater ungulate populations 1986–2005: conservation concerns and the need for ecological research. *Biological Conservation* 131:106–120.
6. Staver, AC, et al. (2019) Grazer movements exacerbate grass declines during drought in an African savanna. *Journal of Ecology* 107:1482–1491.
7. Ogutu, JO; Owen-Smith, N. (2003) ENSO, rainfall and temperature influences on extreme population declines among African savanna ungulates. *Ecology Letters* 6:412–419.
8. Owen-Smith, N; Mills, MGL. (2006) Manifold interactive influences on the population dynamics of a multispecies ungulate assemblage. *Ecological Monographs* 76:73–92.
9. Grange, S, et al. (2004) What limits the Serengeti zebra population? *Oecologia* 140:523–532.

10. Georgiadis, N, et al. (2003) The influence of rainfall on zebra population dynamics: implications for management. *Journal of Applied Ecology* 40:125–136.
11. Owen-Smith, N. (1990) Demography of a large herbivore, the greater kudu *Tragelaphus strepsiceros*, in relation to rainfall. *The Journal of Animal Ecology* 59:893–913.
12. Mduma, SAR, et al. (1999) Food regulates the Serengeti wildebeest: a 40-year record. *Journal of Animal Ecology* 68:1101–1122.
13. Owen-Smith, N (1981) The white rhinoceros overpopulation problem, and a proposed solution. In Jewell, PA et al. (eds) *Problems in Management of Locally Abundant Wild Mammals*. Academic Press, New York, NY, pp. 129–150.
14. Moss, CJ. (2001) The demography of an African elephant (*Loxodonta africana*) population in Amboseli, Kenya. *Journal of Zoology* 255:145–156.
15. Corfield, TF. (1973) Elephant mortality in Tsavo National Park, Kenya. *African Journal of Ecology* 11:339–368.
16. Smit, IPJ; Bond, WJ. (2020) Observations on the natural history of a savanna drought. *African Journal of Range & Forage Science* 37:119–136.
17. Young, TP. (1994) Natural die-offs of large mammals: implications for conservation. *Conservation Biology* 8:410–418.
18. Spinage, CA; Matlhare, JM. (1992) Is the Kalahari cornucopia fact or fiction? A predictive model. *Journal of Applied Ecology* 29:605–610.
19. Walker, BH, et al. (1987) To cull or not to cull: lessons from a southern African drought. *Journal of Applied Ecology* 24:381–401.
20. Owen-Smith, N. (2019) Ramifying effects of the risk of predation on African multi-predator, multi-prey large-mammal assemblages and the conservation implications. *Biological Conservation* 232:51–58.
21. Owen-Smith, N. (2015) How diverse large herbivores coexist with multiple large carnivores in African savanna ecosystems: demographic, temporal and spatial influences on prey vulnerability. *Oikos* 124:1417–1426.
22. Yoganand, K; Owen-Smith, N. (2014) Restricted habitat use by an African savanna herbivore through the seasonal cycle: key resources concept expanded. *Ecography* 37:969–982.
23. Owen-Smith, N; Traill, LW. (2017) Space use patterns of a large mammalian herbivore distinguished by activity state: fear versus food? *Journal of Zoology* 303:281–290.
24. Scheel, D; Packer, C (1995) Variation in predation by lions: tracking a movable feast. In Sinclair, ARE et al. (eds) *Serengeti II: Dynamics, Management, and Conservation of an Ecosystem*. University of Chicago Press, Chicago, IL, p. 299.
25. Sinclair, ARE, et al. (2003) Patterns of predation in a diverse predator–prey system. *Nature* 425:288–290.
26. Harrington, R, et al. (1999) Establishing the causes of the roan antelope decline in the Kruger National Park, South Africa. *Biological Conservation* 90:69–78.
27. Owen-Smith, N; Mills, MGL. (2008) Shifting prey selection generates contrasting herbivore dynamics within a large-mammal predator–prey web. *Ecology* 89:1120–1133.

28. Fryxell, JM, et al. (1988) Why are migratory ungulates so abundant? *The American Naturalist* 131:781–798.
29. Fryxell, JM; Sinclair, ARE. (1988) Seasonal migration by white-eared kob in relation to resources. *African Journal of Ecology* 26:17–31.
30. Cooper, SM, et al. (1999) A seasonal feast: long-term analysis of feeding behaviour in the spotted hyaena (*Crocuta crocuta*). *African Journal of Ecology* 37:149–160.
31. Harris, G, et al. (2009) Global decline in aggregated migrations of large terrestrial mammals. *Endangered Species Research* 7:55–76.
32. Naidoo, R, et al. (2016) A newly discovered wildlife migration in Namibia and Botswana is the longest in Africa. *Oryx* 50:138–146.
33. Morjan, MD, et al. (2018) Armed conflict and development in South Sudan threatens some of Africa's longest and largest ungulate migrations. *Biodiversity and Conservation* 27:365–380.
34. Naidoo, R, et al. (2012) Home on the range: factors explaining partial migration of African buffalo in a tropical environment. *PLoS One* 7:e36527.
35. Hillman, J. (1988) Home range and movement of the common eland (*Taurotragus oryx* Pallas 1766) in Kenya. *African Journal of Ecology* 26:135–148.
36. Fryxell, JM, et al. (2005) Landscape scale, heterogeneity, and the viability of Serengeti grazers. *Ecology Letters* 8:328–335.
37. Cagnacci, F, et al. (2016) How many routes lead to migration? Comparison of methods to assess and characterize migratory movements. *Journal of Animal Ecology* 85:54–68.
38. Kreulen, D. (1975) Wildebeest habitat selection on the Serengeti plains, Tanzania, in relation to calcium and lactation: a preliminary report. *African Journal of Ecology* 13:297–304.
39. Murray, MG. (1995) Specific nutrient requirements and migration of wildbeest. In Sinclair, ARE, et al. (eds) *Serengeti II: Dynamics, Management, and Conservation of an Ecosystem*. University of Chicago Press, Chicago, IL, p. 231.
40. Ogutu, J, et al. (2013) Changing wildlife populations in Nairobi National Park and adjoining Athi-Kaputiei Plains: collapse of the migratory wildebeest. *The Open Conservation Biology Journal* 7:11–26.
41. Ogutu, JO, et al. (2012) Dynamics of ungulates in relation to climatic and land use changes in an insularized African savanna ecosystem. *Biodiversity and Conservation* 21:1033–1053.
42. Owen-Smith, RN. (1988) *Megaherbivores: The Influence of Very Large Body Size on Ecology*. Cambridge University Press, Cambridge.
43. Gaillard, J-M, et al. (2015) Does tooth wear influence ageing? A comparative study across large herbivores. *Experimental Gerontology* 71:48–55.
44. Owen-Smith, N; Ogutu, JO. (2013) Controls over reproductive phenology among ungulates: allometry and tropical-temperate contrasts. *Ecography* 36:256–263.
45. Bigalke, RG. (1970) Observations on springbok populations. *African Zoology* 5:59–70.
46. Mills, MGL. (2017) *Kalahari Cheetahs: Adaptations to An Arid Region*. Oxford University Press, Oxford.
47. Ryan, SJ, et al. (2007) Ecological cues, gestation length, and birth timing in African buffalo (*Syncerus caffer*). *Behavioral Ecology* 18:635–644.

48. Smuts, GL. (1976) Population characteristics of Burchell's zebra (*Equus burchelli* antiquorum. H. Smith, 1841) in the Kruger National Park. *South African Journal of Wildlife Research* 6:99–112.
49. Whyte, IJ; Joubert, CSJ. (1988) Blue wildebeest population trends in the Kruger National Park and the effects of fencing. *South African Journal of Wildlife Research* 18:78–87.
50. Sinclair, ARE, et al. (2000) What determines phenology and synchrony of ungulate breeding in Serengeti? *Ecology* 81:2100–2111.
51. Ogutu, JO, et al. (2015) How rainfall variation influences reproductive patterns of African savanna ungulates in an equatorial region where photoperiod variation is absent. *PloS One* 10: e0133744.
52. Estes, RD. (1976) The significance of breeding synchrony in the wildebeest. *African Journal of Ecology* 14:135–152.
53. Augustine, DJ; McNaughton, SJ. (2004) Regulation of shrub dynamics by native browsing ungulates on East African rangeland. *Journal of Applied Ecology* 41:45–58.
54. Owen-Smith, RN. (1983) Dispersal and the dynamics of large herbivores in enclosed areas: implications for management. In Owen-Smith, N (ed.) *Management of Large Mammals in African Conservation Areas*. Haum Educational Publishers, Pretoria, pp. 127–143.
55. Owen-Smith, N, et al. (2012) Shrinking sable antelope numbers in Kruger National Park: what is suppressing population recovery? *Animal Conservation* 15:195–204.
56. Caughley, G, et al. (1980) Does dingo predation control the densities of kangaroos and emus? *Wildlife Research* 7:1–12.

Chapter 13: How Large Herbivores Transform Savanna Ecosystems

Large herbivores transform savanna ecosystems most fundamentally by selectively consuming plants, thereby altering competitive relations among plant species and the structural balance between trees and grasses. Their impacts on vegetation affect the spread of fires and what proportion of the nutrients contained in the vegetation are recycled via soil organisms rather than through herbivore guts. Through locally concentrating their grazing, browsing, breakage and trampling, they contribute to the spatial heterogeneity that is an inherent feature of Africa's savannas. Elephants have an especially great impact through their capacity to topple or uproot trees and kill even large trees through bark removal. Mega-grazers in the form of white rhinos and hippos transform grasslands by promoting grazing lawns in place of taller grasslands through their broad-lipped cropping actions. Smaller browsers suppress the establishment of tree seedlings by selectively consuming green leaves. Grazers of all sizes affect the ability of grass plants to re-establish their foliage above ground at the start of the wet season and following droughts or other disturbances. By reducing the fuel load, they affect the intensity of fires and hence the effect that fires have in restricting the expansion of woody plants or bush encroachment. Over what time frame can ecosystems be interpreted as stable, given the frequency of disturbances from variable rainfall, fire, and the shifting impacts of large herbivores?[1]

Ecosystems with few large herbivores would look very different from those retaining a full complement of these herbivores. This chapter considers how large herbivores alter the structure, composition and functioning of African savannas.

How Much Grass Do Grazers Consume?

The impacts that grazers have on the grass layer depend on how much of the grass biomass produced they consume annually. The proportion left unconsumed gets either incinerated by fire or deposited as litter for decomposition by soil organisms. This partitioning has important consequences for nutrient cycling. Fires result in nitrogen losses while concentrating inorganic mineral nutrients in ash. Digestion via a warm herbivore gut accelerates the conversion of part of the structural carbon into methane and carbon dioxide gases, thereby concentrating organic and mineral nutrients for more rapid recycling than takes place in the soil.

The total consumption of vegetation by large herbivores depends on the aggregate biomass of these consumers and their relative food requirements. Because of metabolic scaling (see below), larger animals require less food per unit of their biomass than do smaller ones. Aggregate biomass levels depend on the rainfall, governing the annual amount of vegetation produced, and on the soil fertility, governing the proportion of these plants that is potentially edible and digestible.

The highest aggregate biomass density of herbivores in Africa, by a considerable margin, was recorded within protected areas in Uganda, before local warfare intruded (Figure 13.1). The major contribution came from the huge numbers of elephants and hippos that were present then. The Hluhluwe-iMfolozi Park in South Africa, with white rhinos substituting for hippos, pips the Serengeti–Mara ecosystem for second place. A broader region of southern Africa - combining Kruger NP in South Africa, Hwange NP in Zimbabwe, Chobe NP in Botswana, and Ruaha NP in southern Tanzania - supports a herbivore biomass totalling over half of that in Serengeti, largely dominated by elephants. Western Africa, representing moist but dystrophic ecosystems, with Malawi in the east added, exhibits much lower herbivore biomass despite high rainfall. However, elephants were probably more abundant there in the past than they are today. Both Hluhluwe-iMfolozi and Serengeti still have

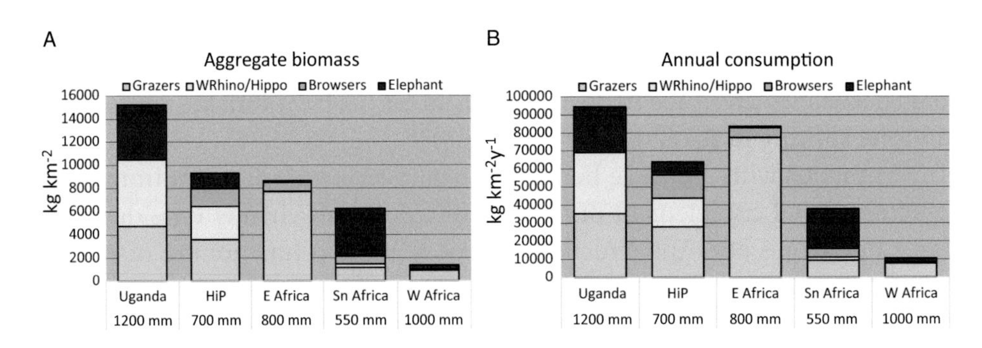

Figure 13.1 (A) Aggregate herbivore biomass and (B) projected combined consumption of vegetation in representative parks retaining largely intact assemblages of large mammals. Uganda is represented by Murchison Falls and Queen Elizabeth National Parks before their elephant and hippo numbers got decimated; HiP is the Hluhluwe-iMfolozi Park in south-eastern South Africa; Eastern Africa plateau grasslands are represented by the Serengeti–Mara ecosystem; southern Africa by Kruger, Chobe, Hwange, and Ruaha national parks combined; and western Africa by a combination of Comoe and Bouba Njida parks in Cameroon and Kasungu National Park in Malawi, all three predominantly miombo woodlands.

growing elephant populations, which will raise their aggregate herbivore biomass levels closer towards the Uganda situation. More broadly across Africa, areas underlain by volcanic deposits or clay-rich sediments generating relatively fertile soils support aggregate herbivore biomass levels 5–10 times greater than found in savannas with similar rainfall underlain by infertile sandy soils.[2,3]

To accommodate the metabolic scaling of food requirements, species-specific biomass contributions need to be transformed into their metabolic mass equivalents. As was outlined in Chapter 10, mass-specific metabolic rates decline with increasing body size according to body mass raised to the power minus one quarter ($M^{-0.25}$). For example, a wildebeest typically consumes grass amounting to around 2.5 percent of its body mass per day, or approaching 10 times its own weight over the course of a year. A white rhino eating 1.5 percent of its body mass per day will consume five times its body mass over a year. As a consequence, Serengeti is closer to Uganda and ahead of Hluhluwe-iMfolozi in terms of amount of vegetation consumed, because a greater proportion of its herbivore biomass is in the form of medium–large grazers rather than megaherbivores. The contribution of browsers, excluding elephants, to aggregate consumption is much smaller than that of grazers in all of these ecosystems.

To relate vegetation consumed to amount produced, an estimate is needed of the total biomass of vegetation generated annually above ground, focusing specifically on grass. Left ungrazed, the standing biomass of grass by the end of the growing season typically amounts to between a half and one gram of dry matter per millimetre of rainfall per square metre.[4,5] However, the amount of grass actually produced could be over twice the peak biomass, because of ongoing turnover of leaves and stems.[6,7] Root growth and turnover below ground can equal or exceed the production of leaves and stems above ground.[8] For Serengeti, the projected annual biomass produced by grasses is about 100 times greater than the total biomass of large herbivores.

Even in Serengeti, the aggregate offtake of grass by large herbivores represents less than 20 percent of the total annual amount of grass produced above ground. This projected offtake is consistent with field measurements in Serengeti, which showed a 30 percent reduction in end-of-season biomass in plots open to grazing compared with fenced exclosures.[9] However, the average estimate obscures wide spatial variation in grass consumption as well as variation between years in response to variable rainfall. Much grass remains ungrazed on hillsides and other localities where few ungulates normally go. Spatially, the total amount of vegetation consumed by large herbivores in Serengeti could entail the removal of nearly half of the plant biomass produced over half of the ecosystem and none over the rest. In drought years when

Figure 13.2 Grazing lawns. (A) An extensive short-grass lawn cultivated by white rhinos in Mfolozi GR; (B) lawn grass cover in Mfolozi GR; (C) lawn grassland cultivated by wildebeest and other grazers in Serengeti plains; (D) lawn grassland generated by wildebeest grazing in Kruger NP; (E) lawn generated by hippos being maintained by puku, Luangwa Valley, Zambia; (F) lawn sustained by grazing in Central Kalahari, Botswana.

rainfall and hence grass growth may be reduced to merely half of the mean, almost all of the grass produced above ground can be consumed over extensive areas (Figures 12.2 and 13.2). Places close to water where grazing ungulates congregate during the dry season routinely become denuded of grass, with hippos adding to the offtake by land-based herbivores. A much smaller fraction of tree foliage produced is consumed by browsers, because most remains out of reach and little of it is retained by trees through the dry season.

Nevertheless, woody plants with evergreen or semi-evergreen leaves may have most of their foliage removed as high as herbivores can reach by late in the dry season.

Spatial variation in grazing impacts was the incentive for my white rhino study in the Hluhluwe-iMfolozi Park. Some areas were being grazed down to bared soil, while elsewhere stands of tall grass were left ungrazed. White rhinos were blamed for locally over grazing the grassland, accentuating soil erosion to the detriment of ecosystem productivity and food for other herbivores. The protected area seemed in danger of becoming a white rhino slum, at the cost of broader biodiversity conservation. My investigations revealed how the nutritious short grasses cultivated by white rhino were preferentially grazed as long as they retained sufficient edible material. Once too little grass remained in these areas, white rhinos shifted their grazing to remaining reserves of taller grass. Instead of being labelled 'overgrazed', areas with short grasses are now recognised as productive 'grazing lawns', benefiting other grazers favouring short grass as well as white rhinos.[10] Grazing lawns cultivated by white rhinos covered 10–30 percent of local landscapes within the Hluhluwe-iMfolozi Park and a similar proportion in parts of Kruger NP.[11,12,13]

Grazing lawns (Figure 13.2) were first recognised in the context of hippo grazing.[14] Short-grass mosaics generated by hippos may extend up to 3 km from water bodies.[14,15,16] In western Africa, kob concentrations can maintain local grazing lawns flanking dambos.[17] Processes involved in the generation and maintenance of lawn grasslands have been especially well studied in Serengeti, with the south-eastern plains representing one vast, although locally variable, lawn grassland (Figure 13.2C).[18] Similar lawn-forming grasslands are cultivated locally by wildebeest herds in Kruger NP on uplands underlain by gabbro bedrock (Figure 13.2D).[19] If grazing is excluded, taller grasses soon replace the short grasses.[20] Repeated episodes of consumption during the course of the growing season favour short grasses, which are promoted by local concentrations of grazers in places where the latter are more secure from predation.[21]

The impact of grazing on future grass growth depends on when consumption takes place during the seasonal cycle. Defoliation of tall grasses during the wet season lowers their competitive superiority for light, opening space for shorter grasses to establish. The latter are better adapted to tolerate defoliation during the wet season through having a larger proportion of their leaves close to the soil and thus not readily removed by grazers. However, grass regrowth depends on an adequate supply of soil nutrients, potentially contributed by the dung and urine deposited by the grazing ungulates. If this is missing, less-palatable grasses increase to the detriment of forage quality. Consumption occurring during the dry season has little impact on future grass growth because grasses are dormant then. Although large areas may be reduced to bare soil during the dry season in drought years, grasses can recover rapidly during the following wet season provided the soil remains intact.

Grazing and Fire

Recurrent fires are intrinsic to savannas because of the accumulation of potential fuel in the grass layer during the dry season. However, by grazing down this dry grass and thereby promoting the spread of lawn-forming grasses, large herbivores suppress the extent of fires.[10] Fires tend not to spread once the proportion of the landscape covered by short grass exceeds around 40 percent.[22] During the early 1960s, about 80 percent of Serengeti grasslands, excluding the short-grass plains, burnt annually.[23] Following the expansion of the wildebeest population to 1.3 million animals, the annual proportion of the ecosystem burnt has been halved to around 30 to 40 percent.[24] The area burnt is reduced further in drought years because less grass remains ungrazed. Plant material that is digested by herbivores escapes the nitrogen loss that occurs with fire. However, both fire and grazing restrict the build-up of decomposing plant material or humus in the soil, reducing its capacity to retain nutrients against leaching. Nevertheless, nutrients are released more slowly through decomposition of plant detritus in the soil than via dung and urine produced by herbivores.

Surface water distribution affects the spatial distribution of grazing. Areas remote from surface water during the dry season escape heavy grazing during the dry season so that most of the grass remains to be burnt periodically. Areas close to few sources of water get heavily grazed and trampled, perhaps locally to bare soil. Hippos add their grazing pressure within a few kilometres of the water bodies where they seek security during the day, leading to rampant soil erosion in the vicinity. White rhinos cultivate grazing lawns further afield, but need to remain within cruising range of water sources. Thus, grazing and fire in combination generate a broad-scale zonation in their relative vegetation impacts.

Browsers and Bush Encroachment

Heavily grazed areas tend to be invaded by woody plants, resulting in the development of quite dense thickets.[25,26] Browsers and mixed feeders like impala[27] can potentially suppress the establishment of woody seedlings and saplings through selective defoliation. Juvenile woody plants emerging in grazing lawns and bared areas are exposed to browsing by the lack of grass cover.[28] Heavy browsing of taller saplings by giraffe can prevent woody

Figure 13.3 Browsing impacts on woody plants. (A) Acacia 'topiaries' heavily pruned by giraffe from above; (B) kudus defoliating acacias at middle height; (C) impala browsing at lower height; (D) acacia sapling with stems heavily clipped by black rhino.

saplings from growing beyond the fire trap (Figure 13.3A).[29] Black rhinos heavily prune small saplings (Figure 13.3D). In Serengeti, wildebeest concentrations contribute additionally to suppressing small woody plants through the breakage they impose. However, the most extreme impacts on woody seedlings came from little dikdiks living in tiny territories at high local densities.[30]

Because of the mechanisms that savanna trees and shrubs have to withstand the effects of fires (Chapter 8), browsing may be needed in addition to suppress the growth of the invading woody plants.[31] Cohorts of establishment by acacia saplings have been related to windows of opportunity following disease outbreaks that temporarily reduced impala numbers.[32]

However, the effectiveness of browsers in suppressing woody plant expansion depends on which species these herbivores choose to eat and where and when their browsing pressure is greatest. These considerations prompted my study on factors influencing diet selection by browsing ruminants undertaken in the Nylsvley Nature Reserve. In this moist/dystrophic savanna, browsing by kudus and impalas actually had little effect on the most common woody species prevalent there, because their leaves generally contained condensed tannins or other chemical deterrents.[33] The spinescent acacias escaped heavy impact by shedding their leaves quite early in the dry season, while browse remained abundant. Moreover, while small browsers like impala, dikdik and other small antelope can have locally severe impacts on woody seedings, they are patchily distributed.[34] They may suppress woody plant invasion locally, but there will be vast areas where they are absent and trees are potentially able to expand, if permitted by fire patterns and other influences. Over wider landscapes, the main agents of woody plant mortality are fire and elephants, separately or in combination.

Elephants and Trees

Elephants are the supreme mixed feeders, breaking branches, toppling trees and uprooting grasses[35] (Figures 13.4 and 13.5). They can push over trees up to 60 cm in trunk diameter.[36] They kill even bigger trees by bark removal, thereby exposing trees to further damage from fires burning into the trunk and from wood-boring beetles. They uproot shrubs and saplings and dig out and consume roots. By persistently breaking leader shoots, they can prevent tree saplings from growing taller, producing browsing hedges, the counterpart of grazing lawns.[37,38] The impacts of elephants on the tree canopy cover can exceed that of fires.[39,40] In Serengeti, increasing numbers of elephants are reversing the expansion by woody plants that had followed the suppression of fires by wildebeest grazing.[23,41,42] In Kruger NP, the expanding elephant population has opened the tree cover on basaltic soils, while granitic soils have shown much less change.[43] Annual rates of tree toppling were amplified sixfold over those shown in fenced areas from which elephants and other herbivores had been excluded.[39]

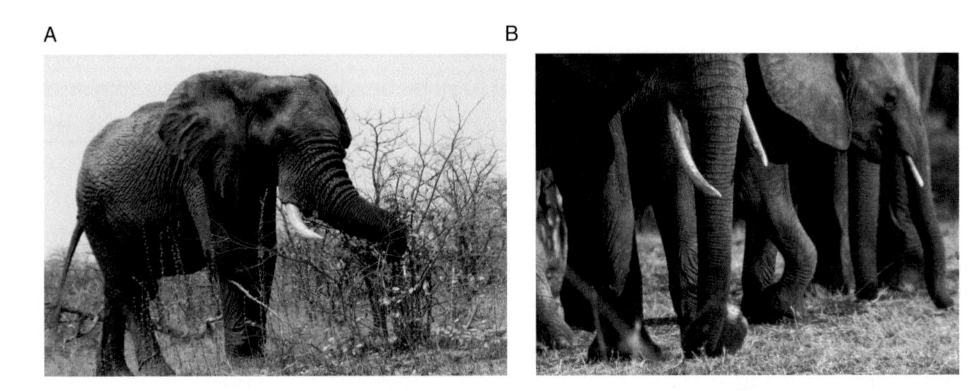

Figure 13.4 Elephants feeding by (A) breaking branchlets of mopane sapling and (B) uprooting grasses.

Figure 13.5 Tree damage inflicted by elephants. (A) Mopane trees felled in northern Botswana; (B) bark stripped from giraffe thorn tree, northern Botswana; (C) mopane tree showing bark response to cumulative damage; (D) debarked dead marula tree, Hluhluwe-Mfolozi Park; (E) baobab tree hollowed, Luangwa Valley, Zambia; (F) elephant yanking acacia sapling, Maasai Mara NR, Kenya; (G) sapling left with top broken off.

The most extreme transformation of the tree cover by elephants took place in Murchison Falls NP in Uganda.[44] A broad-leaved woodland of clusterleaf and bushwillow trees became converted into a tall, mostly treeless grassland with shrinking forest patches within a few decades by elephants hemmed in by surrounding human settlements at a density exceeding 3 elephants per km^2. However, most of these elephants were eliminated during civil warfare in

Figure 13.6 Landscapes transformed by elephants. (A) Woodland that has regenerated in place of grassland following elimination of most elephants from Murchison Falls South NP in Uganda; (B) open grassland generated by elephants and fire in combination in Maasai Mara NR; (C) mopane woodland reduced to stumps in northern Botswana; (D) saplings locally broken by elephants following a fire, northern Botswana; (E) vegetation denuded around Aruba Dam in Tsavo NP, Kenya; (F) dense knobbly bushwillow shrubs replacing acacia woodland, Linyanti, Botswana.

the 1980s. When I visited the park in 2018, the broad-leaved woodland south of the Nile River had become re-established, although the fire regime had not changed (Figure 13.6A). In the Maasai Mara NR in Kenya, elephants and fire in combination transformed an acacia savanna into mostly open grassland, maintained by elephants through their consumption of woody seedlings

(Figure 13.6B).[45,46] The elimination of the trees was ascribed to a succession of very hot fires during years with exceptionally high rainfall. However, I suspect that elephants had actually been the primary agent of woodland removal, toppling trees after lower-level browse had been eliminated by the severe fires. By continuing to break regenerating saplings, elephants are preventing the woodland from regenerating there. In northern Botswana, elephants locally toppled a large proportion of canopy trees in circumstances after either fire or frost had reduced food available in the shrub layer (Figure 13.6D).[36] Woodland destruction of this magnitude may take place episodically where accessible food is lacking, leaving only stumps remaining (Figure 13.6C).

Within Tsavo East NP in Kenya, elephants hemmed in by hunting in surrounding areas uprooted or felled wild myrrh shrubs and umbrella thorn trees and almost entirely eliminated baobabs. The area around a large dam became almost completely denuded of vegetation (Figure 13.6E).[47,48] Nevertheless, much recovery has since taken place in the woodland following the reduction of the elephant population by illegal hunting.[49] In Hwange NP in Zimbabwe where the regional elephant density has reached around 3 animals per km^2, elephants have converted areas of broad-leaved woodland into coppice regrowth. Nevertheless, woodlands dominated by deep-rooted Zimbabwe teak and other large trees not utilised by elephants remain little impacted (personal observations). Furthermore, woody species not eaten by elephants are increasing in compensation. Alongside the Chobe River in northern Botswana, high local elephant densities have converted the formerly dense woodland flanking the river into open shrubland.[50,51] Further west along the Linyanti River, knob thorn and giraffe thorn trees have been mostly eliminated from the riparian woodland and replaced by knobbly bushwillow (*Combretum mossambicensis*), a shrub not utilized by elephants (Figure 13.6F).[52]

Nevertheless, some trees have bark resistant to removal by elephants, like leadwood. Others have very deep roots, and are thus not readily pushed over, like giraffe thorn. Some have foliage rejected by elephants, like African blackwood (*Erythrophleum africanum*). Although not eaten by elephants, knobbly bushwillow is readily browsed by ruminants.[53] Other bushwillow species are consumed by elephants. Such selective utilisation can lead to compositional shifts in the vegetation without overall opening of the woody canopy cover.

However, elephants are water-dependent and need to drink every 2–4 days. This means that the spatial extent of the vegetation transformation that they induce is restricted to <10 km of rivers, lakes or other long-lasting water bodies.[54] In northern Botswana, trees pollarded by elephants occur mostly within 5 km of surface water and elephant impacts diminish further away.[55] Despite the enormous numbers of elephants present in northern Botswana, extensive mopane and sandveld woodlands remote from water remain little damaged by elephants.[56]

The habitat transformation brought about by elephants need not be adverse for other large herbivores. Trees get replaced by lower-growing saplings and shrubs, bringing more food within the height reach of medium-large browsers.[57] Opening of the tree cover promotes grasses and hence more food for grazers. Nutrients contained within trees get cycled more rapidly than if locked up until trees eventually die of old age. However, the structural and compositional diversity of the woodland can be reduced,[52] and birds dependent on tall trees for habitat and for nesting sites, like vultures, may lose out.

Savanna regions were more open than they are today over much of Africa even quite recently. Many wildlife reserves were proclaimed following the resurgence of tsetse flies and the sleeping sickness that they transmit as a consequence of woody thickening, which necessitated the relocation of human settlements. Bush thickening has been widespread throughout southern African over recent decades, except in places retaining elephants.[58] The collapse of large herbivore populations during the rinderpest epizootic during the 1890s, coupled with a reduction in fires lit by people, have both been invoked as the causes of the woody plant expansion elsewhere in Africa. I suspect that the main cause of this vegetation transformation was actually the elimination of elephants by hunting for ivory, rampant during the nineteenth century. The woody vegetation cover in the presence of elephants will inevitably be more open than in places lacking these ecosystem engineers, at least within cruising range of surface water.

A further side benefit of elephants should not be overlooked: the network of trails they develop, many leading to long-lasting water sources. Trails formed by white rhinos greatly facilitated my movements on foot in the Hluhluwe-iMfolozi Park. Elephant trails crossing ridges and valleys have served to guide road construction in many parts of Africa.

The combined roles of elephants and fires in suppressing the tree canopy cover in savannas has generally been regarded as adverse. However, the productive potential of savannas lies mainly in the grass layer, which is promoted by both. Trees are needed somewhere as an essential component of savanna heterogeneity. They persist alongside rivers and drainage lines as well as in areas remote from surface water. Africa's savanna trees and shrubs have evolved along with elephants as well as fires for many millennia. Mopane trees have an amazing capacity to resist elephant impacts by growing back from the stumps of toppled trees. The overall savanna is potentially more productive, and broadly more diverse, in their presence than where their effects are missing. During earlier times, while humans evolved, mega-browsers and mega-grazers were more abundant and diverse than they remained into the

Holocene. Their impacts on savanna structure and composition would have been correspondingly greater, at least locally within reach of water.

Termites and Decomposition

The other major bioengineers in savannas altering structure and function physically, besides elephants, are termites. Their activities are undertaken largely underground and thus out of sight. They are the primary agents of the decomposition of plant biomass not eaten by herbivores and not burnt by fires. The termites that feed mainly on woody material are members of the genus *Macrotermes*. Those dependent mostly on grasses are represented by three genera: *Odontotermes*, *Trinervitermes* and *Hodotermes*. *Macrotermes* spp. build large earth mounds, *Trinervitermes* spp. small domed mounds, and *Odontotermes* spp. low spread-out mounds, while *Hodotermes* colonies are hidden entirely underground (Figure 13.7).[59] *Cubitermes* spp. construct small domes and digest the soil organic matter that they consume via the agency of cellulose-degrading microbes in their gut. Other mound-building termites cultivate symbiotic fungi within chambers inside their mounds, able to degrade the lignocellulose component of woody plant parts. The termites feed mostly on these fungi. Harvester termites contribute to much grass removal, including some green leaves, especially when drought conditions take hold (Figure 13.7E,F). Large *Macrotermes* mounds with basal diameters up to 10 m achieve densities of 2–10 mounds per ha (Figure 13.7C), while the much smaller mounds of other species can reach densities of 1000 or more per ha.[60,61] Mounds are present more frequently on crests than lower down on slopes.

Nutrients contained in plant litter get concentrated in the interiors of termite mounds. Termites play a further role in soil engineering by lifting clay from deeper levels to constitute much of the mound, along with mineral nutrients. Concurrently, stone pebbles gravitate downwards to form stone lines in subsoils.

Low *Odontotermes* mounds support the short grasses favoured by white rhinos and other grazers. The raised eminences of *Macrotermes* mounds provide refuges for woody plant species sensitive to fire.[62,63] Trees growing on mounds attract browsing by black rhinos and elephants because of their relatively palatable foliage compared with the surrounding woodland.[64,65]

The abundance of termites in African savannas is testified by the many mammals that feed on them, some nearly exclusively. These include aardvarks (*Orycteropus afer*), pangolins (*Manis* spp.) and aardwolves (*Proteles cristata*), while numerous birds are attracted to the winged alates, which fly off in vast swarms following rain. In nutrient-poor savannas, where most of the grass is

Figure 13.7 Termite activities. (A) *Macrotermes* mound in grassland, western extension of Serengeti NP; (B) *Macrotermes* mound in woodland, Kruger NP; (C) concentration of *Macrotermes* mounds, Kruger NP; (D) low *Odontotermes* mound on western edge of plains, Serengeti NP; (E) *Hodotermes* gathering grass remains into their underground nest; (F) grass cut by harvesting *Hodotermes.*

too fibrous to be digested by ungulates, the total biomass of termites underground can greatly exceed the combined weight of the mammalian herbivores.[66,67] This situation gets reversed in nutrient-rich savannas, at least locally.

However, the presence of termites of different kinds can vary quite widely spatially. Mound-building termites need clay particles to build these structures, restricting their presence to regions where there is adequate clay in the soil or subsoil. There is also turnover among mounds, as testified by dispersal of the alates seeking sites to establish colonies. The consequences of such dynamics for herbivores, fire and nutrient recycling have yet to be investigated.

Nutrient Cycling

I have emphasised the recycling of the nutrients contained in plant biomass by large herbivores, specifically nitrogen, phosphorus, potassium and calcium. However, there is another mineral nutrient critically important for herbivores that is made available very differently. It is sodium. Most plants actively exclude sodium in their tissues because it is so avidly sought by herbivores.[68] The sodium accumulators include many of the short grasses prevalent in grazing lawns, such as pan dropseed (Figure 6A.4G).[69,70] These grasses benefit indirectly through attracting grazing because they are better able to tolerate defoliation than the taller grasses which are able to shade them out. The bare soil exposed by the reduced grass cover promotes evaporation, which concentrates sodium.[71] Trampling may contribute further by soil compaction, restricting water infiltration. Grazing ungulates tend to be most abundant in drier savannas where sodium is not leached. Notably, white rhinos were absent historically from cool Highveld grasslands. Unpublished studies revealed that grazing lawns promoted by ruminants in the the Suikerbosrand Nature Reserve south of Johannesburg did not contain the sodium-accumulating grasses prevalent elsewhere. In nutrient-deficient miombo woodlands, sodium tends to accumulate downslope on the margins of grassy dambos, attracting grazers such as roan and sable antelope to these regions. Salt licks formed in other localities may also become sources of attraction for herbivores (Figure 13.8). Termite mounds locally concentrate sodium in soils and hence in plant parts. Elephants tended to favour water bodies containing higher sodium levels in regions underlain by nutrient-poor Kalahari sand.[72]

Overview

Large herbivores have fundamental impacts on savanna structure and composition, in interaction with fire. Large grazers cultivate lawn-forming grasses in place of taller tufted grasses, suppressing fires. Browsers of varying sizes help suppress the expansion of woody plants, particularly in the absence of frequent fires. Elephants are major agents of the mortality of trees and shrubs, capable of opening or even removing the tree canopy over large areas, especially in interaction with fire. Concentrations of grazers near surface water can exclude fires through their denudation of the grass cover. Areas remote from

Figure 13.8 Sable antelope herd at a salt lick in northern Botswana.

surface water burn more frequently and intensely because of the absence of water-dependent grazers. All of these features contribute to the spatial heterogeneity that is an inherent feature of Africa's savannas. Grass consumption by grazers plus tree felling by elephants accelerates rates of nutrient recycling. This accentuates landscape heterogeneity in interaction with soil moisture, bedrock geology and local fire intensity. In these various ways, herbivores contribute to the spatial heterogeneity that is integral to African savannas. The outcome is the green, brown and black world mosaic that characterises savanna landscapes,[73] observable from above via satellite, aircraft or drone imagery. The patch mosaic would appear somewhat different in the absence of large herbivores.

However, much of savanna Africa is not thronged with abundant herbivores. The broad-leaved woodlands typifying moist/dystrophic savanna support much lower densities of big herbivores, except locally in floodplains or around dambos. Soil fertility based on mineral nutrients plays a major role in limiting the regional abundance of large herbivores. Sodium availability may serve as a major restriction on herbivore abundance in moister savannas where soils are leached of this mineral.

Elsewhere in the world, the role of herbivores in restricting the tree cover is less apparent, because most were eliminated by human hunters during the late Pleistocene. Prior to these extinctions, mammoths, bisons and horses contributed to maintaining the grassy 'mammoth steppe' that stretched from Siberia through Alaska.[74] Megaherbivores opened localities within deciduous forests, enabling fires to penetrate and promote meadows.[75] Grazing by mammoths would have enhanced the compositional diversity of tall-grass prairies. Gomphotheres and other megaherbivores formerly inhabiting South America's cerrado savannas must have had a similarly great impact on the tree cover there. Nevertheless, the tree-damaging effects of the elephants occupying tropical Asian forests seem quite minor.[76] In India, spotted deer attain local densities matching those of mixed-feeding impalas in Africa, but abundant grazers are lacking.[77] Forest elephants in Africa similarly have little impact on tree saplings, although their grazing and browsing along forest margins may restrict fire penetration by depressing fuel loads.[78]

Before turning attention to human evolution, the somewhat different assemblages of herbivores and carnivores that coexisted with emerging human lineages further back in time need to be recognised. These paleo-faunas will be the subject of the next chapter.

Suggested Further Reading

Owen-Smith, N. (1988) *Megaherbivores. The Influence of Very Large Body Size on Ecology*. Cambridge University Press, Cambridge.

References

1. Briske, DD, et al. (2020) Strategies for global rangeland stewardship: assessment through the lens of the equilibrium–non-equilibrium debate. *Journal of Applied Ecology* 57:1056–1067.
2. Bell, RHV. (1982) The effect of soil nutrient availability on community structure in African ecosystems. In Huntley, BJ; Walker, BH (eds) *Ecology of Tropical Savannas*. Springer, Berlin, pp. 193–216.
3. Fritz, H; Duncan, P. (1994) On the carrying capacity for large ungulates of African savanna ecosystems. *Proceedings of the Royal Society of London Series B: Biological Sciences* 256:77–82.
4. Rutherford, MC. (1980) Annual plant production–precipitation relations in arid and semi-arid regions. *South African Journal of Science* 76:53–57.
5. Deshmukh, IK. (1984) A common relationship between precipitation and grassland peak biomass for east and southern Africa. *African Journal of Ecology* 22:181–186.
6. Deshmukh, I. (1986) Primary production of a grassland in Nairobi National Park, Kenya. *Journal of Applied Ecology* 23:115–123.
7. Cox, GW; Waithaka, JM. (1989) Estimating aboveground net production and grazing harvest by wildlife on tropical grassland range. *Oikos* 54:60–66.

8. Strugnell, RG; Pigott, CD. (1978) Biomass, shoot-production and grazing of two grasslands in the Rwenzori National Park, Uganda. *The Journal of Ecology* 66:73–96.
9. Anderson, TM, et al. (2007) Rainfall and soils modify plant community response to grazing in Serengeti National Park. *Ecology* 88:1191–1201.
10. Waldram, MS, et al. (2008) Ecological engineering by a mega-grazer: white rhino impacts on a South African savanna. *Ecosystems* 11:101–112.
11. Arsenault, R; Owen-Smith, N. (2011) Competition and coexistence among short-grass grazers in the Hluhluwe-iMfolozi Park, South Africa. *Canadian Journal of Zoology* 89:900–907.
12. Cromsigt, JPGM; te Beest, M. (2014) Restoration of a megaherbivore: landscape-level impacts of white rhinoceros in Kruger National Park, South Africa. *Journal of Ecology* 102:566–575.
13. Cromsigt, J, et al. (2017) The functional ecology of grazing lawns – how grazers, termites, people, and fire shape HiP's savanna grassland mosaic. In Cromsigt, JPGM, et al. (eds) *Conserving Africa's Mega-diversity in the Anthropocene: The Hluhluwe-iMfolozi Park Story*. Cambridge University Press, Cambridge, pp. 135–160.
14. Olivier, RCD; Laurie, WA. (1974) Habitat utilization by hippopotamus in the Mara River. *African Journal of Ecology* 12:249–271.
15. Lock, JM. (1972) The effects of hippopotamus grazing on grasslands. *The Journal of Ecology* 60:445–467.
16. Kanga, EM, et al. (2013) Hippopotamus and livestock grazing: influences on riparian vegetation and facilitation of other herbivores in the Mara Region of Kenya. *Landscape and Ecological Engineering* 9:47–58.
17. Verweij, R, et al. (2006) Grazing lawns contribute to the subsistence of mesoherbivores on dystrophic savannas. *Oikos* 114:108–116.
18. McNaughton, SJ. (1983) Serengeti grassland ecology – the role of composite environmental-factors and contingency in community organization. *Ecological Monographs* 53:291–320.
19. Yoganand, K; Owen-Smith, N. (2014) Restricted habitat use by an African savanna herbivore through the seasonal cycle: key resources concept expanded. *Ecography* 37:969–982.
20. McNaughton, SJ. (1985) Ecology of a grazing ecosystem: the Serengeti. *Ecological Monographs* 55:259–294.
21. Anderson, TM, et al. (2010) Landscape-scale analyses suggest both nutrient and antipredator advantages to Serengeti herbivore hotspots. *Ecology* 91:1519–1529.
22. Archibald, S, et al. (2017) Interactions between fire and ecosystem processes. In Cromsigt, JPGM, et al. (eds) *Conserving Africa's Mega-Diversity in the Anthropocene: The Hluhluwe-iMfolozi Park Story*. Cambridge University Press, Cambridge, pp. 233–261.
23. Sinclair, A, et al. (2008) Historical and future changes to the Serengeti ecosystem. In Sinclair, ARE, et al. (eds) *Serengeti III: Human Impacts on Ecosystem Dynamics*. University of Chicago Press, Chicago, pp. 7–46.
24. Eby, S, et al. (2015) Fire in the Serengeti ecosystem: history, drivers, and consequences. In Sinclair, ARE, et al. (eds) *Serengeti IV: Sustaining Biodiversity in a Coupled Human–Natural System*. University of Chicago Press, Chicago, pp. 73–103.

25. Roques, KG, et al. (2001) Dynamics of shrub encroachment in an African savanna: relative influences of fire, herbivory, rainfall and density dependence. *Journal of Applied Ecology* 38:268–280.
26. O'Connor, TG, et al. (2014) Bush encroachment in southern Africa: changes and causes. *African Journal of Range & Forage Science* 31:67–88.
27. Moe, SR, et al. (2009) What controls woodland regeneration after elephants have killed the big trees? *Journal of Applied Ecology* 46:223–230.
28. Voysey, MD, et al. (2021) The role of browsers in maintaining the openness of savanna grazing lawns. *Journal of Ecology* 109:913–926.
29. Pellew, RA. (1984) The feeding ecology of a selective browser, the giraffe (*Giraffa camelopardalis tippelskirchi*). *Journal of Zoology* 202:57–81.
30. Augustine, DJ; McNaughton, SJ. (2004) Regulation of shrub dynamics by native browsing ungulates on East African rangeland. *Journal of Applied Ecology* 41:45–58.
31. Staver, AC; Bond, WJ. (2014) Is there a 'browse trap'? Dynamics of herbivore impacts on trees and grasses in an African savanna. *Journal of Ecology* 102:595–602.
32. Prins, HHT; van der Jeugd, HP. (1993) Herbivore population crashes and woodland structure in East Africa. *Journal of Ecology* 81:305–314.
33. Cooper, SM; Owen-Smith, N. (1985) Condensed tannins deter feeding by browsing ruminants in a South African savanna. *Oecologia* 67:142–146.
34. du Toit, JT; Owen-Smith, N. (1989) Body size, population metabolism, and habitat specialization among large African herbivores. *The American Naturalist* 133:736–740.
35. Owen-Smith, N, et al. (2019) Megabrowser impacts on woody vegetation in savannas. In Scogings, PF; Sankaran, M (eds) *Savanna Woody Plants and Large Herbivores*. Wiley, Oxford, pp. 585–611.
36. Chafota, J; Owen-Smith, N. (2009) Episodic severe damage to canopy trees by elephants: interactions with fire, frost and rain. *Journal of Tropical Ecology* 25:341–345.
37. Cromsigt, JPGM; Kuijper, DPJ. (2011) Revisiting the browsing lawn concept: evolutionary interactions or pruning herbivores? *Perspectives in Plant Ecology, Evolution and Systematics* 13:207–215.
38. du Toit, JT; Olff, H. (2014) Generalities in grazing and browsing ecology: using across-guild comparisons to control contingencies. *Oecologia* 174:1075–1083.
39. Asner, GP; Levick, SR. (2012) Landscape-scale effects of herbivores on treefall in African savannas. *Ecology Letters* 15:1211–1217.
40. Pellegrini, AFA, et al. (2017) Woody plant biomass and carbon exchange depend on elephant-fire interactions across a productivity gradient in African savanna. *Journal of Ecology* 105:111–121.
41. Sinclair, ARE, et al. (2008) Historical and future changes to the Serengeti ecosystem. In Sinclair, ARE, et al. (eds) *Serengeti III: Human Impacts on Ecosystem Dynamics*. University of Chicago Press, Chicago, pp. 7–46.
42. Morrison, TA, et al. (2016) Elephant damage, not fire or rainfall, explains mortality of overstorey trees in Serengeti. *Journal of Ecology* 104:409–418.

43. Trollope, WSW, et al. (1998) Long-term changes in the woody vegetation of the Kruger National Park, with special reference to the effects of elephants and fire. *Koedoe* 41:103–112.
44. Laws, RM, et al. (1975) *Elephants and Their Habitats*. Clarendon Press, Oxford.
45. Dublin, HT, et al. (1990) Elephants and fire as causes of multiple stable states in the Serengeti–Mara woodlands. *The Journal of Animal Ecology* 59:1147–1164.
46. Dublin, HT. (1991) Dynamics of the Serengeti–Mara woodlands: an historical perspective. *Forest & Conservation History* 35:169–178.
47. Agnew, ADQ. (1968) Observations on the changing vegetation of Tsavo National Park (East). *African Journal of Ecology* 6:75–80.
48. Leuthold, W. (1977) Changes in tree populations of Tsavo East National Park, Kenya. *African Journal of Ecology* 15:61–69.
49. Leuthold, W. (1996) Recovery of woody vegetation in Tsavo National Park, Kenya, 1970–94. *African Journal of Ecology* 34:101–112.
50. Mosugelo, DK, et al. (2002) Vegetation changes during a 36-year period in northern Chobe National Park, Botswana. *African Journal of Ecology* 40:232–240.
51. Skarpe, C, et al. (2014) Plant–herbivore interactions. In Skarpe, C, et al. (eds) *Elephants and Savanna Woodland Ecosystems*. Wiley, Oxford, pp. 189–206.
52. Teren, G, et al. (2018) Elephant-mediated compositional changes in riparian canopy trees over more than two decades in northern Botswana. *Journal of Vegetation Science* 29:585–595.
53. Makhabu, SW. (2005) Resource partitioning within a browsing guild in a key habitat, the Chobe Riverfront, Botswana. *Journal of Tropical Ecology* 21:641–649.
54. Chamaillé-Jammes, S, et al. (2007) Managing heterogeneity in elephant distribution: interactions between elephant population density and surface-water availability. *Journal of Applied Ecology* 44:625–633.
55. Sianga, K, et al. (2017) Spatial refuges buffer landscapes against homogenisation and degradation by large herbivore populations and facilitate vegetation heterogeneity. *Koedoe* 59:a1434.
56. Ben-Shahar, R. (1998) Changes in structure of savanna woodlands in northern Botswana following the impacts of elephants and fire. *Plant Ecology* 136:189–194.
57. Rutina, LP, et al. (2005) Elephant *Loxodonta africana* driven woodland conversion to shrubland improves dry-season browse availability for impalas *Aepyceros melampus*. *Wildlife Biology* 11:207–213.
58. Stevens, N, et al. (2017) Savanna woody encroachment is widespread across three continents. *Global Change Biology* 23:235–244.
59. Gosling, CM, et al. (2012) Effects of erosion from mounds of different termite genera on distinct functional grassland types in an African savannah. *Ecosystems* 15:128–139.
60. Goudie, AS. (1988) The geomorphological role of earthworms and termites in the tropics. In Viles, H (ed.) *Biogeomorphology*. Blackwell, Oxford, pp. 43–82.
61. Davies, AB, et al. (2014) Spatial variability and abiotic determinants of termite mounds throughout a savanna catchment. *Ecography* 37:852–862.
62. Mobæk, R, et al. (2005) Termitaria are focal feeding sites for large ungulates in Lake Mburo National Park, Uganda. *Journal of Zoology* 267:97–102.

63. Davies, AB, et al. (2016) Termite mounds differ in their importance for herbivores across savanna types, seasons and spatial scales. *Oikos* 125:726–734.
64. Holdo, RM; McDowell, LR. (2004) Termite mounds as nutrient-rich food patches for elephants. *Biotropica* 36:231–239.
65. Loveridge, JP; Moe, SR. (2004) Termitaria as browsing hotspots for African megaherbivores in miombo woodland. *Journal of Tropical Ecology* 20:337–343.
66. Ferrar, P. (1982) Termites of a South African savanna. IV. Subterranean populations, mass determinations and biomass estimations. *Oecologia* 52:147–151.
67. Deshmukh, I. (1989) How important are termites in the production ecology of African savannas? *Sociobiology* 15:155–168.
68. Borer, ET, et al. (2019) More salt, please: global patterns, responses and impacts of foliar sodium in grasslands. *Ecology Letters* 22:1136–1144.
69. Seagle, SW; McNaughton, SJ. (1992) Spatial variation in forage nutrient concentrations and the distribution of Serengeti grazing ungulates. *Landscape Ecology* 7:229–241.
70. Griffith, DM, et al. (2017) Ungulate grazing drives higher ramet turnover in sodium-adapted Serengeti grasses. *Journal of Vegetation Science* 28:815–823.
71. Stock, WD, et al. (2010) Herbivore and nutrient control of lawn and bunch grass distributions in a southern African savanna. *Plant Ecology* 206:15–27.
72. Weir, JS. (1972) Spatial distribution of elephants in an African National Park in relation to environmental sodium. *Oikos* 23:1–13.
73. Bond, WJ. (2005) Large parts of the world are brown or black: a different view on the 'Green World' hypothesis. *Journal of Vegetation Science* 16:261–266.
74. Zimov, SA, et al. (2012) Mammoth steppe: a high-productivity phenomenon. *Quaternary Science Reviews* 57:26–45.
75. Owen-Smith, N. (1987) Pleistocene extinctions: the pivotal role of megaherbivores. *Paleobiology* 13:351–362.
76. Mueller-Dombois, D. (1972) Crown distortion and elephant distribution in the woody vegetations of Ruhuna National Park, Ceylon. *Ecology* 53:208–226.
77. Karanth, KU; Sunquist, ME. (1992) Population structure, density and biomass of large herbivores in the tropical forests of Nagarahole, India. *Journal of Tropical Ecology* 8:21–35.
78. Cardoso, AW, et al. (2020) The role of forest elephants in shaping tropical forest–savanna coexistence. *Ecosystems* 23:602–616.

Chapter 14: Paleo-faunas: Rise and Fall of the Biggest Grazers

The large mammals that existed during the course of human evolution were different from those around today – not only in species, but also in genera and in size. What we know of them is rather fragmentary. The erosion that levelled Africa's surface after its separation from Gondwana also took away not only the bones of the dinosaurs of the late Jurassic and Cretaceous periods but also those of the early mammals that replaced them during the Palaeocene and Eocene epochs. Most of the fossil record from this dawn age of mammals comes from the Fayum Province of Egypt. This northern region of Africa lay beneath a shallow sea then. Its fossil deposits include early predecessors of whales, affiliated with hippos. Also present are the fossilised bones of a short-legged, rather pig-like swamp-dweller called *Moeritherium*, believed to be the precursor of the elephants.

While Africa was joined with Arabia during this dawn period, it was isolated from the rest of Eurasia by the Tethys Sea. The mammal orders that evolved there in isolation loosely constitute the 'Afrotheria'. Besides the Proboscidea (elephants and relatives), they include the Hyracoidea (hyraxes), Tubulidentata (aardvark), Sirenia (dugongs and manatees), Macroscelidea (elephant shrews) and Chrysochloridea (golden moles). Also entering Africa somehow during this early time were giant rhino-like animals called embrithopods and hippo-like anthracotheres.[1] Although primates originated in Asia, they were represented in Africa by ~30 Ma, as were rodents.

Africa's isolation from other continents ended towards the end of the Oligocene 23 Ma when a land bridge connected Arabia with Eurasia. Elephant-like proboscideans moved out and early ungulates crossed in, along with much else. Most of what we know about the evolution of Africa's large mammals comes from sites located in rift valley troughs, beginning in Ethiopia and extending through eastern Africa (Figure 14.1A–D; see also Chapter 1). Timelines for the fossil sequences are provided by succeeding layers of volcanic ash or lava, which can be dated quite precisely from the radioactive decay of certain isotopes. Most deposition followed the second, Pliocene phase of tectonic uplift, starting around 5 Ma.

Figure 14.1 Fossil sites. (A) Olduvai Gorge in Tanzania, where deposition layers have yielded numerous fossils dated from the early Pleistocene onwards; (B) Lake Turkana region in northern Kenya, which includes several fossil-yielding locations spanning the Pliocene and Pleistocene (photo: Peter Howard); (C) Ledi Guraru region of Ethiopia, one of the sites where sedimentary deposits have yielded crucial fossils of early humans (photo: Erin DiMaggio); (D) fossil bones accumulated in sediments at Olorgesaile, Kenya; (E) Swartkrans limestone cavity entrance situated close to Sterkfontein in the Cradle of Humankind, South Africa; (F) Makapansgat limeworks located in the northern bushveld region of South Africa.

The Highveld region of South Africa provided a rather different context for fossil preservation west of Johannesburg, in the region labelled the 'Cradle of Humankind' (Figure 14.1E,F). Surface erosion following Miocene uplift exposed dolomitic limestone that had been formed over 2000 million years ago when this part of Africa had been under seawater, well before terrestrial life evolved. Subsequent water erosion opened caves and sinkholes, which eventually became exposed to the land surface after the early Pliocene. Animals either fell into the cavities, or were carried in by carnivores, and became encased within the mineral build-up of breccia. Other limestone caves are located at Makapansgat 200 km further north, in a region that is now savanna bushveld. They retain fossils from similarly early times from ~3.5 Ma in the mid-Pliocene.

Elsewhere in southern Africa, coastal deposits have yielded fossils at Langebaanweg in the south-western Cape, dated ~5 Ma, and at sites further north in Namibia. However, a vast region of Africa lacks fossil deposits. The resultant fragmentation of the fossil record in Africa must be borne in mind when interpreting the fossil record. The species and genera represented need to be placed in the context of the major transitions in climate (Chapter 2) and vegetation (Chapter 9) that occurred. They took place (1) between 10 and 6 Ma during the late Miocene when C_4 grasslands spread, (2) after 2.7 Ma when the Pleistocene glaciations developed and grasslands became more extensive, (3) after 1.8 Ma when further drying expanded grasslands, and (4) after 0.9 Ma when glacial cycles lengthened to 100 kyr and cold extremes produced even greater aridity.

Oligocene–Miocene Giants

Following the global extinction of dinosaurs at the end of the Cretaceous period 66 Ma, the small furry mammals that had scurried from the toes of these huge reptiles began to get much larger. Most of the modern orders of mammals originated during the wet and warm environments of the Palaeocene and Eocene epochs, spanning the period from 66 to 34 Ma.[2] Atmospheric levels of carbon dioxide exceeded 2000 ppm, five times current concentrations, promoting lush vegetation growth across the continent. Proboscideans and hyraxes became the predominant herbivores in Africa, with some of the latter matching the largest antelope in size. Proboscideans were especially diverse, including deinotheres with lower incisors curving downward and backward like giant hooks, shovel-tusked gomphotheres and early mastodons.[3]

During the Oligocene epoch (34–23 Ma), a cooling and drying trend caused forests to give way to broad-leaved deciduous woodlands. Following the start of the Miocene 23 Ma, conditions in eastern Africa became cooler and drier. Along Africa's west coast, the Benguela current formed, conveying cold water northward and accentuating dry conditions into the Congo basin. From the Oligocene into the middle Miocene 12 Ma, Africa's fauna was characterised by a predominance of very large animals, many in the megaherbivore size range.[1,4] Enormous claw-bearing chalicotheres, related to tapirs, apparently foraged by pulling down tree branches, like giant ground sloths once did in South America. Short-necked giraffes with bony antlers, called sivatheres, entered Africa from Asia and became widespread, joined by hippopotamuses and diverse rhinoceroses.[2] None of the large herbivores from the early portion of the Miocene showed dental adaptations for a grass diet, although grasses had appeared by then. The stem ruminants were apparently mainly fruit-eaters.

Ruminant Radiations

The earliest ruminant, assigned to the genus *Eotragus*, dispersed from Asia into Africa around 18 Ma early in the Miocene.[5,6] Two distinct subfamilies within the Bovidae (horned ruminants) diverged from this shared ancestor.[7] The Bovinae, encompassing wild cattle (Bovini) and spiral-horned antelope (Tragelaphini), were derived from boselaphine ancestors in Asia, represented today by the Indian nilgai (*Boselaphus tragocamelus*). African bovines split from Asian representatives around 7 Ma, while the tragelaphines diverged slightly later from a boselaphine stem. The lesser kudu (*Tragelaphus imberbis*) is closest genetically to the stem species.[8] The second subfamily, the Antilopinae, encompasses the remaining antelope tribes, plus goats and sheep in the tribe Caprini. The impala, alone in the tribe Aepycerotini, had an origin close to the stem of the Antilopinae. Its closest genetic link is with the dwarf antelope called suni (*Nesotragus moschatus*), similarly ancient in its origin.[9] The gazelle tribe (Antilopini) also had an early origin and is widely represented in Asia as well as Africa. The reduncines (waterbuck, kob, reedbuck, and allied wetland grazers) originated in Asia with early forms represented in fossil deposits in Pakistan dated to the late Miocene. The alcelaphines (wildebeest affiliates) evidently originated in Africa and had radiated into several genera by the late Miocene, all of them now either extinct or replaced by descendants.[10,11] The genus *Beatragus*, represented by the rare hirola (or Hunter's hartebeest, *Beatragus hunteri*), is closest genetically to the alcelaphine stem. Caprines were represented in South Africa by a buffalo-sized form called *Makapania*, now extinct. The hippotragines seemingly had an African origin.[12] By the late Miocene, all of the modern antelope tribes had become distinct.

Giraffes had produced both long-necked and short-necked forms by the late Miocene. Suids (pigs) diversified into various forms, some quite huge.[13] Three-toed hipparion horses (*Eurygnathus* sp.), with an origin in North America, appeared in Africa shortly after 10 Ma and were represented by several species.[14] True elephants, derived from gomphothere ancestors, made their appearance late in the Miocene.[15] White rhinos (*Ceratotherium simum*) diverged from a black rhino (*Diceros bicornis*) ancestor during the later Miocene and were distributed throughout Africa from the Cape to the Mediterranean Sea margin. Two forms of hippo coexisted, including *Hippopotamus gorgops*, with protruding eyes, along with the ancestral form of the modern hippo.

Emergence of Grazers

The tectonic uplift that took place in north-east Africa by 10 Ma deflected rain-bearing winds, accentuating aridity in eastern Africa. Carbon dioxide levels in the atmosphere became reduced towards 400 ppm, further restricting plant growth. Grasslands spread (see Chapter 9), enhanced nutritionally by volcanically derived soils. Shifts towards grass-based diets by large herbivores are revealed by the form of their teeth (see Chapter 10) and from the ratio of stable carbon isotopes in the collagen content of dental enamel or other body parts (see Chapter 7). The transitional period between 10 and 6 Ma is best represented in fossil deposits in the Turkana region of northern Kenya.[4,16,17,18,19] Hipparion horses showed earliest indications of a C_4 grass component in their diet by 9.9 Ma, although their tooth structure remained indistinct from that of browsers until 7.4 Ma. Rhinos showed an increasing proportion of C_4 grasses in their diet after 9.6 Ma, and the grazing white rhino had diverged dentally from the browsing black rhino by 6 Ma. By 7.4 Ma, some bovids had diets consisting mostly of C_4 grasses. Impala remained a browser in dentition and diet.[20] Hippos showed a dietary trend towards an increased proportion of C_4 grasses. By 6.5 Ma, some gomphotheres as well as early elephants had become mainly grazers. Even large suids had adopted C_4-dominated diets by 4.2 Ma, along with short-necked giraffes.

By 4 Ma, in the early Pliocene, grazers dependent on tropical grasses for the bulk of their food had become a major component of the eastern African fauna not only among ruminants, but effectively among all large herbivore families. Tall-necked giraffe and black rhino remained mostly browsers. The tragelaphine antelope, along with the gazelles and other small antelope, consumed a variable mix of grass and browse. Hidden in the forests where few fossils formed, the duikers (Cephalophini) remained browsers on foliage and fruits. In Chad in north-central Africa, white rhinos were still consuming a mix of C_3 and C_4 plants around 6 Ma, but had become purely C_4 grazers by 3.5 Ma after C_4 grasses became dominant.[21] Hipparion horses there were also exclusive grazers by that time, while both the ancestral African elephant and the gomphothere *Anancus* had become mainly grazers. In South Africa at Langebaanweg, dated to 5 Ma, dental adaptations for grazing were evident among early white rhino, hipparion horses and primitive buffalo (*Simatherium*), but stable isotopes indicated that C_4 grasses had yet to make a contribution there.[22,23] The early alcelaphine antelope represented at Langebaanweg (*Damalacra* and *Parmularius* spp.) were browsers or mixed feeders, based on their pointed molar cusps plus pits rather than scratches on their teeth. By 3.3 Ma, stable carbon isotopes indicated that C_4 grass-dominated diets had become prevalent among all grazers inhabiting interior South Africa.[24] Thus, grazers remained predominant throughout savanna Africa through the mid-Pliocene 3.5 Ma.

Plio–Pleistocene Turnover

By the commencement of the Pliocene 5 Ma, Africa's large herbivore fauna was especially diverse, because many of the Miocene browsers still persisted alongside the more recently evolved grazers (Figure 14.2).[25] Browsing ruminants remained prominent in rift valley deposits, especially in the Omo valley in southern Ethiopia, indicating the persistence of woodland mosaics near water (Figure 14.2).[26] Among proboscideans, deinotheres retained a purely C_3 browse diet, while the gomphothere *Anancus* showed a progressive shift

Figure 14.2 Scene representing the African Pliocene in the Lake Turkana basin. Note the presence of grazing Reck's elephants and early equids along with sabretooth cats, as well as the wooded aspect Artwork: Mauricio Anton.

towards consuming more C_4 grasses, despite having low-crowned molars, until it vanished from the fossil record in the mid-Pliocene.[3,27] The upland region around Laetoli in Tanzania supported a mix of grazing alcelaphines, hippotragines and gazelles plus hipparion horses between 3.8 and 3.6 Ma, with browsers somewhat less common.[28] Also present then were several species of proboscidean as well as both species of African rhino. In the Omo Valley of Ethiopia, ruminants expanded at the expense of various monkeys and wild pigs between 4 and 2.8 Ma.[25,29] The most abundant browsers or mixed feeders there were impala plus two species of tragelaphine, resembling bongo and bushbuck, respectively, while the grazers included early waterbuck (*Kobus*) and long-horned buffalo (*Syncerus* (or *Pelerovis*) *antiquus*).[29,30] Early alcelaphine antelope showed greater prominence to the north-west at Hadar in Ethiopia and also near Lake Turkana in Kenya.[29]

Following the climatic cooling that took place between 2.7 and 2.3 Ma during the inauguration of the Pleistocene, several essentially modern ruminant species appeared in the fossil record. They included various forms of kob, along with a large roan antelope.[7] Wildebeest and greater kudu were recorded earliest in the south and exhibited dry savanna adaptations honed in arid Namibia,[31,32] while sable antelope had an exclusively southern African distribution. Grazing zebra entered Africa from Eurasia around 2.6 Ma but remained less common than the coexisting hipparion horses until later. Following the transition into the Pleistocene 2.6 Ma, the species composition of large herbivore assemblages closely resembled that of modern communities. The area around the lake that formed at Olduvai remained quite densely wooded and showed a predominance of reduncine grazers plus browsing tragelaphines in its lowest beds dated to 2 Ma.

After 1.9 Ma, conditions became still cooler and more widely variable in aridity. Ruminant diversity reached its peak, augmented by further grazer radiations.[25] Alcelaphine antelope became prominent everywhere, including giant wildebeest (*Megalotragus*) as well as common wildebeest, a large blesbok or topi (*Damaliscus niro*), and an extinct species in the genus *Parmularius*. The latter became the most common grazer both at Olduvai Gorge in eastern Africa and in cave sites in South Africa.[7,33] Hippotragine grazers were rather uncommon throughout eastern Africa, except at Laetoli. At Olduvai Gorge, grazing equids increased in abundance after 1.7 Ma.[33] Other grazers present then at Olduvai included giant buffalo, white rhino, a springbok (*Antidorcas recki*), short-necked giraffe, giant warthog along with other large suids, hipparion horse, and the gorgops hippo. Reck's elephant (*Elephas recki*) was present, but scarce. Modern wildebeest and impala were rare. Alcelaphine antelope increased in representation in the lake margin habitats of east Turkana until 1.45 Ma, while reduncine grazers along with impala and gazelles became less common, indicating a continuing trend towards dry grassland.[34,35] Certain large pigs became extinct locally but giant warthogs continued to flourish. The browsing deinothere was last recorded in the Omo–Turkana basin around 1.6 Ma.[25] Its disappearance finally ended the tenure of genera representing the Miocene giants.

Fossil assemblages from Sterkfontein and Swartkrans in South Africa's Cradle of Humankind dated after 1.8 Ma contain numerous large grazers, but lack the abundance of reduncine antelope and giant buffalo associated with rift valley sites.[36,37] White rhino and impala are missing from cave deposits there, although they are represented further north at Makapansgat where conditions were more densely wooded.[38] Elephant fossils are scarce at both sites, although frequently recorded elsewhere in South Africa around that time.

After 0.9 Ma, a further climatic shift occurred with glacial cycles lengthening to 100 kyr, making times of glacial advances more extremely cold, dry and prolonged. Around this time, dry country grazers or mixed feeders, represented by the alcelaphine antelope plus gazelles, achieved their greatest dominance in eastern Africa. The representation of mixed-feeding impala declined.[39] Black wildebeest (*Connochaetes gnou*), endemic to treeless Highveld grasslands, originated in South Africa around 1.05 Ma,[40] along with blesbok. During this time, other essentially modern species of antelope made their appearance, including hartebeest and topi in eastern Africa around 0.6 Ma.[41] Numerous grazers predominated in the upland grasslands that developed on the shores of a greatly reduced Lake Victoria up until 36 ka, including forms of wildebeest, blesbok and impala along with long-horned buffalo.[42] The modern African buffalo attained prominence in savanna faunas quite late in the Pleistocene, suggesting that it inhabited mainly forests until the long-horned buffalo faded out. Blesbok and springbok, typical of dry grassland or shrubland, were present further north in Zimbabwe. Lechwe occurred in parts of the Free State and even in the southern Kalahari until near the end of the Pleistocene, indicating the local presence of wetlands.[43] Several grazers typical of savannas further north occurred as far south as Elandsfontein in the south-western Cape, consuming mostly C_3 grasses growing among the fynbos shrubs predominant in this region.[44,45]

The ancestral form of the modern African elephant was the most common proboscidean in fossil deposits in northern Kenya early in the Pliocene, shifting toward a mainly grass diet through time.[46] Later during the Pliocene it was replaced by Reck's elephant, with more extreme dental adaptations for grazing, found throughout Africa until late in the Pleistocene.[27,47] It seems that the modern African elephant became restricted to shrinking forests, occupied by present-day forest elephants (*L. a. cyclotis*), until after the disappearance of Reck's elephant from savanna regions.

Late Pleistocene Extinctions

Starting during the sparsely recorded period between 1 and 0.5 Ma, several of the large grazers that had previously been abundant disappeared from both eastern and southern Africa.[40,48,49,50,51] Among them was Reck's elephant, last recorded in the Omo–Turkana basin at the end of the last-but-one glacial maximum 130 ka.[47] Also last recorded around this time were the hipparion horse, short-necked giraffe and large gorgops hippo.[52] The giant grazing gelada (*Theropithecus oswaldi*), previously abundant in north-eastern Africa, had disappeared from the fossil record a little earlier, after 350 ka. Notably, these were all Pliocene relicts and also all grazers.

A further wave of species extinctions took place in eastern Africa during or shortly after the LGM 20 ka, once again involving solely grazers.[50] This included two forms of topi with exceptionally high-crowned molars (*Damaliscus niro* and *D. hypsodon*), which had both been locally abundant earlier, giant wildebeest and the long-horned buffalo (Figure 14.3). In southern Africa, long-horned buffalo and giant wildebeest became extinct around the same time, along with a big zebra (*Equus capensis*), a small springbok (*Antidorcas bondi*) and two suids: the giant warthog (*Metridiochoerus*) and grazing bushpig (*Kolpochoerus*).[40,43,48] Several of them happened to be the

Figure 14.3 Some of the large grazers that were common through the Pleistocene before becoming extinct towards the end of this epoch. (A) Grazing elephant (*Elephas recki/iolensis*); (B) long-horned buffalo (*Syncerus antiquus*); (C) giant wildebeest or hartebeest (*Megalotragus priscus*); (D) giant gelada (*Theropithecus oswaldi*). Artwork: Roman Uchytel.

largest species in their respective genera. These all represent genuine extinctions, leaving no descendants.

In interior South Africa, grazers that exhibited a substantial dietary contribution from C_3 plants prior to 500 ka had shifted to almost purely C_4 diets by the Holocene.[53] Whether the C_3 component represented a greater presence of C_3 grasses during the cold conditions of the LGM, or low browse in the form of dwarf shrubs spreading during these conditions, remains undetermined.

White rhinos vanished from the swathe of Africa between the Zambezi and Nile rivers, separating its two subspecies, quite late in the Holocene. Cave paintings in central Tanzania,[54] plus the odd tooth,[55] show that white rhinos remained present in eastern Africa merely a few thousand years ago. White rhinos were abundant throughout drier savanna regions of southern Africa into historic times.[56] Cave paintings show that they had been present as far north as Algeria until quite recently. The regional extirpation of this mega-grazer throughout eastern and south-central Africa cannot be ascribed to any apparent unsuitability of habitats, because other dry-country grazers that were associated with white rhinos continued to thrive.

Roan antelope provide another example of a recent local extirpation. Rock art depicts this species in the Drakensberg foothills, several hundred kilometres south of their historic distribution limit, which lay in southern Kruger NP. Roan antelope were present in the southern Cape late in the Pleistocene.[48] The bluebuck (*Hippotragus leucophaeus*), a small version of roan antelope with a narrow distribution in the south-western Cape, persisted until around 1800 CE, when its demise was brought about by hunting plus habitat transformation following European settlement. Its favoured habitat was probably the grassy Agulhas plain before this became inundated by sea-level rise, restricting the species to grassy patches amid fynbos shrubland.

Large Carnivores

The predominant carnivores in Africa through the Oligocene into the early Miocene belonged to an extinct group called the creodonts. The first representatives of the order Carnivora made their appearance during the Miocene in the form of hyenas.[57] During the late Miocene and early Pliocene, five species of hyena coexisted locally at some African sites, including a giant species weighing around 100 kg and a hunting hyena with long limbs for cursorial hunting resembling the modern brown hyena in size. Jaw adaptations for bone-cracking first appeared during the Pliocene and the modern spotted hyena had emerged by 3.5 Ma. Bears were represented by two species during the Pliocene, but they were nowhere common and soon became extinct.[58] The giant hyena went extinct around 1.5 Ma and the hunting hyena around 1 Ma.

The earliest large felids were sabretooth cats (subfamily Machairodontinae), which appeared in Africa during the late Miocene around 7.5 Ma at the time of the bovid radiations. They remained the most abundant carnivores through the Pliocene, represented by three genera.[57,59] *Meganterion* resembled the northern hemisphere *Smilodon* with its robust forelimbs and dagger-like canines, and presumably also specialised in killing young 'pachyderms' like elephants and rhinos. It was never very common in Africa and disappeared there after 1.4 Ma. *Homotherium* slightly exceeded a modern male lion in size, with long legs and extended serrated canines. It may have hunted by ambush like the modern tiger, targeting large ungulates that were not able to run fast in wooded savanna environments.[60] However, its laterally compressed canines with serrated edges suggests that these were adapted to cutting open the carcasses of thick-skinned herbivores rather than applying lethal bites to struggling animals. It persisted until 0.7 Ma. The false sabretooth *Dinofelis* was a little larger than a leopard and had only slightly elongated canines. Its short but robust forelimbs seem adapted for grappling, suggesting it was an ambush predator reliant on vegetation cover and perhaps targeting young animals of the larger herbivores.[61] *Dinofelis* remained widespread until 0.9 Ma. Big cats in the genus *Panthera* and ancestral cheetahs first appeared during the mid-Pliocene around 3.5 Ma, but remained rare until after the demise of the sabretooths. Canids entered Africa after 3.5 Ma in the form of jackals and foxes, but the African wild dog appeared only after 2 Ma.

Thus, during the late Pliocene and early Pleistocene, Africa supported an exceptionally rich assemblage of large carnivores, comprising up to five felids plus five hyenas, double the total that exists today.[58] How did all of these carnivores coexist, and why did half of them become extinct? How did the human ancestors that had adopted a meat-augmented diet manage to persist by hunting when large carnivores that had evolved as hunters disappeared from the African fauna?

Species and Subspecies

Much of the account of faunal changes through time is based on species names. Biologically, species represent morphologically distinct segments of independently evolving lineages, isolated from genetic exchanges by mate recognition cues or by the infertility of hybrids. Recent genetic research reveals that genetic exchanges occur between morphologically distinct species placed in the same genus more frequently than had been imagined, most memorably between modern humans entering Eurasia and the Neanderthal people already present there.[62] However, the potential to interbreed cannot readily be established between populations that are isolated geographically. There is a recent tendency among taxonomists to elevate populations previously regarded as subspecies to full species status, based on divergence in mostly neutral gene mutations.[63] A contentious example concerns the elevation of the forest elephant (*L. a. cyclotis*) to full species status.[64] However, all of the evidence suggests that the two elephant populations have not diverged sufficiently ecologically to coexist without genetic merging. Although the forest buffalo (*Syncerus caffer nana*) is more distinct physically from the savanna buffalo (*S. c. caffer*) than the differences between the forest and savanna elephants, they remain distinguished only at subspecies level.

This taxonomic dilemma is more acute for forms represented only by fossilised remains, too ancient to preserve their DNA. Morphological changes through time ('anagenesis') can justify the assignment of new species names; the former species name is extinguished, while the lineage continues unbroken. All species go extinct, but not necessarily the lineages they represent. The ancestral species can coexist sympatrically with its descendant only if sufficiently distinct ecologically in diet, habitat, or other components of ecological niches. Nevertheless, geographic isolation can form the foundation for ecological divergence.

Among Africa's ungulates, the alcelaphine antelope exhibit numerous subspecies reflecting their patchy distribution across Africa (Table 14.1). The reduncine grazers show a different pattern, with distinct species replacing one another geographically in isolated wetlands. The gazelles are assigned to numerous species, geographically localised through parts of Asia as well as Africa. In contrast, only two impala subspecies are distinguished, from a very minor distinction in face colour in Namibia, despite the long evolutionary history of their genus. Greater kudu from the southern and north-eastern ends of Africa are distinguished only subspecifically, despite their distant geographic separation.

Do the alcelaphine subspecies indicate incipient splitting into species? Or are the criteria used to differentiate them - horn shape and coat colour - too trivial to be of any weight ecologically? I raise these issues here because they will come to the fore later when we confront the taxonomic proliferation among early hominin specimens.

Overview

Africa's diverse assemblage of grazing ruminants originated during the late Miocene when savanna vegetation formations dominated by C_4 grasses spread. Other large herbivores also shifted their diets towards C_4 graminoids during this period, even some of the Miocene giants. Further diversification among the grazers, especially the alcelaphines, occurred when open savanna

Table 14.1 Bovid subspecies recognised (from species accounts in Kingdon & Hoffmann (2013) *The Mammals of Africa*)

Tribe	Common name	Subspecies	Distribution range
Alcelaphini	Blue wildebeest	*Connochaetes taurinus*	South Africa into S Angola and SW Zambia
	Cookson's wildebeest	*C. t. cooksoni*	Luangwa Valley, Zambia
	Eastern white-bearded wildebeest	*C. t. albojubatus*	NE Tanzania and S Kenya
	Western white-bearded wildebeest	*C. t. mearnsi*	NW Tanzania
	White-banded wildebeest	*C. t. johnstoni*	N Mozambique and S Tanzania; extinct
	Black wildebeest	*Connochaetes gnou*	South African Highveld
	Bubal hartebeest	*Alcelaphus buselaphus*	Morocco; extinct
	Swayne's hartebeest	*A. b. swaynei*	Somalia, N Ethiopia
	Lelwel hartebeest	*A. b. lelwel*	S Sudan, S Ethiopia, Uganda, Chad
	Western hartebeest	*A. b. major*	Western Africa
	Coke's hartebeest or kongoni	*A. b. cokei*	S Kenya, N Tanzania
	Lichtenstein's hartebeest	*A. b. lichtensteini*	S Tanzania, Zambia
	Red hartebeest	*A. b. caama*	Southern Africa

Table 14.1 (cont)

Tribe	Common name	Subspecies	Distribution range
	Hunter's hartebeest or hirola	*Beatragus hunteri*	S Somalia, N Kenya
	Tsessebe	*Damaliscus lunatus*	Southern Africa
	Topi	*D. l. jimela*	Eastern Africa
	Tiang	*D. l. tiang*	NW Kenya, Ethiopia, S Sudan, Chad
	Korrigum	*D. l. korrigum*	Western Africa
	Bontebok	*D. pygargus pygargus*	SW Cape
	Blesbok	*D. pygargus phillipsi*	South Africa
Reduncini	Common waterbuck	*Kobus ellipsiprymnus*	Southern to eastern Africa E of Rift
	Defassa waterbuck	*K. e. defassa*	W of Eastern Rift into western Africa
	Southern lechwe	*K. leche*	Botswana to Zambia
	Black lechwe	*K. l. smithemani*	N Zambia
	Bufffon's kob	*K. kob kob*	Western Africa
	Uganda kob	*K. k. thomasi*	Uganda
	White-eared kob	*K. k. leucotis*	S Sudan
	Puku	*K. vardoni*	Zambia
	Nile lechwe	*K. megaceros*	S Sudan, W Ethiopia

continues

Table 14.1 (cont)

Tribe	Common name	Subspecies	Distribution range
	Bohor reedbuck	*Redunca redunca*	Eastern and western Africa
	Common reedbuck	*R. arundinum*	Southern Africa
	Mountain reedbuck	*R. fulvorufula*	Southern Africa
	Chanler's mountain reedbuck	*R. f. chanleri*	Eastern Africa
Hippotragini	Roan antelope	*Hippotragus equinus*	From Sudan through central and southern Africa
	Northern roan antelope	*H. e. koba*	Western Africa
	Sable antelope	*Hippotragus niger*	Africa south of the Zambezi River
	Giant sable antelope	*H. n. variani*	Central Angola
	Roosevelt's sable	*H. n. roosevelti*	Coastal Tanzania and Kenya
	Gemsbok or South African oryx	*Oryx gazella*	Southern Africa in arid regions
	Fringe-eared or beisa oryx	*O. beisa*	North-eastern Africa
	Scimitar-horned oryx	*O. dammah*	Sahara desert margins
	Addax	*Addax nasamaculata*	Chad and Mauritania

Table 14.1 (cont)

Tribe	Common name	Subspecies	Distribution range
Tragelaphini	Greater kudu	*Tragelaphus strepsiceros*	Southern, central and NE Africa
	Lesser kudu	*T. imberbe*	Eastern Africa
	Nyala	*T. angasi*	SE Africa
	Bushbuck	*T. sylvaticus*	Africa-wide
	Mountain nyala	*T. buxtoni*	Ethiopia
	Situtunga	*T. spekii*	Central, eastern and western Africa
	Eland	*T. oryx*	Eastern and southern Africa
	Derby eland	*T. o. derbianus*	Western Africa
	Bongo	*T. eurycerus*	West-central and western Africa
Aepycerotini	Impala	*Aepyceros melampus*	Eastern and southern Africa
	Black-faced impala	*A. m. petersi*	Namibia

conditions expanded further, around 2.7 Ma and, less markedly, around 1.8 Ma. Browsing ruminants remained more conservative in species than grazers, while mixed-feeding impala retained a single lineage from the late Miocene through to the present. The greatest diversity of large herbivores was manifested during the early Pleistocene, before the last of the Miocene relicts faded out.

Comparable radiations of grazing ungulates did not occur on other continents, despite the global expansion in C_4 grasslands. North America developed a diversity of grazing and browsing equids during the Miocene, later spreading into South America, along with various gomphotheres.[65,66] All of America's very large herbivores disappeared during the wave of extinctions that followed the arrival of modern human hunters in both continents during the late

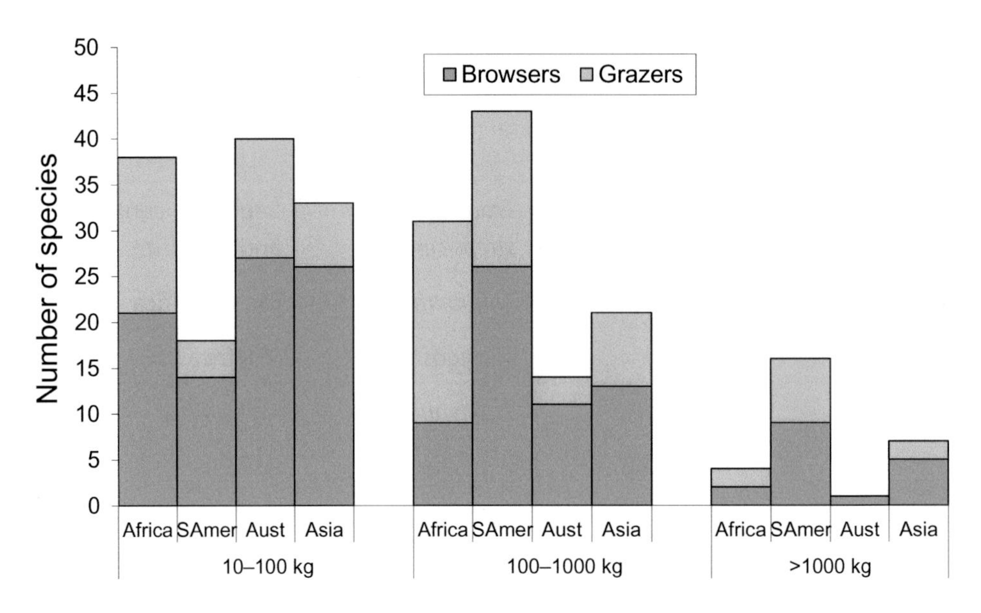

Figure 14.4 Browser–grazer distributions of large mammalian herbivores extant on different continents during the late Pleistocene within different body size ranges (excluding African forest duikers). The two-way division is between species that fed solely or mainly on woody plants and herbs and those that solely or mainly grazed on grasses and sedges (adapted from Owen-Smith (2013) *Journal of Biogeography* 40:1215–1224).

Pleistocene (Figure 14.4).[67] The only specialist grazer that survived in South America was the vicuna, a smallish camelid restricted to grassy meadows high in the Andes ranges. No deer is an obligate grazer in its diet or anatomy, although several species include grasses in a mixed diet. Bovines including gaur (*Bos gaurus*), bisons (*Bison bonasus* and *B. bison*), yak (*Bos grunniens*), and auroch (*Bos primigenius*, the ancestor of domestic cattle) are the only ruminants phenotypically and dietary adapted as grazers present outside of Africa today. All of the gazelles found in Asia are mixed feeders. Australia houses a diverse assemblage of medium-small marsupials on the kangaroo theme, but with no obligate grazers among them, even before the arrival of humans. The predominance of very large herbivores in the faunal assemblages found in the Americas and northern Eurasia prior to the late Pleistocene extinctions resembles the faunal features shown in Africa during the early Miocene.[67]

Africa did not completely escape the extinctions that decimated large mammal diversity on other continents around the end of the Pleistocene. Reck's elephant, dentally specialised for grazing, disappeared, while the modern African elephant with a more diverse diet expanded in its place. Both of the mega-grazers surviving into modern times – white rhino and the hippo – have wide mouths facilitating cropping short grass. Extinctions involved several species that were the largest of their kinds: long-horned buffalo, giant wildebeest, big zebra, gorgops hippo and giant warthogs. Last appearance dates in southern Africa cluster shortly after the last glacial maximum ~20 ka,[48,50] although some forms faded from the fossil record around the end of the preceding glacial maximum. Several medium-sized grazers with exceptionally high-crowned teeth that had previously been abundant in dry grasslands also went extinct. Large carnivore extinctions, involving various sabretooths and several hyenas, took place earlier, around the time when most of the relicts of the Miocene giants faded out. The grazers that did survive into modern times, like wildebeest and white rhino, have contracted distribution ranges compared with those they manifested historically.[68] Why were the environmental conditions that prevailed during the last glacial extreme so inimical for large grazers that had previously thrived?

The last glacial maximum was just one in a series of glacial oscillations with 100-kyr periods between peaks that were established after 0.8 Ma. The progressive downward trend in global temperatures continued, with each glacial maximum attaining unprecedented extremes of cold, most especially the last one ~20 Ma and that which preceded it ~140 ka (Figure 19.1). Global cold means less moisture evaporated from oceans and hence less falling as rain in tropical and subtropical savannas. Lower rainfall means less grass produced and shorter grass remaining through the dry season. Abundant medium–large grazers further reduced the height of the grass remaining while human hunters contributed by setting fire to ungrazed grass to improve visibility and passage.

Reductions in amount and height of grass remaining through the critical dry season months would be especially inimical for the largest grazers.[69] This seems a tenable explanation for the apparent size bias of the extinctions. Lacking resource buffers of sufficiently tall grass, the larger grazers ran out of food soonest. Their populations became reduced to zero throughout Africa, largely synchronous in time. Reck's elephants, using their trunks to yank out tall grass tufts, would have been especially compromised. Presciently, modern African elephants turned to bark and roots to tide them through dry seasons when they extended their range from forests into savannas.

However, some contribution from early humans to these extinctions cannot be ruled out, through hunting in addition to the deployment of fire. This contentious issue will be addressed in the final set of chapters, forming Part IV. Somehow, early humans surmounted these tough times, while several larger grazers that had survived through multiple climatic extremes became extinct. What were the crucial adaptations that enabled these comparatively puny omnivores to make it through into the benign Holocene while numerous large herbivores fell by the wayside?

Suggested Further Reading

Bobe, R. (2011) Fossil mammals and paleoenvironments in the Omo–Turkana Basin. *Evolutionary Anthropology* 20:254–263.

Elliot, MC; Berger, LR. (2018) *A Handbook to the Cradle of Humankind.* Reach Publishers, Wandsbeck.

Werdelin, L; Sanders, WJ (eds) (2010) *The Cenozoic Mammals of Africa.* University of California Press, Berkeley.

References

1. Abbate, E, et al. (2014) The East Africa Oligocene intertrappean beds: regional distribution, depositional environments and Afro/Arabian mammal dispersals. *Journal of African Earth Sciences* 99:463–489.
2. Janis, CM. (1993) Tertiary mammal evolution in the context of changing climates, vegetation, and tectonic events. *Annual Review of Ecology and Systematics* 24:467–500.
3. Sanders, WJ, et al. (2010) Proboscidea. In Werdelin, L; Sanders, WJ (eds) *Cenozoic Mammals of Africa.* University of California Press, Berkeley, pp. 161–252.
4. Leakey, M, et al. (2011) Faunal change in the Turkana Basin during the late Oligocene and Miocene. *Evolutionary Anthropology: Issues, News, and Reviews* 20:238–253.
5. Bibi, F. (2009) The fossil record and evolution of Bovidae. *Palaeontologia Electronica* 12:1–11.
6. Bibi, F. (2013) A multi-calibrated mitochondrial phylogeny of extant Bovidae (Artiodactyla, Ruminantia) and the importance of the fossil record to systematics. *BMC Evolutionary Biology* 13:166.
7. Gentry, AW (2010) Bovidae. In Werdelin, L; Sanders, WJ (eds) *Cenozoic Mammals of Africa.* University of California Press, Berkeley, pp. 741–796.
8. Willows-Munro, S, et al. (2005) Utility of nuclear DNA intron markers at lower taxonomic levels: phylogenetic resolution among nine *Tragelaphus* spp. *Molecular Phylogenetics and Evolution* 35:624–636.
9. Marcot, JD.(2007) Molecular phylogeny of terrestrial artiodactyls. In Prothero, DR; Foss, SE (eds) *The Evolution of Artiodactyls.* Johns Hopkins University Press, Baltimore, pp. 4–18.
10. Vrba, ES. (1997) New fossils of Alcelaphini and Caprinae (Bovidae: Mammalia) from Awash, Ethiopia, and phylogenetic analysis of Alcelaphini. *Palaeontologica Africana* 34:127–198.
11. Leakey, MG, et al. (1996) Lothagam: a record of faunal change in the Late Miocene of East Africa. *Journal of Vertebrate Paleontology* 16:556–570.
12. Bibi, F. (2011) Mio–Pliocene faunal exchanges and African biogeography: the record of fossil bovids. *PLoS One* 6:e16688.
13. Bishop, LC. (1999) Suid paleoecology and habitat preferences at African Pliocene and Pleistocene hominid localities. In Bromage, TG; Schrenki, F (eds) *African Biogeography, Climate Change, and Human Evolution.* Oxford University Press, Oxford, pp. 216–225.

14. Gentry, AW. (2000) The ruminant radiation. In Vrba, ES; Schaller, GB (eds) *Antelopes, Deer, and Relatives: Fossil Record, Behavioral Ecology, Systematics, and Conservation*. Yale University Press, New Haven, pp. 11–25.
15. Todd, NE. (2010) New phylogenetic analysis of the family Elephantidae based on cranial-dental morphology. *The Anatomical Record: Advances in Integrative Anatomy and Evolutionary Biology* 293:74–90.
16. Cerling, TE, et al. (2010) Stable carbon and oxygen isotopes in East African mammals: modern and fossil. In Werdelin, L; Sanders, WJ (eds) *Cenozoic Mammals of Africa*. University of California Press, Berkeley, pp. 941–952.
17. Cerling, TE, et al. (2015) Dietary changes of large herbivores in the Turkana Basin, Kenya from 4 to 1 Ma. *Proceedings of the National Academy of Sciences* 112:11467–11472.
18. Uno, KT, et al. (2011) Late Miocene to Pliocene carbon isotope record of differential diet change among East African herbivores. *Proceedings of the National Academy of Sciences* 108:6509–6514.
19. Geraads, D; Bobe, R. (2020) Ruminants (Giraffidae and Bovidae) from Kanapoi. *Journal of Human Evolution* 140:102383.
20. Schubert, BW, et al. (2006) Microwear evidence for Plio-Pleistocene bovid diets from Makapansgat Limeworks Cave, South Africa. *Palaeogeography, Palaeoclimatology, Palaeoecology* 241:301–319.
21. Zazzo, A, et al. (2000) Herbivore paleodiet and paleoenvironmental changes in Chad during the Pliocene using stable isotope ratios of tooth enamel carbonate. *Paleobiology* 26:294–309.
22. Franz-Odendaal, TA, et al. (2002) New evidence for the lack of C_4 grassland expansions during the early Pliocene at Langebaanweg, South Africa. *Paleobiology* 28:378–388.
23. Stynder, DD. (2011) Fossil bovid diets indicate a scarcity of grass in the Langebaanweg E Quarry (South Africa) late Miocene/early Pliocene environment. *Paleobiology* 37:126–139.
24. Lee-Thorp, JA, et al. (2007) Tracking changing environments using stable carbon isotopes in fossil tooth enamel: an example from the South African hominin sites. *Journal of Human Evolution* 53:595–601.
25. Bobe, R. (2011) Fossil mammals and paleoenvironments in the Omo-Turkana Basin. *Evolutionary Anthropology: Issues, News, and Reviews* 20:254–263.
26. Andrews, P; Humphrey, L. (1999) African Miocene environments and the transition to early hominines. In Bromage, TG; Schrenk, F (eds) *African Biogeography, Climate Change, and Human Evolution*. Oxford University Press, Oxford, pp. 282–300.
27. Sanders, WJ. (2020) Proboscidea from Kanapoi, Kenya. *Journal of Human Evolution* 140:102547.
28. Su, DF; Harrison, T. (2007) The paleoecology of the Upper Laetolil Beds at Laetoli. In Bobe, R, et al. (eds) *Hominin Environments in the East African Pliocene: An Assessment of the Faunal Evidence*. Springer, Dordrecht, pp. 279–313.
29. Bobe, R, et al. (2002) Faunal change, environmental variability and late Pliocene hominin evolution. *Journal of Human Evolution* 42:475–497.

30. Geraads, D, et al. (2013) New ruminants (Mammalia) from the Pliocene of Kanapoi, Kenya, and a revision of previous collections, with a note on the Suidae. *Journal of African Earth Sciences* 85:53–61.
31. Pickford, M. (2004) Southern Africa: a cradle of evolution. *South African Journal of Science* 100:205–214.
32. Lorenzen, ED, et al. (2012) Comparative phylogeography of African savannah ungulates 1. *Molecular Ecology* 21:3656–3670.
33. Bibi, F, et al. (2018) Paleoecology of the Serengeti during the Oldowan-Acheulean transition at Olduvai Gorge, Tanzania: the mammal and fish evidence. *Journal of Human Evolution* 120:48–75.
34. Patterson, DB, et al. (2017) Ecosystem evolution and hominin paleobiology at East Turkana, northern Kenya between 2.0 and 1.4 Ma. *Palaeogeography, Palaeoclimatology, Palaeoecology* 481:1–13.
35. O'Brien, K, et al. (2020) Ungulate turnover in the Koobi Fora Formation: Spatial and temporal variation in the Early Pleistocene. *Journal of African Earth Sciences* 161:103658.
36. Brain, CK. (1983) *The Hunters or the Hunted? An Introduction to African Cave Taphonomy.* University of Chicago Press, Chicago.
37. Elliott, MC; Berger, LR. (2018) *A Handbook to the Cradle of Humankind.* Reach Publishers, Wandsbeck.
38. McKee, JK. (1999) The autocatalytic nature of hominid evolution in African Plio–Pleistocene environments. In Bromage, TG; Schrenk, F (eds) *African Biogeography, Climate Change, and Human Evolution.* Oxford University Press, Oxford, pp. 57–75.
39. Bobe, R. (2006) The evolution of arid ecosystems in eastern Africa. *Journal of Arid Environments* 66:564–584.
40. Codron, D, et al. (2008) The evolution of ecological specialization in southern African ungulates: competition or physical environmental turnover? *Oikos* 117:344–353.
41. Arctander, P, et al. (1999) Phylogeography of three closely related African bovids (tribe Alcelaphini). *Molecular Biology and Evolution* 16:1724–1739.
42. Faith, JT, et al. (2015) Paleoenvironmental context of the Middle Stone Age record from Karungu, Lake Victoria Basin, Kenya, and its implications for human and faunal dispersals in East Africa. *Journal of Human Evolution* 83:28–45.
43. Klein, RG. (1984) The large mammals of southern Africa: late Pliocene to Recent. In Klein, RG (ed.) *Southern African Prehistory and Paleoenvironments.* Balkema, Rotterdam, pp. 107–146.
44. Stynder, DD. (2009) The diets of ungulates from the hominid fossil-bearing site of Elandsfontein, Western Cape, South Africa. *Quaternary Research* 71:62–70.
45. Lehmann, SB, et al. (2016) Stable isotopic composition of fossil mammal teeth and environmental change in southwestern South Africa during the Pliocene and Pleistocene. *Palaeogeography, Palaeoclimatology, Palaeoecology* 457:396–408.
46. Cerling, TE, et al. (1999) Browsing and grazing in elephants: the isotope record of modern and fossil proboscideans. *Oecologia* 120:364–374.

47. Manthi, FK, et al. (2020) Late Middle Pleistocene elephants from Natodomeri, Kenya and the disappearance of *Elephas* (Proboscidea, Mammalia) in Africa. *Journal of Mammalian Evolution* 27:483-495.
48. Faith, JT. (2011) Ungulate community richness, grazer extinctions, and human subsistence behavior in southern Africa's Cape Floral Region. *Palaeogeography, Palaeoclimatology, Palaeoecology* 306:219–227.
49. Faith, JT, et al. (2012) New perspectives on middle Pleistocene change in the large mammal faunas of East Africa: *Damaliscus hypsodon* sp. nov.(Mammalia, Artiodactyla) from Lainyamok, Kenya. *Palaeogeography, Palaeoclimatology, Palaeoecology* 361:84–93.
50. Faith, JT. (2014) Late Pleistocene and Holocene mammal extinctions on continental Africa. *Earth Science Reviews* 128:105–121.
51. Brink, JS. (2016) Faunal evidence for Mid- and Late Quaternary environmental change in southern Africa. In Knight, J; Grab, SW (eds) *Quaternary Environmental Change in Southern Africa: Physical and Human Dimensions*. Cambridge University Press, Cambridge, pp. 286–307.
52. Faith, JT, et al. (2018) Plio-Pleistocene decline of African megaherbivores: No evidence for ancient hominin impacts. *Science* 362:938–941.
53. Codron, D, et al. (2008) Functional differentiation of African grazing ruminants: an example of specialized adaptations to very small changes in diet. *Biological Journal of the Linnean Society* 94:755–764.
54. Leakey, M. (1983) *Africa's Vanishing Art. The Rock Paintings of Tanzania*. Doubleday, New York.
55. Hooijer, DA. (1969) Pleistocene East African rhinoceroses. In Leakey, LSB (ed.) *Fossil Vertebrates of Africa*. Vol. 1. Academic Press, New York, pp. 71–98.
56. Delegorgue, A. (1990) *Adulphe Delegorgue's Travels in Southern Africa*. University of Natal Press, Durban.
57. Werdelin, L; Peigné, S. (2010) Carnivora. In Werdelin, L; Sanders, WJ (eds) *Cenozoic Mammals of Africa*. University of California Press, Berkeley, pp. 603–657.
58. Werdelin, L; Lewis, ME. (2005) Plio-Pleistocene Carnivora of eastern Africa: species richness and turnover patterns. *Zoological Journal of the Linnean Society* 144:121–144.
59. Lewis, ME; Werdelin, L. (2007). Patterns of change in the Plio-Pleistocene carnivorans of eastern Africa. In Bobe, RA, et al. (eds) *Hominin Environments in the East African Pliocene: An Assessment of the Faunal Evidence*. Springer, Dordrecht, pp. 77–105.
60. Turner, A; Antón, M. (2004) *Evolving Eden: An Illustrated Guide to the Evolution of the African Large-mammal Fauna*. Columbia University Press, New York.
61. Kuhn, BF, et al. (2016) The carnivore guild circa 1.98 million years: biodiversity and implications for the palaeoenvironment at Malapa, South Africa. *Palaeobiodiversity and Palaeoenvironments* 96:611–616.
62. Prüfer, K, et al. (2014) The complete genome sequence of a Neanderthal from the Altai Mountains. *Nature* 505:43–49.
63. Groves, C; Grubb, P. (2011) *Ungulate Taxonomy*. Johns Hopkin University Press, Baltimore.
64. Rohland, N, et al. (2010) Genomic DNA sequences from mastodon and woolly mammoth reveal deep speciation of forest and savanna elephants. *PLoS Biology* 8:e1000564.

65. Webb, SD. (1978) A history of savanna vertebrates in the New World. Part II: South America and the Great Interchange. *Annual Review of Ecology and Systematics* 9:393–426.
66. Sánchez, B, et al. (2004) Feeding ecology, dispersal, and extinction of South American Pleistocene gomphotheres (Gomphotheriidae, Proboscidea). *Paleobiology* 30:146–161.
67. Owen-Smith, N. (2013) Contrasts in the large herbivore faunas of the southern continents in the late Pleistocene and the ecological implications for human origins. *Journal of Biogeography* 40:1215–1224.
68. Rowan, J, et al. (2015) Taxonomy and paleoecology of fossil Bovidae (Mammalia, Artiodactyla) from the Kibish Formation, southern Ethiopia: implications for dietary change, biogeography, and the structure of the living bovid faunas of East Africa. *Palaeogeography, Palaeoclimatology, Palaeoecology* 420:210–222.
69. Shrader, AM, et al. (2006) How a mega-grazer copes with the dry season: food and nutrient intake rates by white rhinoceros in the wild. *Functional Ecology* 20:376–384.

Part III: Synthesis: Movers of Savanna Dynamics: Grazers, Elephants and Fires

The distinguishing feature of Africa's large mammal fauna is its diversity of grazing ruminants. Relative security from predation contributes importantly to niche separation among herbivore species of similar size, coupled with distinctions in grass height grazed. Grazers present on other continents prior to the late Pleistocene extinctions tended to be very large and mostly non-ruminants. Although some deer consume much grass, none is specialised in dentition and digestive anatomy for an exclusively grass diet. This is perhaps because the C_3 grasses prevalent in higher northern latitudes are more readily digested than the C_4 grasses prevalent through tropical and subtropical Africa, especially during the season of plant dormancy. Grasses growing under higher rainfall elsewhere in the tropics seemingly require hindgut fermentation to handle their high fibre contents. The prevalence of volcanic soil substrates under moderately low rainfall regimes in Africa contributes to the high local abundance levels attained by some of the grazing ruminants. Equids are widely distributed, but do not reach the biomass densities shown by grazing ruminants like buffalo and wildebeest under favourable conditions. Browsers, apart from elephants with their exceptionally broad dietary range, remain much less abundant than similar-sized grazers.

The local abundance of grazers is enhanced further by their concentrations around remaining sources of surface water during the dry season. Migratory populations concentrate additionally in nutritious but ephemeral grasslands during the wet season. Locally intense grazing restricts the spread of fires. The outcome is a mosaic interspersion of cropped, burnt and ungrazed areas. Regions remote from surface water burn frequently while those in close proximity to water can become grazed down to bared soil, especially in drought years, leading to starvation-induced mortality among grazers.

Elephants make a major contribution to the openness of the tree canopy in Africa's savannas, in interaction with fire and smaller browsers. They induce mortality among quite tall trees by toppling and debarking and uprooting or severely breaking regenerating saplings and shrubs. Nevertheless, in some situations the damage they impose keeps woody plants at shrub height, enhancing food availability for them and smaller browsers. The latter restrict the woody plant cover through the defoliation that they impose locally on seedlings, counteracting bush encroachment. Nevertheless, all herbivores, even elephants, are selective among tree species and size classes in the damage they impose. Certain woody species may gain a selective advantage as a consequence, while others get excluded. Thus, compositional changes may result rather than a general opening of the woody canopy, especially in broad-leaved miombo woodlands. Areas further than 10 km from water sources can provide a refuge for vulnerable tree species because elephants are water-dependent and feed on woody plant parts mostly during the dry season. Trees with deep roots are resistant to felling, but not necessarily to bark damage. The grazing and browsing impacts of various herbivores contribute in addition to geology, water redistribution and fire spread to the spatial heterogeneity that is a striking feature of Africa's savannas.

The form of the dynamics generated by rainfall variation disrupts equilibrium tendencies within savannas. If wet-season conditions persisted for multiple years rather than months, savannas would become transformed into woodland or forest, with far fewer large herbivores. If dry-season conditions precluded plant growth for long enough, savannas would come to resemble deserts. Animals must cope with contrasting aridity not only seasonally, but also over successive years in response to multi-year fluctuations in rainfall. Primary production can be halved in drought years, correspondingly reducing the ecosystem capacity to support herbivores. Megaherbivores cruise through the dry years with little reduction in their abundance, unless drought conditions last several years or animals become crowded around few remaining water sources. Grazing ruminants can incur substantial population collapses when the grass cover becomes depleted within range of water sources. Browsers may have the effects of rainfall failure alleviated by the capability of savanna trees to regenerate their foliage from stored water reserves. Carnivores constrict the spatial distribution of their prey rather than local abundance in relatively secure habitats. Migrants largely escape this limitation.

Africa's large herbivore diversity originated in the later Miocene when grazing ruminants diversified and the earliest hominins made their appearance. Large carnivores were also more diverse then. The large Miocene relicts faded out during the course of the Pliocene and Pleistocene, while a pulse of extinctions took place among some of the largest grazers during the late Pleistocene and subsequent transition into the Holocene.

This represents the ecosystem context that our hominin predecessors entered when they colonised spreading savannas back in the Miocene. While landscapes became acutely depauperate of plant resources during dry seasons, they were locally packed with herbivores representing a wide range in body sizes. How did early humans use the opportunities provided to exploit animal rather than plant resources to make a living under conditions that became increasingly arid as the Pleistocene advanced? This is the story to be taken up in the following chapters forming the last section of this book.

These are the notable points to carry forward from Part III:

1. Africa's large herbivore fauna is unrivalled in its diversity of medium-large grazers, especially ruminants.

2. These grazers can attain exceptionally high population levels through exploiting relatively nutritious, volcanically enriched savanna grasslands.

3. Grazers concentrate in the vicinity of remaining sources of surface water during the dry season.

4. Drought-related die-offs among large herbivores can be substantial in drought years.

5. Conditions were prevalently drier and hence more widely fertile during the course of the Pleistocene than experienced during the current interglacial interlude.

6. The herbivores that became extinct during the most recent glacial extremes in aridity were all grazers and among the largest species of their type.

Figure III.3 Migratory wildebeest concentration in Serengeti NP.

Part IV: Evolutionary Transitions: From Primate Ancestors to Modern Humans

Figure IV.1 Evolutionary transitions from ape to human (from Wikimedia commons).

Primates (order Primata) are basically herbivores, but are mostly not so large and mainly forest-dwelling. They are represented by a dazzling diversity of species throughout tropical forests where fruits are produced and trees retain foliage almost year-round. The platyrrhine ('flat-nosed') monkeys found in South America split off from the haplorhine ('simple-nosed') suborder inhabiting the 'Old World' tropics early in the Eocene ~50 Ma. Prosimians (bushbabies, lemurs and allies) diverged even earlier during the late Cretaceous 74 Ma, based on genetic evidence, and are most diverse in Madagascar. The monkeys inhabiting Africa and Asia (Cercopithecoidea) are separated from the apes (Hominoidea), which lack tails, at superfamily level. Among apes, the gibbons, found solely in Asia, are placed in a different family (Hylobatidae) from the remainder (Hominidae). Humans are allied with chimpanzees (*Pan troglodytes* and *P. paniscus*) and gorillas (*Gorilla gorilla*) in the latter family, but allocated to a distinct subfamily, the Homininae. Ecologically, we have diverged from other apes in habitat, locomotion and diet.

Among the African monkeys, baboons are largely ground-dwelling savanna inhabitants, penetrating only the margins of forest blocks. Some other monkeys also spend quite a lot of time on the ground, most notably the gelada (*Theropithecus gelada*), but remain mainly quadrupedal. Most monkeys augment their vegetarian diet with insects, baby birds or other animals when opportunities present. While chimpanzees kill monkeys and baby duikers, gorillas remain strictly vegetarian.

The first two chapters in this concluding section of the book establish the adaptive foundations laid by our primate ancestry, covering evolutionary linkages and ecological relationships. Thereafter, attention will be focused on the origins of our lineage and how our adaptations in locomotion, diet and culture relate to features of Africa's savannas. In the concluding two chapters, I attempt to make sense of the taxonomic complexity of early hominins and look ahead to the place of Africa's diverse ungulate fauna in changing human lifestyles.

Chapter 15: Primate Predecessors: From Trees to Ground

Primates originated in Asia and became recognisable in Africa around 55 Ma, quite early in the Eocene, although still not very distinct from other mammals of that time.[1] During the Oligocene, a split developed between the colobine subfamily of monkeys, which are primarily leaf-eaters, and the more omnivorous cercopithecines. Colobine monkeys have sacculated stomachs, facilitating the fermentation of leaves along with unripe fruits but less complex than those of ruminants. Strangely, colobus monkeys have their thumbs reduced to stubs. The cercopithecine monkeys, or guenons, have simpler stomachs and consume a variable mix of fruits and young leaves, augmented by insects. They characteristically possess cheek pouches where food can be stuffed and chewed later in more secure places. Apes split off from the cercopithecine monkeys around 15 Ma during the Miocene, when forests were becoming more open.[1,2,3] Besides lacking tails, they possess shoulder adaptations for brachiating through treetops. A putative stem to the hominids is a monkey named *Proconsul*, found both in Africa and Asia during the early Miocene.[2] Another candidate for ape ancestor is *Kenyapithecus*, which consumed hard fruits and nuts gathered from forest floors in seasonally dry woodlands in Africa.[4] Both showed adaptations for walking bipedally. Partly terrestrial monkeys, like the vervet (*Chlorocebus pygerythrus*), were also around at that time. The emergence of ground-dwelling primates was associated with the opening of the forest cover taking place in parts of equatorial Africa during the early portion of the mid-Miocene.

The great apes are placed in a separate family, the Hominidae. Very little is known about their ancestry because the wet forests that they inhabit are not conducive to fossil formation. Rather large teeth recovered from the Afar formation in north-eastern Ethiopia, dated to 8 Ma when forests were still present there, have been claimed to represent an ancestor of the gorilla.[5] For chimpanzees, the only fossils known are a few teeth discovered in the rift valley in northern Kenya, along with hominin fossils, in deposits dated to the mid-Pleistocene. The split of the hominin lineage from its shared ancestor with the great apes evidently occurred during the obscure period in the late Miocene between 10 and 6 Ma.

By the start of the Pliocene 5 Ma, largely terrestrial baboons with their eclectic food habitats had emerged from monkey-like ancestors, first in southern Africa then spreading northward.[1] In eastern Africa, ground-dwelling geladas made their appearance early in the Pliocene. One giant form (*T. oswaldi*) became primarily a grazer, like the modern gelada found in Ethiopia. It remained abundant in fossil faunas throughout Africa, as far south as the Cape, until quite late in the Pleistocene. Baboons became common in eastern Africa only after the demise of the giant gelada, suggesting competition for food between these two terrestrial primates.

Primates currently rival the ungulates in diversity of species, but remain associated primarily with tropical evergreen forests (Figure 15.1). They include generalist guenons, leaf-eating colobines, seed-crunching mangabeys, ground-dwelling baboons, baboon-like mandrills and the modern gelada. Mandrills (*Mandrillus sphinx*) are largely terrestrial scavengers moving in large herds, seeking fruits and other plant parts plus invertebrates on rainforest floors within central Africa. They are more closely related to mangabeys than to baboons. The modern gelada is narrowly restricted to high-elevation grasslands in Ethiopia, where it lives in large aggregations. Vervet monkeys are represented by various local species or subspecies throughout Africa from the Cape to Sudan and western Africa. They inhabit savanna woodlands and even strips of riverine thicket within karoo shrublands. The long-limbed, largely terrestrial patas monkey (*Erythrocebus patas*) has a restricted distribution in open grassy savanna in equatorial regions. Among the forest-inhabiting guenons, the most widely distributed is the blue monkey (*Cercopithecus mitis*), present in forest patches in subtropical South Africa in the form of the local subspecies known as the samango.

Baboons (*Papio* spp.) are widely distributed throughout Africa. Their broad habitat tolerance stretches from savanna woodlands to high-altitude grasslands in the Drakensberg foothills, semi-arid shrublands in the Cape Province, and river corridors through the Namib Desert. Geographically distinct populations connecting the chacma baboon of South Africa with the hamadryas baboon in Ethiopia have been assigned to distinct species, but interbreed and produce fertile offspring.[6] Their lineage diverged in southern Africa ~2 Ma and has since shown repeated range shifts, fragmentation, isolation and reconnection of populations.

All three of Africa's great apes are restricted to equatorial forests or woodlands (Figure 15.2). Chimpanzees are distributed from the forests of western Africa through the Congo Basin into Uganda, while bonobos (or pygmy chimpanzees) are found in tropical deciduous forest in a restricted region south of an upper tributary of the Congo River. Gorillas are represented by three distinct subspecies. The shaggy mountain gorilla is found in the region near the Virunga volcanoes stretching between Rwanda, Uganda and Congo

Figure 15.1 A miscellany of monkeys. (A) Blue monkey (*Cercopithecus mitis*); (B) red-tailed monkey (*Cercopithecus ascanius*); (C) red colobus monkey (*Procolobus badius*); (D) black-and-white colobus monkey (*Colobus guereza*); (E) Patas monkey (*Erythrocebus patas*); (F) Chacma baboon (*Papio ursinus*).

DRC. It is replaced by eastern lowland gorillas in eastern regions of Congo DRC, and by western lowland gorillas more widely from Gabon through adjoining countries in west-central Africa.

Primates are numerous and diverse in tropical forests and woodlands in other continents. In Asia, various macaques and langurs spend much time on the ground, while apes, represented by the orang-utan and various gibbons,

Figure 15.2 The great apes. (A) Chimpanzee brachiating in Kigale Forest, Uganda; (B) mountain gorilla on forest floor in Bwindi Forest, Uganda.

remain arboreal forest inhabitants. The monkeys occupying South and Central America have retained mostly a treetop existence in forests, although capuchins do spend time on the ground cracking nuts. None of them inhabits the grassy cerrado savannas prevalent there.

Suggested Further Reading

Jablonski, N; Frost, S. (2010) Cercopithecoidea. In Werdelin, L; Sanders, WJ (eds) *Cenozoic Mammals of Africa*. University of California Press, Berkeley, pp. 393–428.

References

1. Jablonski, NG; Frost, S. (2010) Cercopithecoidea. In Werdelin, L; Sanders, WJ (eds) *Cenozoic Mammals of Africa*. University of California Press, Berkeley, pp. 393–428.
2. Harrison, T. (2010) Dendropithecoidea, proconsuloidea, and hominoidea. In Werdelin, L; Sanders, WJ (eds) *Cenozoic Mammals of Africa*. University of California Press, Berkeley, pp. 429–469.
3. MacLatchy, LM. (2010) Hominini. In Werdelin, L; Sanders, WJ (eds) *Cenozoic Mammals of Africa*. University of California Press, Berkeley, pp. 471–540.
4. Benefit, BR. (1999) Biogeography, dietary specialization, and the diversification of African Plio-Pleistocene monkeys. In Bromage, TG; Schrenk, F (eds) *African Biogeography, Climate Change, and Human Evolution*. Oxford University Press, Oxford, pp. 172–188.
5. Suwa, G, et al. (2007) A new species of great ape from the late Miocene epoch in Ethiopia. *Nature* 448:921–924.
6. Fischer, J, et al. (2019) The natural history of model organisms: insights into the evolution of social systems and species from baboon studies. *Elife* 8:e50989.

Chapter 16: Primate Ecology: From Forests into Savannas

Monkeys seem to have an easy life. Up in the trees, they are surrounded by leaves – no need to wait for them to fall as is the case for ground-based ungulates. Succulent fruits ripen periodically, loaded with easily digested carbohydrates. At night, they can perch above the reach of most ground-based carnivores and climb a little higher than the leopards that may ascend after them. The smaller monkeys do need to watch out above for attacks that come from large eagles. In some places big apes, in the form of chimpanzees, hunt them. On shaded branches, it is rarely too hot or too cold. But this easy lifestyle is not viable in temperate regions where broad-leaved trees lose their foliage through winter and fruit production is narrowly pulsed at the end of summer.

Nevertheless, there are temptations drawing monkeys down from the trees. Many trees drop their ripe fruits, which accumulate on the ground, eaten by duikers if monkeys do not get there first. Insects may be extracted from the litter, adding protein spikes to a carbohydrate-rich diet. If trees are too far apart, terrestrial travel is necessary to move in between, or from one cluster bearing fruit to another. But as woodlands were transformed into savannas the tree spacing widened. Baboons successfully established a lifestyle as ground-based primates traversing savanna grasslands. They rely on being in large groups to detect or even cooperatively chase off threatening predators. There is usually a tree somewhere nearby to be ascended for safety from lions. Baboons ascend tall trees for nocturnal security. Where there are no trees, they employ their tree-climbing capabilities to ascend cliffs. Patas monkeys also occupy savannas, relying on their lengthened limbs for a speedy departure to the nearest tree, but are restricted to the tropics.

Actually, subsisting as a primate is not so idyllic. Primate ecologists raised the concept of fallback foods, eaten only when little else remains, because fruits are restricted seasonally in their availability, even in the tropics.[1,2] Most tree leaves contain various secondary chemicals, which can be toxic and disrupt digestive enzymes. Evergreen foliage tends to be reinforced by fibre requiring bacterial fermentation to release its energy content. Rainforests retain seasonal variation dependent on rainfall. There is a period towards the end of the dry season when trees divert their resources to new leaf production and fruits become scarce. Miombo woodlands produce a wider range of succulent fruits than acacia savannas (Chapter 7),[3] but still have a shortage of ripe fruits around the transition from the dry season into the wet season.

Africa's apes currently remain forest dwellers restricted to the tropics. The chimpanzees rely on finding fruits through most of the year, falling back on new leaves and other plant parts during the times when fruits are gone. The gorilla is largely folivorous, eating fruits when available but able to survive on a diversity of herbaceous plants (but not tree leaves) in places lacking fruits. It is the colobine monkeys, not the big apes, that have taken the adaptive step of accommodating microbial fermentation to release energy even from mature leaves.

In this chapter, I will explore the consequences of dietary shifts from fruit-seeking or leaf-eating monkeys and apes to savanna-exploiting baboons. The aspects of ecology covered include areas traversed as home ranges and diurnal versus nocturnal tactics restricting the risks of predation. For whatever combination of reasons, primates as a group have evolved relatively larger brains than other herbivores and even carnivores. Large brains have costs, and the major one is to slow development and progression through life-history stages. Slower growth to reproductive maturity means that reproductive lifespans must be extended to compensate, making it especially crucial that mortality from predation is curtailed.

Primate Diets: Fruits Versus Foliage

All primates, even most humans, draw their diets largely from plant parts (Figure 16.1). Most monkeys are rather omnivorous, supplementing leaves, fruits and seeds with insects, eggs, and occasionally smaller mammals when opportunities arise. The lean time for fruit production in rainforests is around the beginning of the wettest months when trees flush new leaves and flowers. The flowers yield ripe fruits several weeks later. Tree species partition the times when their fruits ripen to span much of the year, including the drier months. Fig trees are especially valuable because they bear fruits through most of the year.[4] Nevertheless, fruit crops can vary widely from one year to the next and trees bearing fruits are localised spatially. This generates lean periods when ripe fruits become scarce.[5]

Colobus monkeys get through lean times through their capacity to digest even mature leaves. Nevertheless, they are selective among the tree species that they exploit, probably due to the varying secondary chemical contents presented. They even choose unripe fruits over ripe ones, because fermenting sugars would make their stomach contents too acid for microbes.[6,7]

Figure 16.1 Primates feeding. (A) Blue monkey eating berries up in a tree; (B) vervet monkey feeding on the ground (photo: Peter Henzi); (C) baboon feeding on the ground (photo: Peter Henzi); (D) gorilla eating fig fruits from the ground.

Mangabeys rely on hard seeds, fruits and pods, frequently unripe, supplemented by insects, especially ants and caterpillars.[8,9] Vervet monkeys inhabiting woodland strips within savannas consume acacia gum along with young leaves, flowers, unripe pods, and some insects when ripe fruit is unavailable.[10] Patas monkeys inhabiting savannas also obtain much of their energy from acacia gum during the dry season, supported by insects, especially grasshoppers, for protein.[11,12,13] They also eat the ants that live inside swollen thorns on ant-gall acacias. Geladas living in mountainous grasslands retain their dietary concentration on the leaves and seeds of grasses during lean periods, supplemented by underground corms.[14] They depend on a high digestive turnover supported by efficient chewing to extract sufficient nutrition from grass blades. Fruits and insects make a small contribution opportunistically when available.

Savanna-dwelling baboons are particularly eclectic in their food habitats. Fruits are sought, but constitute only about a third of the diet of baboons, averaged across studies.[15,16] Fruits are supplemented by leaves, seeds, flowers, corms, and invertebrates hidden under stones.[17,18,19,20] In grassy savannas with few trees, baboons fall back on underground corms of graminoids (grasses plus sedges) and forbs along with grass rhizomes. Vertebrates such as hares and newly born antelope are caught and eaten opportunistically, but mostly by adult males and make little or no nutritional contribution to other classes. The C_4 plant component in the diet of baboons, indicated by stable carbon isotopes in droppings, ranges from 20 percent to 40 percent in different study areas, but part of this comes from CAM succulents (see Chapter 7) rather than grasses.[21]

Chimpanzees seek out fleshy fruits as long as these remain available and gather at fruiting trees in noisy aggregations.[22,23,24] When fruits become scarce, chimpanzees turn to young leaves, pithy stems, flowers, and other plant parts. Chimpanzees also probe for animal matter by poking termites from mounds and spear bushbabies hiding in tree cavities, using sharp sticks. Monkeys and baby antelope are hunted when opportunities present and are eaten mainly by males. Chimpanzees occupying wooded savanna in Senegal retain a mainly fruit diet, including some consumed even while still unripe. They supplement fruits with seeds and pods, underground plant storage organs and even some bark.[25,26] Among the fruits sought are the large pods produced by baobab trees. Chimpanzees seek easily digestible fruits during the morning, switching to eating more leaves in the afternoon.[27] Bonobos are somewhat more reliant on herbaceous plants than chimpanzees, converging on the diet of gorillas, which are absent from the bonobo distribution range.[28]

Gorillas subsist on an entirely vegetarian diet, consisting mostly of leaves and pithy stems of herbaceous plants common in the forest understorey. Fruits make a variable contribution, depending on their local availability.[29,30,31,32] Lowland gorillas eat more fruit than mountain gorillas, but only while these are ripe, and pass seeds undigested in their droppings. Termites and other insects are eaten occasionally, but not any vertebrate flesh. Among lowland gorillas, males consume relatively more leaves and females more herbs and insects.[30] The leafy diet of gorillas can provide more protein than these large animals actually need for maintenance.[33]

Hence most primates seek out fruits, some even still unripe, and fall back on vegetative plant parts, including underground storage organs or acacia gum, when fruit is unavailable. They include insects or small vertebrates in their diet opportunistically. Only gorillas and various colobus monkeys are strictly vegetarian.

Primates also need ready access to surface water for drinking, especially when ripe fruits are unavailable during the savanna dry season. Baboons drink regularly, travelling to pools to do so generally around midday.[34] Colobus monkeys occupy home ranges next to rivers where they can readily obtain water.[6,35]

Home Range Extents

Home ranges encompass the places where animals find food, security, shelter and reproductive requirements. Bigger animals need larger areas to fulfil their greater individual food requirements. The home range extent depends also on the number of animals in the group sharing the food resources. Home ranges are smaller if leaves provide the main food source than if fruits are sought.

Colobus monkeys feeding on leaves can live within tiny home ranges covering only 2 ha, defended by pairs or family groups (Table 16.1).[6,35] Forest-inhabiting guenons with more varied diets require much larger home ranges, typically around 50 ha.[36,37] Vervet monkeys inhabiting savanna regions generally traverse home ranges exceeding 1 km^2, but spend most of their time in smaller core areas.[38,39] Home ranges of mangabeys extend up to 4 km^2 to provide the fruits and seeds they need.[40] Patas monkeys living in open grassy savannas cover especially large home ranges, encompassing around 4 km^2 in Cameroon,[41] but perhaps even as much as 50 km^2 in Uganda.[42]

Baboon troops, which sometimes include over 100 individuals, move over home ranges varying in extent from 5 to 20 km^2, smallest in the Okavango Delta and largest in Kenya (Table 16.1). Chimpanzees communities typically containing around 30 individuals inhabit home ranges covering around 15 km^2 in forests,[32] expanded to 45 km^2 in savanna vegetation in Senegal.[26] Home ranges occupied by gorilla groups including 5–15 animals cover 20–40 km^2, with core areas contributing about half of this.[31]

Daily distances traversed are much less than a kilometre for colobus monkeys and around 1 km for forest-inhabiting guenons and mangabeys (Table 16.1). Savanna-inhabiting vervets travel 1–2 km per day, but patas monkeys up to 5 km. Baboon troops typically move 3–5 km per day, lengthened to 10 km in the dry savanna at Amboseli in southern Kenya. Chimpanzees at Gombe in western Tanzania travel about 4 km per day on average while foraging. Gorillas are more sedentary, travelling on average less than a kilometre per day, with a daily range varying between 0.25 and 2 km depending on the amount of fruit obtained.[43] Bonobos with a more herbaceous diet than chimpanzees exhibit short daily ranges like gorillas.

Thus, home ranges traversed relative to group size increase with the fruit component in the diet and in savanna compared with forest habitats. Daily movement distances vary correspondingly. Home ranges covered by savanna-living monkeys and forest-dwelling apes are similar in size to those occupied by non-migratory ungulates, like kudu and sable antelope in Kruger NP (Chapter 10).

Table 16.1 Comparative space use among primates

Species	Place	Home range (km^2)	Daily range (km)	Population density (no./km^2)	Reference
Black-and-white colobus	Ethiopia	0.02–0.2			35
Black-and-white colobus	E Africa	0.15	0.5	104	Clutton-Brock 1977
Red colobus	Gombe	0.67	0.7	186	Clutton-Brock 1977
Black colobus	Equatorial Guinea	0.6	0.3	30	Clutton-Brock 1977
Blue monkey	Kenya	0.14		120	36
Blue monkey	Cape Vidal	0.15		200	37
Moustached monkey	Gabon	0.35	0.9	25	Clutton-Brock 1977
De Brazza's monkey	Gabon	0.15	0.3	42	Clutton-Brock 1977
Putty-nosed monkey	Gabon	0.67	1.5	30	Clutton-Brock 1977
Crowned monkey	Gabon	0.78	1.75	23	Clutton-Brock 1977
Vervet monkey	E Africa	0.27	1.3	112	Clutton-Brock 1977
Vervet monkey	Amboseli	8	1.2		34
Vervet monkey	Loskop	1.4–2.2			38
Vervet monkey	Samara, E Cape	0.6–1.8	1.0		39
Vervet monkey	Cameroon	0.4–0.9	1.3–2.5		41
Patas monkey	Cameroon	4.0	5		41
Grey-cheeked mangabey	E Africa	4.1	1.2	2.7	Clutton-Brock 1977

Table 16.1 (cont)

Species	Place	Home range (km^2)	Daily range (km)	Population density (no./km^2)	Reference
Agile mangabey	Gabon	1.3	1.3	60	Clutton-Brock 1977
Baboon	Amboseli	17	4.5		Clutton-Brock 1977
Baboon	Uganda	24	3.6		Clutton-Brock 1977
Baboon	South Africa	15	4.8		Clutton-Brock 1977
Baboon	Kekopey-Gilgil	20			20
Baboon	Nairobi NP	23		3.9	20
Baboon	QENP, Uganda	4.6		11	20
Baboon	Ivory Coast	4–15		1.2	20
Baboon	Botswana	5		24	44
Chimpanzee	Gombe	12.5	3.9	2.7	27
Chimpanzee	Kahuzi-Biega	16			43
Chimpanzee	Ivory Coast	25		3.5	51
Chimpanzee	Senegal	45			26
Bonobo	Congo DRC	8	0.5	2.0	28
Gorilla	Bwindi	21–40	0.4		31
Gorilla	Kahuzi-Biega	42			43

Additional reference: Clutton-Brock, TH, ed. (1977) *Primate Ecology: Studies of Feeding and Ranging Behaviour in Lemurs, Monkeys and Apes*. Academic Press, New York.

Evading Predation

By living in cohesive groups, primates reduce the risk of being killed by a predator. As is the case for ungulates, more eyes increase the chance of detecting a lurking predator before it gets too close. There is also a smaller chance of each individual being killed if it is a member of a group. Male monkeys and baboons may break away from groups temporarily, scouting opportunities to transfer between groups, but females and their offspring are never encountered alone.

During daylight, baboons usually detect lions at distances up to 100 m and either move away or ascend trees.[44] Their main threat comes from leopards, nearly as adept at tree-climbing as they are. Baboons may respond to a threatening leopard with a coordinated attack led by adult males, which can even result in the leopard being killed.[44,45] Hence leopards infrequently threaten baboons during daylight. Smaller monkeys must watch out for eagles swooping from the sky. Pythons can kill and eat monkeys, as can crocodiles when rivers are crossed.

Risks of predation increase greatly at night, particularly from leopards. Monkeys are somewhat safer than baboons because they can climb up onto smaller branches than those supporting the predator.[46] In the absence of sufficiently tall trees, baboons seek nocturnal security on cliffs, as also do geladas. Chimpanzees and gorillas construct nests of branches high up in trees each evening, providing support for their large bodies. Only adult male gorillas, big enough to be a formidable opponent for a leopard, sleep frequently on the ground at night.[47]

More than half of the adult mortality incurred by vervet and patas monkeys in Kenya was inflicted by leopards.[48] However, for red colobus monkeys about 20 percent of the mortality they experience can be due to chimpanzees.[49] In Kruger NP, baboons killed by predators, including leopards, contributed only 0.1 percent of all found carcasses.[50] Lions seldom eat baboons they have killed. Although about 40 percent of chimpanzee deaths in the Tai Forest in Ivory Coast were due to predation by leopards, some of this mortality was on ageing adults likely to have died soon anyway.[51] All of the observed attacks by leopards on adult male gorillas were unsuccessful.[52] Mortality rates incurred from predation amounted to only 5 percent per year for vervets, and 3 percent per year for baboons, during a long-term study in Kenya.[46]

Life Histories and Population Dynamics: Costs of Bigger Brains

Larger animals have their life-history stages prolonged (see Chapter 12). Gestation periods become longer, reproductive maturity is delayed, intervals between births get lengthened, and lifespans are extended. Among primates, the progression through life-history stages is slower than for other mammals of similar size, especially for apes and humans. Delayed maturity has been ascribed to the costs of building and maintaining relatively larger brains.[53,54]

Gestation generally lasts 5–6 months among African monkeys, including baboons, while chimpanzees show a 7.5-month gestation and gorillas an 8.5-month gestation. This closely matches the gestation periods of medium-large ungulates, which range between 6 and 9 months. However, baby ungulates are able to run shortly after birth, while primates are born at a feebler stage of development. Female monkeys generally delay the birth of their first offspring until they have attained 4–7 years of age and inter-birth intervals generally span 2–3 years, unless the previous offspring dies in infancy.[48,55,56,57] In order for populations to be viable, the slower reproductive rates of primates must be coupled with lower rates of mortality and lengthened reproductive lifespans. Adult monkeys incur annual mortality rates ranging between 5 and 15 percent, while from 10 to 40 percent of infants die before they reach one year of age. Ground-dwelling patas monkeys exhibit higher mortality rates than other guenons, around 20–30 percent annually for adults coupled with 13–30 percent losses of infants during their first year. In compensation, female patas monkeys first reproduce at 3 years of age and typically give birth annually thereafter.[48,56] Monkeys can live up to 35 years in the wild, although lifespans of around 18 years seem more typical. This is considerably longer than the lifespans of most ungulates. However, patas monkeys rarely lived longer than 10–15 years in places where they have been studied. My demographic modelling indicated that, in order to maintain a viable population, a female monkey reproducing only every other year and losing half of her offspring before they attain maturity at 5 years of age must expect to survive for 8 years or longer beyond her age at first reproduction (Table 16.2).

The life-history patterns of baboons are similar to those of savanna-inhabiting vervets, despite the difference in body size. Female baboons first give birth at a median age of 6 years and show inter-birth intervals typically around 2 years.[44,58,59,60] Juvenile baboons remain dependent on their mothers until they reach 2 years of age, and around 30–45 percent of them die during this juvenile stage.[61] Much of the infant mortality is caused by male baboons.[44,60] Generally, fewer than half of newborn baboons survive to reach adulthood. Annual mortality rates among adult female baboons during their early prime years may be as low as 3–4 percent annually,[60] but rise as high as 9 percent annually in places where predation by leopards is more common. Baboons can live 26 years or longer in the wild. Under idealised conditions with zero mortality among both young and adult animals until the end of the lifespan, the maximum rate of population growth that could be achieved by baboons is about 12.5 percent per year, around half of the growth rate possible for ungulates (Table 16.2). Note that this takes into account terminal mortality at the end of the lifespan. Male mortality rates can be double those incurred by females.

Table 16.2 Modelled combinations of vital rates yielding either maximum or zero population growth for baboons, chimpanzees and gorillas. Zero growth has been achieved by density-dependent increases in mortality between birth and first reproduction affecting recruitment into the adult segment without much change in age at maturity, adult survival or birth intervals

Species	Scenario	Population growth rate (% pa)	Maturity (years)	Longevity (years)	Birth interval (years)	Prime adult mortality (% pa)	Overall adult mortality (% pa)	Recruitment (%)
Baboon	Maximum	12.8	5	25	2.0	0	1.5	100
	Zero	0	5	25	2.2	2.8	6.5	55
Chimpanzee	Maximum	5.7	11	50	4	0	1.0	100
	Zero	0	14	50	5	0.5	3.1	80
Gorilla	Maximum	6.3	10	45	4	0	0.8	100
	Zero	0	10	45	5	0.8	3.2	70

Chimpanzees first reproduce as late as 11–14 years of age and exhibit interbirth intervals averaging over 5 years in the wild but 4 years in captivity. Mortality rates estimated compositely from five studied populations averaged 21 percent during the first year, dropping to 5 percent annually among females during their prime stage between 15 and 30 years of age.[62] However, in combination such demographic rates yield a declining population with a replacement rate ('lambda') of only 0.8, reflecting mortality imposed by viral epidemics as well as poaching. A specific assessment for the slowly increasing Kanyawara subpopulation in Kibale Forest, Uganda, indicated much lower mortality rates: 11 percent over the first year decreasing to 3 percent between 10 and 15 years and as low as 1 percent through the prime age range from 15 to 30 years.[63] Male chimpanzees consistently incur greater mortality than females at all stages from birth through adulthood. A maximum longevity of 55 years has been recorded for a chimpanzee in the wild. The maximum potential growth rate for a chimpanzee population, assuming zero mortality throughout the adult lifespan, is around 5.5 percent per year (Table 16.2), similar to that of elephants and slower than that of white rhinos.[64] Small increases in predation, infanticide or mortality among adults could pull the population growth rate below zero. In western Africa, targeted predation by leopards caused chimpanzee numbers to decline.[51]

Despite being larger, gorillas show a somewhat faster life history than chimpanzees. In the Virunga volcanoes, females first gave birth at 10 years and maintained birth intervals averaging 4 years.[65] Annual mortality rates were 11 percent for infants, 2.5 percent for juveniles and 3.5 percent among prime-aged females. Most of the infant mortality was due to infanticide by adult males taking over groups, while some older juveniles were injured by snares. Gorillas in the Bwindi forest showed longer birth intervals averaging almost 5 years (with a minimum of 2.5 years recorded), but without any difference in offspring mortality.[66] Maximum longevity attained by a gorilla is 45 years. A gorilla population could potentially grow at 6 percent per year in the absence of mortality prior to the end of the lifespan (Table 16.2).

Thus both of these great apes resemble elephants, rather than similar-sized antelope, in their life-history schedules. How can these and other primates keep predation low enough among both adults and offspring to make populations viable? This seems readily achieved in forests where there are few predators, but how could this be accomplished by savanna-inhabiting hominins?

How Is Primate Abundance Regulated?

Monkeys and baboons can attain high numerical densities locally within occupied home ranges, while densities of chimpanzees and gorillas are much lower and resemble those of low-density ungulates like kudu and sable antelope (Table 16.1). While food supplies constitute the basic limiting factor, predation may restrict the habitats that can be occupied, as noted for ungulates (Chapter 12).

The Amboseli baboon population in Kenya decreased from a density of 73 animals per km^2 during the 1960s[67] to under 2 animals per km^2 eventually, following habitat loss. This crash was brought about by the disappearance of most of the fever tree woodlands they had depended on as their major food source, due to a rise in water table elevating soil salinity. The disappearance of these large trees also eliminated the treetop refuges they had occupied at night. Fifty-one separate troops present during the 1960s became reduced to eight over a period of 5 years. The median group size fell from 43 to 27 animals, with the largest and smallest groups being most reduced. The troops that survived did so by shifting their home ranges towards a region where umbrella thorn trees still remained. The subsequent recovery of this baboon population from a low of 123 animals in 1979 to somewhere between 1000 and 1200 animals around 2015[60] is consistent with a sustained annual growth rate of 6.5 percent, half of the potential maximum rate. However, an influx by baboons of a different subspecies contributed to the recovery in numbers.[68] Fertility rates decrease in the largest troops and may lead to fissioning into smaller groups.[69]

Population declines among monkeys have been associated with drought conditions and habitat deterioration, causing elevated mortality particularly among juveniles.[70] At Amboseli, vervet monkeys responded to the progressive disappearance of umbrella thorn trees as well as fever trees by expanding their territories initially, but later whole groups disappeared.[71] In Cameroon, drought conditions led to substantially increased mortality among both infant and adult patas monkeys, producing a population decline.[56]

Although primate populations are limited fundamentally by food availability, regulation generally operates through social mechanisms effective within groups. More dominant females secure a greater share of resources than subordinate females. In both Botswana and Amboseli, high-ranking female baboons had shorter birth intervals than low-ranking ones.[44] In Tanzania, infant mortality was greater in larger troops and among low-ranked females.[61] However, larger groups of baboons are able to displace smaller groups into less-suitable habitat, imposing greater movement costs on the latter. Because competition gets intensified within larger groups of baboons, animals in intermediate-sized groups numbering 40–80 individuals seem to be most advantageously placed.[72] When troops become much larger, low-ranked individuals may split off to form a new group, provided vacant habitat exists. Among chimpanzees, much of the mortality incurred is inflicted by adult males on juveniles, rising as density and hence social strife and inequality increases.

Overview

Primates depend to varying degrees on leaves, fruits and other plant parts as food resources. Their dietary protein content may be augmented by insects, although leaves alone can supply adequate protein. Both vervet and patas monkeys occupying savannas still obtain most of their food from trees. Chimpanzees are strongly dependent on forest fruits, while gorillas and bonobos rely largely on herbaceous plant parts obtained from the forest floor. Chimpanzees and baboons augment their carbohydrate-loaded diets with animal flesh obtained opportunistically, with males gaining most of the benefit.

Only baboons and geladas subsist substantially on the grass parts and forbs prevalent in savannas. To survive through the dry season, baboons fall back on carbohydrates stored in corms and bulbs underground. Their digging capabilities for deeper bulbs and roots are outdone by porcupines and undermined by mole-rats. Chimpanzees occupying savanna woodlands must cover substantially larger ranges than traversed by forest-inhabiting communities in order to support their frugivorous diet. Densities attained by the great apes in tropical forests are no greater than those shown by browsing ungulates in open savannas, despite the lush plant growth.

All non-human primates rely on their climbing capabilities to restrict predation, especially at night. They must keep mortality due to predation lower than incurred by ungulates because of their slow life-history progression, ascribed to building relatively bigger brains than other animal forms. Mortality losses incurred by gorillas and chimpanzees are indeed lower than those exhibited by most ungulates and resemble that of megaherbivores like rhinos and elephants, but these great apes inhabit forested regions where leopards are sparse and other large carnivores absent. Keeping mortality rates this low in savanna regions thronged with large carnivores is a more formidable undertaking.

Suggested Further Reading

Boesch, C; Boesch-Achermann, H. (2000) *The Chimpanzees of the Taï Forest: Behavioural Ecology and Evolution*. Oxford University Press, Oxford.

Hohmann G, et al. (eds). (2006) *Feeding Ecology in Apes and Other Primates*. Cambridge University Press, Cambridge.

Schaller, GB. (1963) *The Mountain Gorilla*. University of Chicago Press, Chicago.

References

1. Copeland, SR. (2009) Potential hominin plant foods in northern Tanzania: semi-arid savannas versus savanna chimpanzee sites. *Journal of Human Evolution* 57:365–378.

2. Marshall, AJ, et al. (2009) Defining fallback foods and assessing their importance in primate ecology and evolution. *American Journal of Physical Anthropology: The Official Publication of the American Association of Physical Anthropologists* 140:603–614.
3. O'Brien, EM; Peters, CR (1999) Landforms, climate, ecogeographic mosaics, and the potential for hominid diversity in Pliocene Africa. In Bromage, TG; Schrenk, F (eds) *African Biogeography, Climate Change and Human Evolution*. Oxford University Press, Oxford, pp. 115–137.
4. Uwimbabazi, M, et al. (2019) Influence of fruit availability on macronutrient and energy intake by female chimpanzees. *African Journal of Ecology* 57:454–465.
5. Van Schaik, CP; Pfannes, KR. (2005) Tropical climates and phenology: a primate perspective. In Brockmann, DK; Van Schaik, CP (eds) *Seasonality in Primates*. Cambridge University Press, Cambridge, pp. 23–54.
6. Oates, JF. (1977) The guereza and its food. In Clutton-Brock, TH (ed.) *Primate Ecology: Studies of Feeding and Ranging Behaviour in Lemurs, Monkeys, and Apes*. Academic Press, London, pp. 275–321.
7. Danish, L, et al. (2006) The role of sugar in diet selection in redtail and red colobus monkeys. In Hahmann, MM, et al. (eds) *Feeding Ecology in Apes and Other Primates*. Cambridge University Press, Cambridge, pp. 473–487.
8. Poulsen, JR, et al. (2001) Seasonal variation in the feeding ecology of the grey-cheeked mangabey (*Lophocebus albigena*) in Cameroon. *American Journal of Primatology: Official Journal of the American Society of Primatologists* 54:91–105.
9. Doran-Sheehy, DM, et al. (2006) Sympatric western gorilla and mangabey diet: re-examination of ape and monkey foraging strategies. In Hohmann, G, et al. (eds) *Feeding Ecology in Apes and Other Primates*. Cambridge University Press, Cambridge, pp. 49–72.
10. Wrangham, RW; Waterman, PG. (1981) Feeding behaviour of vervet monkeys on *Acacia tortilis* and *Acacia xanthophloea*: with special reference to reproductive strategies and tannin production. *The Journal of Animal Ecology* 50:715–731.
11. Isbell, LA. (1998) Diet for a small primate: insectivory and gummivory in the (large) patas monkey (*Erythrocebus patas pyrrhonotus*). *American Journal of Primatology* 45:381–398.
12. Nakagawa, N. (2000) Seasonal, sex, and interspecific differences in activity time budgets and diets of patas monkeys (*Erythrocebus patas*) and tantalus monkeys (*Cercopithecus aethiops tantalus*), living sympatrically in northern Cameroon. *Primates* 41:161.
13. Nakagawa, N. (2003) Difference in food selection between patas monkeys (*Erythrocebus patas*) and tantalus monkeys (*Cercopithecus aethiops tantalus*) in Kala Maloue National Park, Cameroon, in relation to nutrient content. *Primates* 44:3–11.
14. Dunbar, RIM. (1977) Feeding ecology of gelada baboons: a preliminary report. In Clutton-Brock, TH (ed.) *Primate Ecology*. Academic Press, London, pp. 251–273.
15. Norton, GW, et al. (1987) Baboon diet: a five-year study of stability and variability in the plant feeding and habitat of the yellow baboons (*Papio cynocephalus*) of Mikumi National Park, Tanzania. *Folia Primatologica* 48:78–120.
16. Whiten, A, et al. (1991) Dietary and foraging strategies of baboons. *Philosophical Transactions of the Royal Society of London Series B: Biological Sciences* 334:187–197.

17. Whiten, A, et al. (1987) The behavioral ecology of mountain baboons. *International Journal of Primatology* 8:367–388.
18. Barton, RA; Whiten, A. (1994) Reducing complex diets to simple rules: food selection by olive baboons. *Behavioral Ecology and Sociobiology* 35:283–293.
19. Alberts, SC, et al. (2005). Seasonality and long term change in a savannah environment. In Brockmann, DK; Van Schaik, CP (eds) *Seasonality in Primates*. Vol. 16. Cambridge University Press, Cambridge, pp. 157–195.
20. Kunz, BK; Linsenmair, KE. (2008) The disregarded west: diet and behavioural ecology of olive baboons in the Ivory Coast. *Folia Primatologica* 79:31–51.
21. Codron, D, et al. (2008) What insights can baboon feeding ecology provide for early hominin niche differentiation? *International Journal of Primatology* 29:757–772.
22. Wrangham, RW, et al. (1998) Dietary response of chimpanzees and cercopithecines to seasonal variation in fruit abundance. I. Antifeedants. *International Journal of Primatology* 19:949–970.
23. Newton-Fisher, NE. (1999) The diet of chimpanzees in the Budongo Forest Reserve, Uganda. *African Journal of Ecology* 37:344–354.
24. Morgan, D; Sanz, CM. (2006) Chimpanzee feeding ecology and comparisons with sympatric gorillas in the Goualougo Triangle, Republik of Congo. In Hohmann, G, et al. (eds) *Feeding Ecology in Apes and Other Primates*. Cambridge University Press, Cambridge, pp. 97–122.
25. McGrew, WC, et al. (1988) Diet of wild chimpanzees (*Pan troglodytes verus*) at Mt. Assirik, Senegal: I. Composition. *American Journal of Primatology* 16:213–226.
26. Pruetz, JD. (2006) Feeding ecology of savanna chimpanzees (*Pan troglodytes verus*) at Fongoli, Senegal. In Hohmann, G, et al. (eds) *Feeding Ecology in Apes and Other Primates*. Cambridge University Press, Cambridge, pp. 161–182.
27. Wrangham, RW. (1977) Feeding behaviour of chimpanzees in Gombe national park, Tanzania. In Clutton-Brock, TH (ed.) *Primate Ecology: Studies of Feeding and Ranging Behaviour in Lemurs, Monkeys and Apes*. Academic Press, London, pp. 503–538.
28. Malenky, RK; Stiles, EW. (1991) Distribution of terrestrial herbaceous vegetation and its consumption by *Pan paniscus* in the Lomako Forest, Zaire. *American Journal of Primatology* 23:153–169.
29. Rogers, ME, et al. (2004) Western gorilla diet: a synthesis from six sites. *American Journal of Primatology: Official Journal of the American Society of Primatologists* 64:173–192.
30. Doran-Sheehy, D, et al. (2009) Male and female western gorilla diet: preferred foods, use of fallback resources, and implications for ape versus old world monkey foraging strategies. *American Journal of Physical Anthropology: The Official Publication of the American Association of Physical Anthropologists* 140:727–738.
31. Robbins, MM, et al. (2006) Variability of the feeding ecology of eastern gorillas. In Hohmann, G, et al. (eds) *Feeding Ecology in Apes and Other Primates*. Cambridge University Press, Cambridge, pp. 25–48.
32. Yamagiwa, J; Basabose, AK. (2006) Effects of fruit scarcity on foraging strategies of sympatric gorillas and chimpanzees. In Hohmann, G, et al. (eds) *Feeding Ecology in Apes and Other Primates*. Cambridge University Press, Cambridge, pp. 73–96.

33. Rothman, JM, et al. (2011) Nutritional geometry: gorillas prioritize non-protein energy while consuming surplus protein. *Biology Letters* 7:847–849.
34. Altmann, SA. (2009) Fallback foods, eclectic omnivores, and the packaging problem. *American Journal of Physical Anthropology: The Official Publication of the American Association of Physical Anthropologists* 140:615–629.
35. Dunbar, RIM. (1987) Habitat quality, population dynamics, and group composition in colobus monkeys (*Colobus guereza*). *International Journal of Primatology* 8:299–329.
36. De Vos, A; Omar, A. (1971) Territories and movements of Sykes monkeys (*Cercopithecus mitis kolbi* Neuman) in Kenya. *Folia Primatologica* 16:196–205.
37. Lawes, MJ. (1991) Diet of samango monkeys (*Cercopithecus mitis erythrarchus*) in the Cape Vidal dune forest, South Africa. *Journal of Zoology* 224:149–173.
38. Barrett, AS, et al. (2010) A floristic description and utilisation of two home ranges by vervet monkeys in Loskop Dam Nature Reserve, South Africa. *Koedoe* 52:1–12.
39. Pasternak, G, et al. (2013) Population ecology of vervet monkeys in a high latitude, semi-arid riparian woodland. *Koedoe* 55:1–9.
40. Waser, P. (1977) Feeding, ranging and group size in the mangabey (*Cercocebus albigena*). In Clutton-Brock, TH (ed.) *Primate Ecology*. Academic Press, London, pp. 183–221.
41. Nakagawa, N. (1999) Differential habitat utilization by patas monkeys (*Erythrocebus patas*) and tantalus monkeys (*Cercopithecus aethiops tantalus*) living sympatrically in northern Cameroon. *American Journal of Primatology: Official Journal of the American Society of Primatologists* 49:243–264.
42. Hall, KR. (1966) Behaviour and ecology of the wild patas monkey, *Erythrocebus patas*, in Uganda. *Journal of Zoology* 148:15–87.
43. Yamagiwa, J; Basabose, AK. (2009) Fallback foods and dietary partitioning among *Pan* and *Gorilla*. *American Journal of Physical Anthropology: The Official Publication of the American Association of Physical Anthropologists* 140:739–750.
44. Cheney, DL, et al. (2004) Factors affecting reproduction and mortality among baboons in the Okavango Delta, Botswana. *International Journal of Primatology* 25:401–428.
45. Cowlishaw, G. (1994) Vulnerability to predation in baboon populations. *Behaviour* 131:293–304.
46. Isbell, LA, et al. (2018) GPS-identified vulnerabilities of savannah-woodland primates to leopard predation and their implications for early hominins. *Journal of Human Evolution* 118:1–13.
47. Anderson, JR. (1998) Sleep, sleeping sites, and sleep-related activities: awakening to their significance. *American Journal of Primatology* 46:63–75.
48. Isbell, LA, et al. (2009) Demography and life histories of sympatric patas monkeys, *Erythrocebus patas*, and vervets, *Cercopithecus aethiops*, in Laikipia, Kenya. *International Journal of Primatology* 30:103–124.
49. Stanford, CB. (2002) Avoiding predators: expectations and evidence in primate antipredator behavior. *International Journal of Primatology* 23:741–757.

50. Owen-Smith, N; Mills, MGL. (2008) Shifting prey selection generates contrasting herbivore dynamics within a large-mammal predator–prey web. *Ecology* 89:1120–1133.
51. Boesch, C; Boesch-Achermann, H. (2000) *The Chimpanzees of the Taï Forest: Behavioural Ecology and Evolution*. Oxford University Press, Oxford.
52. Fay, JMR. (1995) Leopard attack on and consumption of gorillas in the Central African Republic. *Journal of Human Evolution* 29:93–99.
53. Ross, C. (1998) Primate life histories. *Evolutionary Anthropology: Issues, News, and Reviews* 6:54–63.
54. Ross, C. (2003). Life history, infant care strategies, and brain size in primates. In Kappeler, P, et al. (eds) *Primate Life Histories and Socioecology*. University of Chicago Press, Chicago, pp. 54–63.
55. Harvey, PH; Clutton-Brock, TH. (1985) Life history variation in primates. *Evolution* 39:559–581.
56. Nakagawa, N, et al. (2003) Life-history parameters of a wild group of West African patas monkeys (*Erythrocebus patas patas*). *Primates* 44:281–290.
57. Cords, M; Chowdhury, S. (2010) Life history of *Cercopithecus mitis stuhlmanni* in the Kakamega Forest, Kenya. *International Journal of Primatology* 31:433–455.
58. Altmann, J, et al. (1977) Life history of yellow baboons: physical development, reproductive parameters, and infant mortality. *Primates* 18:315–330.
59. Hill, RA, et al. (2000) Ecological and social determinants of birth intervals in baboons. *Behavioral Ecology* 11:560–564.
60. Alberts, SC. (2019) Social influences on survival and reproduction: insights from a long-term study of wild baboons. *Journal of Animal Ecology* 88:47–66.
61. Rhine, RJ, et al. (1988) Eight-year study of social and ecological correlates of mortality among immature baboons of Mikumi National Park, Tanzania. *American Journal of Primatology* 16:199–212.
62. Hill, K, et al. (2001) Mortality rates among wild chimpanzees. *Journal of Human Evolution* 40:437–450.
63. Muller, MN; Wrangham, RW. (2014) Mortality rates among Kanyawara chimpanzees. *Journal of Human Evolution* 66:107–114.
64. Owen-Smith, RN. (1988) *Megaherbivores: The Influence of Very Large Body Size on Ecology*. Cambridge University Press, Cambridge.
65. Robbins, MM; Robbins, AM. (2004) Simulation of the population dynamics and social structure of the Virunga mountain gorillas. *American Journal of Primatology: Official Journal of the American Society of Primatologists* 63:201–223.
66. Robbins, MM, et al. (2009) Population dynamics of the Bwindi mountain gorillas. *Biological Conservation* 142:2886–2895.
67. Altmann, J, et al. (1985) Demography of Amboseli baboons, 1963–1983. *American Journal of Primatology* 8:113–125.
68. Samuels, A; Altmann, J. (1991) Baboons of the Amboseli basin: demographic stability and change. *International Journal of Primatology* 12:1.

69. Dunbar, RIM, et al. (2018) Trade-off between fertility and predation risk drives a geometric sequence in the pattern of group sizes in baboons. *Biology Letters* 14:20170700.
70. Isbell, LA, et al. (1990) Costs and benefits of home range shifts among vervet monkeys (*Cercopithecus aethiops*) in Amboseli National Park, Kenya. *Behavioral Ecology and Sociobiology* 27:351–358.
71. Lee, PC; Hauser, MD. (1998) Long-term consequences of changes in territory quality on feeding and reproductive strategies of vervet monkeys. *Journal of Animal Ecology* 67:347–358.
72. Dunbar, RIM; Mac Carron, P. (2019) Does a trade-off between fertility and predation risk explain social evolution in baboons? *Journal of Zoology* 308:9–15.

Chapter 17: How an Ape Became a Hunter

Humans are the product of the ape lineage that launched out into expanding savannas. Several adaptive challenges needed to be overcome. Increased competence in bipedal locomotion was required to traverse the widening gaps between trees. Then there was the problem of what to eat during intensifying dry seasons when trees were mostly leafless, while the leaves that did remain were tough to digest and fruits were few and far between. However, savanna plants must store carbohydrates in roots, bulbs and tubers to regenerate foliage at the start of each wet season. To extract them from underground, the evolving ape-men needed to dig and to chew, like porcupines do. Above ground, there remained another potential source of food in the bodies of the herbivores that were adapted to digest the leaves of expanding grasslands as well as those on trees. Scraps of their flesh and bones became available on the carcass remains of animals killed by carnivores, particularly during the dry season when plant resources were sparsest. Exploiting bits of meat and bone marrow required stone tools to be deployed to scrape off flesh and break open bones. Further adaptations in locomotion, dentition, thermal tolerance and stone tool technologies led evolving humans to become hunters in their own right, exploiting time windows to reduce competition from hunting and scavenging carnivores. Of course, they also needed to avoid becoming food themselves for the carnivores.

In this chapter, I consider first the evolutionary transitions that took place in physical features preserved in fossilised bones, skulls and teeth. Four major adaptive shifts can be recognised. The first entailed the development of bipedal locomotion, during the late Miocene. This was followed by trends in dentition towards greater chewing capacity during the Pliocene. The dental trend then became reversed as humans evolved longer legs and larger brains, with a surge in cranial capacity associated with the climatic transition into the Pleistocene. Lastly, following amplified climatic oscillations defining the start of the Middle Pleistocene, brain volume surged to modern levels and facial features converged on those of modern humans. My focus in this chapter is on the ecological processes driving adaptive changes in physical forms. Supporting

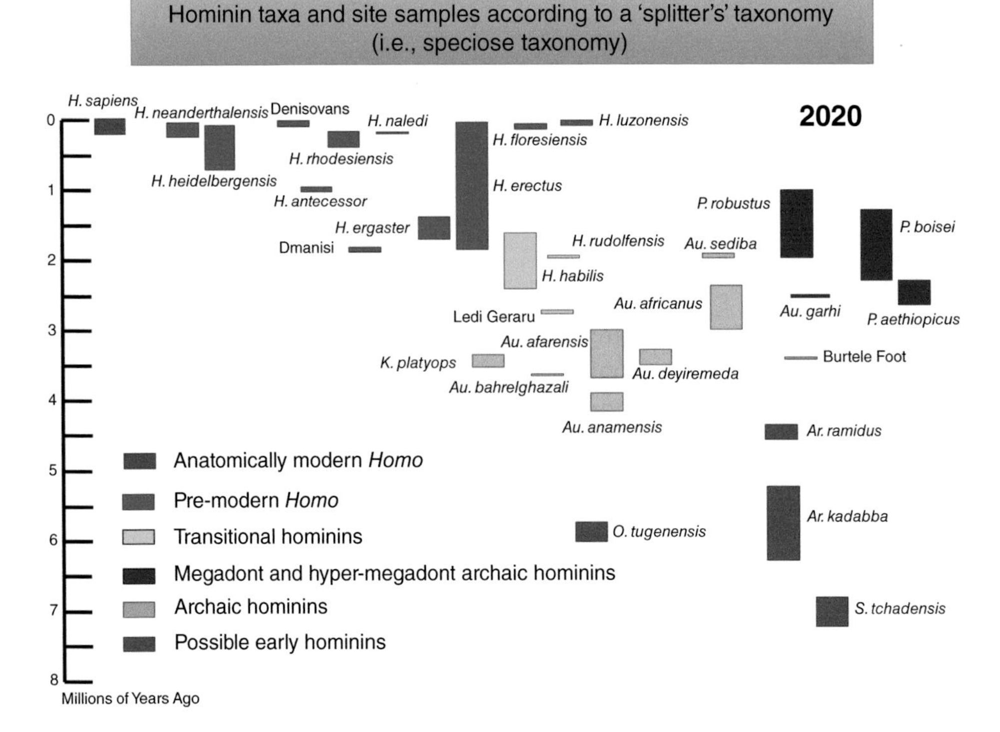

Figure 17.1 Timeline of hominin species recognised with colours indicating adaptive grades from early ape-men to modern humans (supplied by Bernard Wood, updated from Wood & Boyle (2016) *American Journal of Physical Anthropology* 159:37–78).

changes in cultural artefacts in the form of stone tools and, increasingly, other technologies will be covered in the following chapter.

The fragmentary nature of the fossil record in time and space must be acknowledged. Sites providing hominin fossils in eastern Africa are located mostly alongside rivers or lakes within or adjacent to the Eastern Rift Valley. Those in southern Africa are located mostly within cavities formed in dolomitic limestone in the high central interior. Figure 17.1 shows timelines linking named species representing the various hominin forms distinguished, indicating advances in adaptive grades.

Emergence of Ape-men

The connections between the earliest hominins and the common ancestor shared with chimpanzees have yet to be revealed and may never be. The crucial period between 12 and 6 Ma, from mid to late Miocene, has yielded few fossil deposits. So far, only two enigmatic fossil finds from this time

A

B

Figure 17.2 Competence in bipedal locomotion. (A) *Ardipithecus* depicted walking among trees in a wooded environment (artwork: Jay Matternes); (B) *Australopithecus afarensis* family depicted walking bipedally across a volcanic ash surface at Laetoli, where their footprints are preserved (artwork: Marco Anson).

interval have been linked with hominin divergence.[1] The 'Toumai' skull, scientifically named *Sahelanthropus tchadensis*, was discovered on the land surface in Chad in north-central Africa and dated approximately to around 7 Ma. Its foramen magnum connecting the brain to the spinal cord had shifted anteriorly, consistent with an upright posture, while its dental features appear intermediate between those of chimpanzees and later australopiths (which I will colloquially label 'ape-men', while recognising that there were women as well). However, its cranial volume remained no bigger than that of a chimpanzee. The second set of fossils, given the name *Orrorin tugenensis*, consists of a mandible plus limb and foot bones extracted from sediments in the Tugen Hills situated in west-central Kenya, dated to ~6 Ma. These limb bones show greater support for upright balancing than is typical of apes.

The earliest more firmly hominin fossils come from Ethiopia, dated to the period between 5.7 and 4.4 Ma and assigned to the genus *Ardipithecus* (Figure 17.2A).[1] They show features of leg and foot bones that would facilitate upright walking, but retain grasping big toes that would help with tree climbing. The oldest fossils placed in the genus *Australopithecus*, named *Australopithecus anamensis*, come from deposits dated between 4.2 and 4.0 Ma found from north-eastern Ethiopia into the Turkana region of northern Kenya, but no further south.[1,2] They show further advances in hind-limb and ankle bones for bipedal walking, but no change in cranial capacity. Dentally, these ape-men deviated from modern apes by having smaller incisor and canine teeth coupled with larger molars bearing thicker enamel, supportive of a dietary shift from soft fruits towards tougher plant parts. *Au. anamensis* is abundantly represented in fossil assemblages through this northern region of Africa, contributing as much as 5 percent of fossil specimens, similar to early baboons. The lakeshore or riparian vegetation present around the fossil sites during that period took the form of a varying mosaic of quite dense woodland and quite open grassland.[3,4] Other animals present included numerous proboscideans along with frugivorous monkeys, while among the ungulates browsers were more common than grazers. The diet of these early australopiths was restricted entirely to C_3 plants.[5]

Following on chronologically, *Au. afarensis* was distributed more widely throughout eastern Africa in deposits dated to the period between 3.85 and 2.95 Ma.[6] A nearly complete skeleton discovered in Ethiopia was labelled 'Lucy', from the Beatles' song about diamonds in the sky played during the celebration following its finding. The bipedal walking capabilities of this species are documented by footprints of three individuals etched on volcanic tuff at Laetoli in Tanzania, dated to 3.6 Ma (Figure 17.2B). A cranium dated to 3.4 Ma found west of Lake Turkana with a flattened face and somewhat smaller molars was named *Kenyanthropus platyops*; but the generic distinction is questionable given the range of variation among specimens assigned to *Australopithecus*. Nevertheless, it might represent an initial stage of the shift towards the more 'gracile' features shown by early *Homo*. After 2.9 Ma, two hominin lineages diverged. Fossils with massive jaws and especially large cheek teeth, colloquially named 'nutcracker man', appear earliest in the Omo valley of south-western Ethiopia, dated to 2.7 Ma, and shortly thereafter in the Turkana basin in adjoining Kenya.[7] They are generally assigned to a distinct genus, *Paranthropus*. These robust ape-men were far rarer in local faunas than *Au. anamensis* had been, contributing less than 1 percent of fossils. *Paranthropus boisei* persisted in eastern African fossil deposits with little morphological change until around 1.3 Ma.

Within South Africa, an almost complete skeleton named 'Little Foot', from the bones that were first exposed in the enclosing breccia at Sterkfontein, was given the name *Au. prometheus* and dated to 3.67 Ma,[8,9] although this early time has been contested.[10] The physical features of its skull and dentition resemble those of *Au. anamensis* from Ethiopia dated around 4 Ma. *Au. africanus*, its likely descendant, is abundantly represented in various cave sites in the Cradle region from 2.8 Ma until around 2.3 Ma. Its type specimen is the Taung skull of a child, apparently killed by an eagle. A nearly complete skull found in Sterkfontein Cave was named 'Mrs Ples' (Figure 17.3; but 'she' might have been a male). *Au. africanus* persisted somewhat later in South Africa than its eastern African counterpart *Au. afarensis*, overlapping in time with the transition to early *Homo* in the north. Fossils dated more recently to 1.98 Ma, found

Figure 17.3 Skulls of robust ape-man (*Paranthropus robustus*) (left) and more gracile ape-man (*Australopithecus africanus*) (right) from cave deposits in the Cradle of Humankind, South Africa.

in a single cave in South Africa, were given the name *Au. sediba*.[11] They show human-like molar teeth and pelvic girdle, but brain size and facial shape similar to chimpanzees.[12] Their long thumbs coupled with short fingers could have facilitated the precision grip needed for manipulating stone tools, while bone structures in the hands and wrist would have aided arboreal climbing. It has been suggested that they are closely linked to the ancestry of *Homo*,[11] but I wonder whether they might be ancestral to the strange *Homo naledi* found likewise in a single cave and dated to 260 ka (see below). The local robust ape-man, *Paranthropus robustus*, made its appearance in South African cave sites around 2 Ma and became the most abundant hominin represented there until around 1.0 Ma. Its limbs and hand bones suggest that it was less effectively bipedal than coexisting forms of *Homo*, but somewhat better at tree-climbing.

Becoming Human

The earliest fossils placed in the genus *Homo* come from Ledi-Geraru in the Afar region of north-eastern Ethiopia, in the form of a mandible with teeth dated to almost 2.8 Ma.[13] They exhibit features transitional between *Au. afarensis* and later specimens from Ethiopia assigned more firmly to early *Homo*, dated to around 2.4 Ma. Vertebrate fossils suggest that the habitat had become prevalently more open with a mixture of grassland and riparian forest and expanded C_4 component around that time.[14,15] Around one-third of the bovid taxa identified at the site made their first appearance then, including the earliest wildebeest (*Connochaetes* sp.).

The Ledi-Geraru mandible is followed after an interval of 400 kyr by fossils assigned to early *Homo* recorded not only in Ethiopia, but also further south in Kenya and even Malawi.[7] This gap in time happens to span the climatic transition into the start of the recurrent ice ages typifying the Pleistocene.

Further specimens assigned more firmly to *H. habilis*, including a nearly complete skull, come from both the Omo–Turkana Basin and Olduvai Gorge, dated ~1.9 Ma. This name, meaning 'handy man', was given because the fossils were associated with stone tools defining the Oldowan culture at Olduvai. Features distinguishing this species from earlier australopiths include a larger body size and cranial capacity, less-protruding face, longer legs, smaller jaws, and smaller molar teeth with thinner enamel. Hominins somewhat larger in size discovered to the west of Lake Turkana, likewise dated to ~2.0 Ma, were named *H. rudolfensis*, but may not be sufficiently distinct from *H. habilis* (or *H. ergaster*) to justify the species distinction. Fossils from the South African cave sites allied with early *Homo* are generally too fragmented to be assigned reliably to any particular species. These very early humans were rare components of local faunas, making up less than 0.1 percent of all large mammal fossils, which is less than contributed by *Paranthropus* spp. and similar to that formed by large carnivores. The scarcity of fossils representing early *Homo* in South African cave sites suggests that early humans were either rare in this Highveld plateau region, or much less susceptible to falling (or being dragged) into cavities than *Paranthropus*.

The earliest fossils assigned to *H. ergaster* come from the Omo–Turkana Basin, dated to ~1.9 Ma, but became common elsewhere only after 1.7 Ma. *H. ergaster* is commonly merged with *H. erectus*, named from fossils found in Java and elsewhere in south-east Asia. It made its appearance during the time when glacial oscillations defining the Pleistocene, coupled with local earth movements, further accentuated aridity in eastern Africa. These proto-humans exhibited a cranial volume a little larger than that of *H. habilis*, stood as tall as modern humans, and walked as competently.[16,17] Their molar teeth were relatively smaller than those of the gracile australopiths, and incisors somewhat larger, thereby reverting towards dental features of the great apes. Cranially, they retained a sloping forehead and prominent brow ridges. Surprisingly, a juvenile cranium recently recovered in Drimolen Cave in the Cradle region of South Africa, ascribed to *H. ergaster*, has been dated as early as 2.0 Ma, preceding the first appearance of this species in eastern Africa.[18] The next earliest record of *H. ergaster* in the Cradle region is a specimen from Swartkrans dated shortly after 1.8 Ma.[12]

Specimens allied with *H. habilis* apparently persisted alongside *H. ergaster* until ~1.4 Ma. Whether these two forms segregated ecologically or represent a single polytypic species remains debatable. Distinctions between them are based mainly on skull shape.[17] *H. ergaster* showed a slow expansion in brain size and progressive reduction in molar surface area through time, although rather poorly documented. It was replaced by *H. heidelbergensis*, clearly a descendent, around 800 ka, both in Europe and in Africa.[19,20] This was around the time when the period between glacial extremes lengthened to around 100 kyr, defining the transition from early to middle Pleistocene when seasonal extremes of aridity intensified in Africa. By that stage the cranial volume of these early humans was almost equal to that of modern humans. *H. heidelbergensis* was clearly ancestral to the European Neanderthals (*H. neanderthalensis*), but how it connected with early humans inhabiting Africa remains obscure. Genetic evidence suggests that the Neanderthals separated from the human lineage in Africa around 500 ka.[21] They differed from modern humans in their stocky build, prominent brow ridges, and sloping rather than domed foreheads.

The earliest fossils sufficiently similar to modern humans to be assigned to the species *Homo sapiens* come from a cave in Morocco dated to between 300 and 350 ka.[22] These craniums show a reduction in face size, like later humans, but retain fairly prominent brow ridges and lack the globular shape of the braincase characteristic of modern humans. The Kabwe skull, found in a limestone cave in Zambia, has also been dated to ~300 ka.[23] It has been affiliated with *H. heidelbergensis* due to its lack of more modern features. A partial cranium from the Lake Ndutu region near Olduvai also represents this time period.[24] Slightly more recent is the Florisbad skull from interior South Africa, dated approximately to ~260 ka and regarded as representing early *H. sapiens*. Partial human skulls from Herto and Omo in Ethiopia and Turkana in Kenya, dated to between 160 and 195 ka, are more firmly accepted as modern humans.

Strangely out on an evolutionary limb are the remains of diminutive humans with small but complexly shaped brains named *Homo naledi*, which were found deep in a cave system in South Africa. Although their features resemble those of early humans living around 2 Ma, their hardly fossilised bones date from as recently as ~280 ka.[25] While their leg and foot bones were adapted for efficient walking, their hands retained long fingers, which would have facilitated climbing trees. They had smaller but higher-crowned molars with greater wear-resistance than the dentition of the australopiths, suggesting that abrasive particles were prominent in their diet.[26] They apparently coexisted alongside *H. ergaster* for well over a million years without showing up elsewhere in fossil deposits.

Around 130 ka, humans with modern features appeared in the Levant region of the Middle East, but did not persist there for very long. This was around the beginning of the relatively moist interglacial period that defines the transition into the late Pleistocene. The major dispersal of modern humans beyond Africa through the Middle East and onwards occurred ~60 ka. These people evidently followed a coastal route through tropical Asia, arriving in Australia ~55 ka. By 47 ka they had spread westward into Europe, displacing the Neanderthals within a few thousand years.

The Locomotor Transition: From Brachiating Arms to Legs for Walking

During the late Miocene, the space between trees occupied by grassland widened as temperatures cooled and precipitation became more seasonal, promoting the radiation of browsing and grazing ruminants. For the stem hominins, the conversion of forest or woodland into savanna meant more time spent in terrestrial travel and less time foraging in tree canopies. If the tree canopy cover thinned from over 90 percent, representing a forest, to under 30 percent as is typical of savanna, this would approximately triple the home range needed by chimpanzees to encompass sufficient fruit-bearing trees, towards ~50 km^2. Covering such a large area could not readily be accomplished by waddling clumsily on two legs, or by knuckle-walking. More likely, fruit-dependent apes became confined to localities where the tree cover remained closer to 50 percent, or where forest strips traversed savanna vegetation formations, as found where chimpanzees occupy moist savanna in Senegal.

Whichever situation prevailed, selective pressures would favour hind-limb adaptations facilitating bipedal walking, while forelimbs retained their capabilities for ascending trees to access fruits and to escape from ground-based carnivores at night. Bipedal locomotion also provides an elevated perspective for detecting predators lurking in the grass and frees arms for carrying food or other things to safer sites.

Food supplies for herbivores diminish drastically during the dry season when trees shed their leaves and herbaceous plants whither and become dormant. This is illustrated by the foraging ecology of kudus, which consume a foliage diet augmented by fruits and flowers when available (Chapter 10). Towards the end of the wet season while trees retained abundant foliage, kudus needed only 8 steps to be taken for each minute of feeding time; but by the late dry season their movement rate became tripled to 24 steps per minute of feeding, i.e. only a third as much food gained per step taken, despite concentrating where most food remained.[27,28] Kudus restrict their dry season foraging to localities near rivers or around the bases of hills where most evergreen foliage remains. Fruits contribute rather little to their diets during the dry season, after fallen acacia pods had been depleted (Figure 10.5).[29] Kudus are absent from savanna regions where trees retain insufficient leaves, for example from the open umbrella thorn savanna prevalent in Serengeti NP.

Adaptations for bipedal locomotion were apparent in *Ardipithecus*, the earliest unquestionably hominin fossils, after the start of the Pliocene ~5 Ma. They include lengthened legs, shifts in how the hind-limbs articulated with the pelvis, and stiffened rather than dextrous ankle joints. Furthermore, the skull opening connecting the brain to the spine ('foramen magnum') had shifted centrally to support an upright head posture. By 3.6 Ma in the mid-Pliocene, australopiths had become competent bipedal walkers, as shown by the footprints preserved at Laetoli, despite retaining offset big toes (Figure 17.2B). By 1.7 Ma in the early Pleistocene, long-legged *H. ergaster* was as effectively equipped for bipedal walking, and perhaps also for running, as modern humans are.

Dietary Shifts: From Fleshy Fruits to Tough Tubers

As aridity intensified during the Pliocene, trees potentially providing fruits became more widely spaced, and the herbaceous layer between became dominated by grasses. Grass leaves may be sufficiently nutritious while young, even for primates with simple stomachs like geladas, but become more fibrous and resistant to digestion after they mature and turn brown during the dry season. Hence Pliocene hominins needed to look for food below the soil surface, especially during the dry season.

Many of the plants grouped as forbs store carbohydrates underground in the form of bulbs (compacted stem bases) or tubers (swollen roots). Grasses and sedges relocate nutrients in corms (swollen stems) and rhizomes (connecting roots) during the dry season. Underground plant parts form the staple food resource for mole-rats, which forage in tunnels beneath the soil surface, and for porcupines, which dig up bulbs and tubers. Warthogs root for grass rhizomes during the dry season and baboons scratch for corms. Mole-rats first appeared in the fossil record during the Miocene, while porcupines are first recorded late in the Miocene. This indicates that underground storage organs of plants had become sufficiently abundant for the needs of these large rodents by the late Miocene, along with the expansion of grasses. Omnivorous pigs, which use their snouts to dig, diversified a little later during the early Pliocene. The australopiths needed to join this 'rhizophage' guild.[30]

Fossils document the dental trend that australopiths showed towards enlarged molars of low relief with thickened enamel, enabling them to chew and chew, while their canine teeth became reduced so as not to obstruct grinding.[19] *Au. anamensis* had initiated this trend by 4 Ma and it became accentuated in later australopiths. Molar enlargement reached its epitome in the *Paranthropus* forms, which possessed massive jaws with huge molars. These dental features seem adapted especially for processing the 'fallback' foods providing subsistence through the dry season, rather than the staples forming the bulk of the diet during the wet season, which were probably still fruits, seeds and leaves. Nevertheless, greater chewing capabilities might also have been helpful for breaking open fruits with hard shells, like monkey oranges and baobab pods.

Plants with large underground tubers are particularly abundant in sandy savanna regions and less common on volcanic soils.[31] Sedges are especially abundant in wetlands, although dryland forms do exist. Sandy soils intermingled by wetlands are typical of broad-leaved miombo woodlands and coupled there with seasonally abundant fruits. It has been suggested that this dietary mix could best have been obtained within Zambia in south-central Africa, rather than in the rift valley regions of eastern Africa,[32] but quite possibly miombo-type woodlands extended as far north as Ethiopia during the late Miocene and Pliocene while rift valley troughs were still forming.

Besides the form of the teeth, information about the diets consumed by extinct hominins can be gleaned from (1) the form of scratches, pits and other micro-wear on tooth surfaces[33,34] and (2) stable carbon isotope ratios in tooth enamel.[35] From micro-wear abrasions, it can be deduced that *Paranthropus* consumed hard-brittle corms, tubers and seeds. *Paranthropus* fossils are frequently associated with wetland grasslands, which contain abundant sedges, in eastern Africa.[36] The comparatively gracile australopiths, which were around earlier than the robust-jawed ape-men, evidently ate tough but less-hard bulbs mixed with some animal matter.[37,38]

The difference between the heavy and light carbon isotopes ^{13}C and ^{12}C in body tissues indicates the relative portion of the diet derived from C_4 grasses, or from animals eating these grasses (like some rodents, grasshoppers and termites), versus fruits or other plant parts obtained from trees and non-grassy herbs (see Chapter 10). Stable carbon isotope ratios in the teeth of *Au. africanus* indicate that food derived directly or indirectly from C_4 grasses constituted 30–40 percent of its diet on average, very similar to that shown subsequently by *Paranthropus robustus* despite their dental differences.[5,39] More remarkable is the wide individual variation in the C_4 contribution to the diets of both species of ape-man in southern Africa.[5,40] There is also a surprising divergence in the C_4 component between the robust ape-men inhabiting eastern African and those in southern Africa, with a maximum of 80 percent recorded for *P. boisei* at Olduvai.[5,36] The dental micro-wear shown by *Paranthropus* in eastern Africa is also less complex than shown by the robust ape-men from South Africa's Cradle region, suggesting a more leafy diet. The two *Au. sediba* skeletons from around 2 Ma had both consumed solely C_3 plants, although large mammals associated with them had eaten large amounts of C_4 grasses at that time.[41] A reversal of the trend towards robust dentition from the australopiths to *Paranthropus* is a distinguishing feature of the lineage that become *Homo*. Following the emergence of *H. ergaster* there is a dietary shift towards increased consumption of C_4 resources around 1.65 Ma.[42]

Robust-jawed *Paranthropus* spp. persisted alongside forms of early *Homo* for around a million years, fading out after 1.3 Ma in eastern Africa but only by

~1 Ma in southern Africa.[43] It seems that progressive dryness eventually made the 'nutcracker' lifestyle no longer tenable. Nevertheless, porcupines survived into modern times consuming a mix of fruits and underground plant parts in savanna regions; but being smaller they require less food per day than would a hominin.

Homo naledi retained morphological features typical of hominins dated around 2 Ma, and hence must have coexisted alongside the robust australopiths and *H. ergaster* since that time. How might it have differed in its diet? Perhaps it exploited the forb component in Highveld grasslands, supplemented by grass seeds plus grassland insects and rodents as a protein supplement. Large mammals are less abundant in high-altitude grasslands, so making a living as a hunter or scavenger in this savanna variant does not seem a viable proposition. Further insights are needed to resolve its mysterious niche.

Scavenging for Marrow or Meat

Life in savanna became increasingly tough as global cooling intensified dry season aridity. Another opportunity was there to be exploited: gaining from the energy and nutrients contained in the bodies of the large herbivores that were able to eat and digest tough plant parts. Helpfully, these herbivores are most likely to die in the dry season when food and water run short, especially in drought years.

While digging for bulbs and tubers, the early australopiths would have encountered burrows of mole-rats, mice and other rodents, and likely extracted the inhabitants using wooden or bone tools like modern San people do for springhares. Reptiles like tortoises and lizards would also have been found, perhaps singed by the passage of dry season fires, and added to the protein larder. Thus 'faunivory' in some form probably contributed protein to the diet of the australopiths, like it does for chimpanzees and baboons.

While nutritious plant parts became scarcer and harder to process for a primate with the passage of the Pliocene, grazing ruminants became more abundant and diverse. These animals might have seemed tantalisingly out of reach to a primate not much more robust physically than a chimp. However, as noted in Chapter 10, all animals die eventually, either through the agency of a predator or simply from old age or starvation. While searching for food through dry-season landscapes bereft of much vegetation cover, the ape-men would have encountered the carcass remains of large herbivores that had been killed by carnivores, or had starved to death during droughts.[44,45] Large carnivores were especially diverse during the late Pliocene, including 2–3 species of sabretooth cat along with several species of hyenas, some adapted for bone-crushing and others for pursuing prey (Chapter 11). Lions appeared late in the

Pliocene, but did not become common until after the sabretooth cats had faded out in the early Pleistocene. Leopards and cheetahs were, however, represented during the Pliocene. The danger for ape-men from encountering such large carnivores during daylight is not as great as is commonly feared. As ambush predators, the sabretooth cats probably hunted mainly nocturnally as also would have the cursorial hyenas. As long as the ape-men retained capabilities for ascending trees following encounters, they should have been sufficiently secure, especially if they moved in groups.

Modern lions leave little flesh behind on the carcasses of the medium–large ungulates that they preferentially kill, and what is left gets rapidly removed by vultures (Figure 17.4A,B), but limb bones containing fat-rich marrow remain, while skulls enclose fatty brain tissues. Sabretooth cats probably used their enormous canine daggers to kill and dismember thick-skinned animals ('pachyderms'), not only young elephants, rhinos and hippos, but also short- and long-necked giraffes and giant long-horned buffalos, etc. Leopard kills stashed up in trees could also be stolen while the owners were absent. Hyenas can crush limb bones, but would be active mainly nocturnally. Modern spotted hyenas seem not to hunt much in woodlands, perhaps because of the difficulty of prolonged chases there, and the extinct hyenas may also have shown this pattern. Thus, more bones along with some bits of flesh might have remained particularly on carcasses in wooded areas, with climbable trees nearby, and be less readily detected by vultures under tree canopies.[46]

Securing bone marrow, brain tissue, and perhaps meat fragments from carcasses need not entail 'confrontational' scavenging, i.e. driving carnivores from their kills.[44] Left-overs could be gathered towards midday while big cats rested nearby (Figure 17.4C). Hyenas would be long gone back into their dens. Hominins could thus have exploited this temporal window around midday without incurring much risk. Marrow contained within bones would not be putrid, so there was no requirement for meat-eating primates to cope with bacterial toxins. Large herbivores are most readily killed by carnivores during the dry season when plant resources run short, facilitated by concentrations of water-dependent animals around remaining water sources. During extreme droughts when ape-men most desperately needed something else to eat because of the scarcity of plant foods, more carcasses would become available from animals dying of starvation.

How ample would the supply of carcasses have been for the food needs of hominin scavengers getting only what was left behind by large carnivores? Scavenging typically entails long distance travel to find sufficient remains of animals[47] (see Chapter 11). Only birds able to fly far, like vultures, are obligate scavengers. Mammalian scavengers like brown and striped hyenas

Figure 17.4 Carcass remains. (A) Lion feeding on the remains of a zebra, Kruger NP; (B) carcass left behind from lion kill of a buffalo, Luangwa Valley, Zambia; (C) early humans depicted scavenging for meat and bone marrow from a buffalo carcass – although it is unlikely that females and young would have been present at a kill site while hyenas would be in their dens during daylight (© O-M. Nadel 2020).

traverse distances averaging 20–30 km nightly, and they kill mostly small animals encountered opportunistically.[48] Jackals supplement meat obtained from carcasses or from their own kills of small mammals and birds with various fruits.

Based on ungulate population totals in the Serengeti ecosystem and their annual mortality rates estimated during the 1970s, the annual production of large herbivore carcasses was then around 2000 kg/km^2.[47] Medium–large wildebeest and zebra contributed 60 percent of this, of which 30 percent were juveniles. Large carnivores killed 30 percent and scavenged 6 percent of the animals that died. The remaining two-thirds of animal deaths resulted from either starvation or illegal hunting within the national park. Over half of the mortality took place during the three months of the late dry season from August to October. During this time of the year, migratory ungulates were concentrated in the north within about a quarter of the ecosystem. Hence the rate of carcass production during the late dry season would amount to around 5000 kg/km^2 locally, equivalent to 3 wildebeest-sized carcasses per km^2 per week within the dry season range. A practical exploration of the carcass remains available to hominin scavengers that was undertaken in the northern Serengeti woodlands during the mid-dry season yielded five carcasses of animals killed by lions or found dead within a travel distance effectively representing about 10 days of searching, i.e. about one every second day.[49] Two injured or sick animals were also encountered and could have been dispatched. Based on these projections, it seems that the production of large herbivore carcasses could indeed constitute an economically exploitable resource for hominins through the critical dry-season months in parts of the Serengeti ecosystem. Back in the Pliocene, the larger herbivore fauna would have been different, with a greater contribution from very large herbivores and less from smaller ones. This implies that the total herbivore biomass at that time would have been greater than at present, but with less relative turnover of this biomass through mortality. Comparable seasonal concentrations of ungulates also occurred in parts of the Eastern Rift Valley historically.[50] Water limitations would have drawn local concentrations of ungulates more widely through Africa.

Kruger NP in South Africa supports only about half of the herbivore biomass found in Serengeti (Figure 13.1), most of this in the form of elephants and other megaherbivores rarely killed by lions. From herbivore population totals and mortality rates recorded in Kruger NP, potential food for lions, represented by carcasses of potential prey animals dying annually, amounts to 100–150 kg/km^2, equivalent to one wildebeest-sized carcass per km^2 per year.[51] However, most of this mortality is incurred during the dry season when mobile grazers are confined to the proximity of perennial water sources. This raises the local rate of carcass production at least 10-fold, to 10 or more wildebeest equivalents through the last 3 months of the dry season, or one carcass per week within each square kilometre.

Scavenging for marrow and meat would initially have been a fallback response to enable survival through the critical dry-season months when little plant food remains and what remains is mostly of low nutritional value. It was viable because of the locally high biomass of large herbivores sustained by fertile volcanic soils and the seasonal concentrations of grazers near water during the dry season. Furthermore, the bones of the large ungulates that predominated in the Pliocene would have contained more marrow than those of the smaller antelope found today. Scavenging opportunities during the wet season would have been meagre, because carnivore kills during that time of the year are mostly of young animals that get consumed completely. Hominins would have been dependent largely on plant food during this time of the year.

Scavenging for marrow and flesh from carcasses would have had ramifying consequences. The broad, flat-crowned molars of the australopiths had become adapted for chewing tough plants, not macerating raw meat. Early *Homo* initially retained quite large mandibles, but showed a progressive reduction in surface area of its molar teeth and in the thickness of the dental enamel through time. Its dental features became adapted to handle food with a wide range of fracture properties, potentially including bone marrow and flesh.[34,52] No specially worked tools were needed to break open bones to get at the marrow, merely large hammerstones.[53] The fatty tissue obtained from within long bones and inside skulls would have been especially valuable as an energy source to balance the protein provided by the scraps of meat that remained on carcasses.[54] The dental needs for chewing marrow and fragments of meat were less demanding than those required to macerate plant storage organs, reversing the trends towards larger jaws and molar teeth that produced the robust-jawed ape-men.

A further adaptation, not preserved in fossils, was crucial to make scavenging a viable niche for these ape-men. To enable foraging excursions to be carried out during the heat of midday, when hyenas were back in their dens and the big cats were sleepy, a reduction in body hair was necessary to allow sweating to counteract body heat build-up.

Upright walking frees hands to carry bones and other carcass pickings back to secure sites, plus of course also fruits and tubers, but the location of carcass remains would have differed from the places where plant products were most readily gathered. This situation would have promoted a division of labour between males, taking on the danger of confronting carnivores still lurking near carcasses, and females plus their offspring, choosing less-risky places to gather fruits, seeds and roots. Males would need to be in groups to detect carnivores and deter attacks, while females would need to move in groups to be secure. A shared home base would be required to exchange fruits, bones and meat obtained. This central base would need to be located in woodland patches providing trees for shade and for nocturnal sleeping platforms.

It would also need to be within walking distance of year-round surface water, and therefore amid dry-season concentrations of water-dependent ungulates.

This is the ecologically consilient scenario that I propose for the trophic divergence of the lineage leading to *Homo* from that ending in *Paranthropus*, taking place during the Pliocene/Pleistocene transition. Making a living as a facultative scavenger was crucially dependent on a sufficiency of large herbivore carcasses to provide sustenance through intensifying dry seasons.

Becoming a Hunter

The amount of food obtained by scavenging is somewhat meagre for the energy invested, although essential as a supplement enabling ancestral humans to survive through the dry seasons. Could not animals be dispatched before they died from whatever cause? There would surely have been opportunities to do so during severe droughts, when starving herbivores may exceed the consumption capacity of large carnivores.[55] Sharp sticks and big rocks might be sufficient to kill these weakened animals. But how could animal flesh be obtained more reliably over a wider portion of the year? How could ungulates able to outrun hyenas and wild dogs be captured?

The pre-adaptations facilitating the transition from scavenging to hunting were those enabling endurance running capabilities.[56] Lengthened legs and bared skin allowed scavenging excursions to take place during the heat of midday when big cats were lethargic and hyenas had not yet arrived to clean up bones and remaining flesh. With a bit of coordination among group companions, early humans equipped with these features could succeed in chasing down living ungulates by persistently running after them until the targeted animal reached potentially lethal levels of over heating, particularly during midday heat that sparsely hairy hominins could better tolerate (Figure 17.5). Soils bared of much grass during the dry season would permit them to follow the tracks of selected animals until these herbivores came to a standstill. Anatomical adaptations in the shoulder region of early *Homo* became evident around 2.0 Ma, which would assist in the launching of projectiles, probably made from wood, at high speed.[57] At close quarters, these primitive spears could be driven through ribs into vital tissues to dispatch animals that could run no more. If females lacking horns were chosen to chase, there would be little risk of injury to the hunters at this final stage. Hunting in this fashion would have been productive particularly under dry-season conditions when herbivores become weakened and grazers concentrate near remaining water sources. This does not preclude ambushing ungulates from the cover provided by termite mounds or tree clumps alongside trails.[58] These conditions enabled relatively puny upright primates, lacking the teeth,

A

B

Figure 17.5 The transition to hunting by *Homo ergaster*. (A) Chasing after an antelope around midday, enabled by bare skin facilitating cooling by sweating; (B) thrusting spears to bring about the death of the antelope once it can run no further due to overheating (original artwork by Marco Anson).

claws and muscle power of true carnivores, to become effective cursorial hunters.

The composition of *H. habilis* diets in eastern Africa estimated from stable carbon isotope ratios in tooth enamel indicated 25–50 percent to be of C_4 plant origin.[34,36] Because grass seeds and leaves would not provide adequate food for these large omnivores, most of this isotopic signal probably came from tissues of large grazers they had scavenged or, at a later stage, killed.[42,59] A diet consisting almost solely of plant resources eventually failed to suffice for the robust-jawed ape-men, which dropped out of the fossil record after ~1.4 Ma in eastern Africa. The capacity of early *Homo* to exploit animal flesh as a dry-season fallback surely contributed crucially to their persistence.

Whether early humans hunted or scavenged is revealed by whether cut marks on bones overlay tooth marks from carnivores, or vice versa, as well as the kinds of bones assembled and the age profiles of the ungulates represented.[60,61] Butchery marks on bones associated with Oldowan tools at Olduvai dated 1.84 Ma suggest that smaller ungulates had primarily been hunted, whereas larger species were either hunted or scavenged.[62,63] Subsequent bone assemblages show a high frequency of cut marks and other signs indicating prior access by early humans rather than carnivores, and thus that the animals had mostly been hunted.[59,64,65] In addition, marrow had apparently been extracted from the large limb bones of elephants, rhinos and hippos unlikely to have been killed by the early humans.[66] Further evidence of the butchery of large vertebrates comes from Swartkrans Cave in South Africa's Cradle of Humankind dated ~2.0 Ma.[67,68] Where the ungulate species were identified, they mostly showed a predominance of mainly medium to large grazers.[69,70,71] A site at Olduvai where ungulates as large as giant buffalo and short-necked giraffe had been butchered evidently represented circumstances where these animals had been trapped in mud.[72]

Sites in the Afar region of Ethiopia dated 2.6–2.1 Ma have yielded limb bones showing cut marks interpreted as signs of butchery by humans, although whether these were products of scavenging or hunting could not be firmly established.[73,74] Cut and percussion marks from Dikika in Ethiopia dated much earlier at 3.4 Ma[75] are controversial because these marks could have been imposed by crocodiles.[76] No fossil sites exist spanning the period between 2.8 Ma, when the first humans appeared, and 1.9 Ma, when *H. ergaster* emerged, retain adequate bone preservation to indicate whether human rather than carnivore damage was primary.

Increased dependence on hunting would have accentuated the division in food procurement between hunting by males and foraging for plant products primarily by females plus young. Food gathered by each sex would need to be carried back to the home base to be shared at least within families. Signs of a central gathering place where worked bones and stone artefacts accumulated are evident at Olduvai in layers dated ~1.8 Ma.[77] More effective communication would be needed to plan foraging excursions, based on recent or longer experience. The extra social competence required must have contributed to the expansion in cranial capacity from *H. habilis* to *H. ergaster*. However, fossils continued to be assigned to *H. habilis* contemporaneous with *H. ergaster*, until ~1.4 Ma.[7,78] Whether the morphological distinctions between them indicate ecological segregation or a polytypic species will be addressed in Chapter 19.

Evading Predation

The australopiths retained arboreal competence in climbing, as shown by hand and foot bones, coupled with more effective terrestrial locomotion. This meant that they could readily ascend trees at night to gain sufficient security from predation by large cats and hyenas. Both chimpanzees and gorillas routinely spend time constructing new nests in trees around sunset each night. Only mature male gorillas, but neither females nor young, dare to doze at ground level during darkness. All other primates, apart from modern humans, restrict their vulnerability to predation by ascending trees or cliff faces at night.

Arboreal capabilities became compromised with the emergence of *H. ergaster*, generally bigger in height and weight and with relatively longer legs than its predecessor, *H. habilis*. Although modern humans can still climb trees when confronted by lions, we do not ascend very adeptly compared with other primates. During my white rhino study, I experienced the discomfort of being perched in a tree beside a waterhole to observe what animals came down to drink during the night. After squirming there for a few hours, I abandoned this quest and walked back to camp through the darkness despite the danger. Of course, the early hominins would have constructed some form of platform upon which to sleep, whether nightly or permanently can only be guessed. But an essential requirement was trees sufficiently tall, yet still climbable, to elevate them beyond the reach of lions and leopards. This would have restricted their habitat occupation to river margins where such big trees grow.

The metabolic resources needed to grow larger brains delay reproductive maturity and perhaps restrict how much energy is available for digestive processes.[79,80] Dental eruption patterns plus cementum lines show how the life-history stages of hominins became prolonged as brain size expanded.[81] The age when females first gave birth was retarded from 14 years among chimpanzees to between 18 and 20 years, as shown by women living as modern hunter-gatherers.[82] In compensation, humans extend their reproductive period by living much longer than any other primate, and even than elephants. Furthermore, human females cease reproducing at menopause around 40–45 years of age, two or more decades before they die. Other large mammals keep reproducing until the end of their life.

To counteract their reduced reproductive period, women in modern hunter-gatherer societies exhibit a 2–3-year inter-birth interval, shorter than that shown by both chimpanzees and gorillas. The survival of human females beyond the age at which they cease contributing offspring has been ascribed to the substantial benefits provided by grandmothers.[83,84] Rather than raising their own offspring, older females shift their nurturing contributions towards the progeny of their daughters, alleviating the costs incurred by younger females arising from the briefer inter-birth interval.

The age at which reproductive maturity is attained, coupled with maximum longevity, determine the minimal mortality loss that can be sustained by the adult segment of the population. For human hunter-gatherers maturing at 20 years and living beyond 60 years of age, this minimal adult mortality is 2 percent annually. Their relatively short inter-birth interval allows human mothers to produce potentially 8–10 offspring between 20 and 40 years of age, exceeding the seven offspring that chimpanzee mothers contribute during a reproductive lifespan enduring nearly twice as long. This means that human populations can support greater mortality among offspring prior to their incorporation into the adult segment than chimpanzees can.

The population growth rate that could be achieved is restricted by mortality taking place during the adult stage. Annual mortality rates recorded for modern human hunter-gatherers[82] prevent population growth from much exceeding 3 percent per year (Table 17.1). In order for populations to be maintained, predation rates on adult hominins must be half as great as the mortality losses incurred by chimpanzees living in forested regions, despite occupying savanna environments thronged with large ungulates and associated carnivores.[85]

Large ungulates spending every night exposed to predation on the ground, but with sharp senses and honed running capabilities, incur annual mortality rates among prime-aged females amounting to 8 percent or more annually (Chapter 12).[86] How might early humans too big to be secure perched in trees have restricted their adult mortality rate to less than a third of this?

Early humans would have been especially vulnerable to predation in two circumstances: (1) while foraging for left-overs from carcasses not yet abandoned by large carnivores, or when protecting carcasses of their own kills, and (2) while sleeping during the night, once their competence in tree climbing had become compromised in favour of walking. Risks of predation during daylight are not that great. The most threatening carnivores, i.e. lions and leopards, are most active nocturnally, and so probably were sabretooth cats

Table 17.1 Life-history parameters for growing and stable populations of humans, chimpanzees, elephants and white rhinos. For human hunter-gatherers,[82] elephants and white rhinos, values from near-maximally growing populations were adjusted to obtain a likely combination generating zero population growth. For chimpanzees, values derived from a stable population in Kibale Forest, Uganda were adjusted to generate near-maximum population growth. Kudu data are mine. For humans, reproduction terminates at the age of senescence, while for other species reproduction continues at a reduced rate until maximum longevity is reached.

Species	Body mass (kg)	Population growth rate (% pa)	Age 1st reproduction (years)	Age of senescence (years)	Longevity (years)	Inter-birth interval (years)	Annual mortality rate					Survivorship until reproductive age
							Infant	Juvenile	Immature	Prime	Old	
Human	60	3.0	18	40	60	3.0	0.11	0.03	0.02	0.01	0.02	0.58
		0	20	40	60	3.5	0.22	0.06	0.04	0.02	0.04	
Chimpanzee	40	3.8	12	40	50	4	0.11	0.05	0.03	0.02	0.04	0.63
		0	14	40	50	5	0.11	0.025	0.125	0.04	0.08	
Elephant	2800	5.7	10	45	55	4	0.05	0.01	0.005	0.005	0.12	
		0	16	45	55	6	0.3	0.04	0.02	0.02	0.05	0.68
White rhino	1800	9.0	6	35	45	2.5	0.08	0.035	0.015	0.005	0.05	
		0	10	35	45	4	0.24	0.08	0.06	0.03	0.08	0.68
Greater kudu	180	0.27	2	10	15	2	0.10	-	0.05	0.02	0.03	
		0	3	6	15	2	0.52	-	0.15	0.085	0.12	

and hunting hyenas. During the 3.5 years I spent wandering on foot after rhinos in the Mfolozi GR, I was never directly threatened by a carnivore on any of the occasions when I met one. Scientists studying baboons spend countless days on foot among troops, without any of them being injured or killed by a predator, so far as I know. Modern hunter-gatherers like the San and Hadza seldom fall prey to lions, leopards or hyenas, despite commonly sleeping outside the protection of huts at night (see Chapter 18).

Risks are greatly elevated at night. Human refugees from Mozambique traversing Kruger NP to enter South Africa, and forced to sleep out overnight during the journey, were all too frequently eaten by lions. When other prey runs short, lions may switch their hunting to humans.[87] How did members of *H. ergaster* bands avoid being killed by lions or leopards during the hours of darkness? Did they construct durable sleeping platforms in trees at each temporarily occupied home base? Or were they able instead to build impenetrable barriers of thorny branches around overnight abodes at ground level each night? Notably, fossils ascribed to *Au. sediba* retained long curved fingers and wrist bones that would aid arboreal climbing. Specimens assigned to *H. habilis* also had long arms that would facilitate climbing. The robust-jawed *Paranthropus* also retained greater adaptive competence for tree-climbing than that shown by *H. ergaster*.

Carnivores sneaking up at night could be deterred by throwing rocks at them; but it would help to know where to aim. Could these earliest humans have had the benefit of campfires to illuminate the darkness? The contentious issue I am raising is whether the active deployment of fire to light the darkness, render meat and tough plant parts chewable, and keep sufficiently warm in the absence of much fur, helped foster the emergence of *Homo ergaster* after 1.8 Ma, despite the lack of much evidence for the use of fire until thousands of years later.[88,89,90]

The earliest evidence claimed to show the deployment of fire comes from burnt bones found in Swartkrans Cave in South Africa[91] and from baked surfaces of volcanic soils associated with stone tools in northern Kenya,[92] both dated ~1.6 Ma.[93] Burnt bones have also been recorded at Koobi Fora near Lake Turkana dated around 1.5 Ma. The earliest evidence of fire used for cooking in hearths is in Wonderwerk Cave in the Northern Cape of South Africa, dated to ~1 Ma,[94] and from numerous sites in Eurasia from around the same time.[93] Most likely, debris from open-air fires would be indistinct from that left by the grassland fires that recur within savannas. Lightning-ignited fires would have presented opportunities to transport smouldering logs back to home bases. Over time, humans should have learnt how to ignite fires by rotating sticks within cavities drilled in soft wood, even during the wet season when wood is mostly wet.

Overview

The ecologically compatible narrative covering the transition from frugivorous forest-dwelling ape to partially meat-eating, savanna-inhabiting human takes this form. The initial adaptive changes shown by evolving hominins in response to expanding savanna vegetation during the late Miocene were towards bipedal locomotion. Primates occupying savannas needed to walk further to cope with the widened spacing between trees and seasonally restricted production of fruits and young leaves on these trees. This also freed hands for carrying resistant plant parts to secure places where they could be processed for consumption.

Following closely in time were dental adaptations for coping with tough plant parts sought underground between the trees once fruits were no longer available year-round. Molar teeth for chewing expanded in surface area while incisor teeth used for biting into fruits became reduced. This adaptive trend, initiated during the Pliocene, ultimately culminated in the robust-jawed ape-men around the start of the Pleistocene. They continued to exist for well over a million years, disappearing later in southern Africa than recorded in eastern Africa. Hence falling back on tough plant parts for bridging dry-season shortages eventually faded as a viable niche, except among baboons and rodents able to get by on lesser amounts of food.

Around the commencement of the Pleistocene ice ages, a diverging lineage reversed the trend towards greater dental robustness by adding marrow-containing bones scavenged from carnivore kills to its diet, promoting greater dietary versatility. While leaves contract in edibility and fruits in availability during the dry season, carcasses of herbivores killed or dying of starvation are concentrated during this lean period. Furthermore, a temporal window could be exploited around midday when heat inhibited predator activity. Hyenas had gone to dens and the big cats had ceased feeding, or at least defending what remained of their kills. Consequently, emerging humans incurred a reduction in hair cover over most of the body except on top of the head, facilitating evaporative cooling. Enlarging brains and lengthening legs supported the transition from ape-men to earliest humans. Multi-mode foraging favoured the expansion in brain size to cope with the organisational complexity. When and how these adaptive shifts took place during the period between 2.9 and 1.8 Ma, spanning the climatic transition into the early Pleistocene, remains obscure.

These adaptations for scavenging opened a new window of opportunity, leading early humans to become superior to any ungulate in endurance running capabilities. The latter remained handicapped by their fur cover and resultant over heating. Scavenging became augmented by active hunting of increasingly large prey, enabling greater reliability in food procurement through the dry season. However, lengthened legs and larger body size handicapped tree-climbing capabilities and thereby safety from predators. It seems that the deployment of fire became necessary around this time to deter predators, as well as providing warmth and tenderising meat and plant tubers.

The ecological circumstances that enabled hominins to become scavengers on animal tissues and later function as hunters were (1) high abundance of large mammalian herbivores, found especially in dry/eutrophic savannas in semi-arid regions adjoining the African rift valleys where volcanic influences are pervasive; (2) predominance of medium to large grazers in these faunal assemblages, providing bones of manageable size for carrying back from kills; (3) restricted surface water sources, generating concentrations of water-dependent grazers during the dry season when plant foods are most deficient in their availability; (4) thermal tolerance, enabling temporal partitioning from scavenging hyenas; and (5) endurance running capabilities, enabling humans to chase more densely furry ungulates to a standstill. Still unresolved is the stage at which fire became deployed, crucially not only for cooking but also to illuminate lurking carnivores. The fossil dating and ecological contexts seem consistent with the evolutionary divergence of the lineage leading to *Homo* taking place earliest in eastern Africa and dispersing southward from there to form sister-species in South Africa.

This transition from obligate rhizophage to facultative meat-eater was both necessary and opportunistic: necessary for survival through dry seasons, without specialised chewing and digestive adaptations; and opportunistic, exploiting the concentrations of large grazing ungulates that developed around remaining water sources during the late dry season. This lifestyle enabled *H. ergaster* to persist with only minor changes in adaptive morphology for around a million years, from 1.9 to 0.9 Ma, before the surge in brain capacity into our immediate human predecessors took place. While features of bones and teeth changed little, important adaptive changes were taking place in the artefacts supporting this hunting and gathering lifestyle. This cultural evolution will be the subject of the chapter that follows.

Suggested Further Reading

Klein, RG. (2009) *The Human Career: Human Biological and Cultural Origins*, 3rd ed. University of Chicago Press, Chicago.

Thompson, JC, et al. (2019) Origins of the human predatory pattern: the transition to large-animal exploitation by early humans. *Current Anthropology* 60:1–23.

Wood, B; Leakey, M. (2011) The Omo–Turkana basin fossil hominins and their contribution to our understanding of human evolution in Africa. *Evolutionary Anthropology* 20:264–292.

References

1. MacLatchy, LM. (2010) Hominini. In Werdelin, L; Sanders, WJ (eds) *Cenozoic Mammals of Africa*. University of California Press, Berkeley, pp. 471–540.
2. Bobe, R, et al. (2020) The ecology of *Australopithecus anamensis* in the early Pliocene of Kanapoi, Kenya. *Journal of Human Evolution* 140:102717.
3. Wynn, JG. (2000) Paleosols, stable carbon isotopes, and paleoenvironmental interpretation of Kanapoi, Northern Kenya. *Journal of Human Evolution* 39:411–432.
4. Reed, KE. (2008) Paleoecological patterns at the Hadar hominin site, Afar regional state, Ethiopia. *Journal of Human Evolution* 54:743–768.
5. Sponheimer, M, et al. (2013) Isotopic evidence of early hominin diets. *Proceedings of the National Academy of Sciences of the United States of America* 110:10513–10518.
6. Wood, B; Leakey, M. (2011) The Omo–Turkana Basin fossil hominins and their contribution to our understanding of human evolution in Africa. *Evolutionary Anthropology: Issues, News, and Reviews* 20:264–292.
7. Bobe, R; Carvalho, S. (2019) Hominin diversity and high environmental variability in the Okote Member, Koobi Fora Formation, Kenya. *Journal of Human Evolution* 126:91–105.
8. Clarke, RJ; Kuman, K. (2019) The skull of StW 573, a 3.67 Ma *Australopithecus prometheus* skeleton from Sterkfontein caves, South Africa. *Journal of Human Evolution* 134:102634.
9. Crompton, RH, et al. (2018) Functional anatomy, biomechanical performance capabilities and potential niche of StW 573: an *Australopithecus skeleton* (circa 3.67 Ma) from Sterkfontein Member 2, and its significance for the last common ancestor of the African apes and for Hominin origins. *bioRxiv*:481556.
10. Pickering, R, et al. (2019) U–Pb-dated flowstones restrict South African early hominin record to dry climate phases. *Nature* 565:226–229.
11. Berger, LR, et al. (2010) *Australopithecus sediba*: a new species of *Homo*-like australopith from South Africa. *Science* 328:195–204.
12. de Ruiter, DJ, et al. (2017) Late australopiths and the emergence of *Homo*. *Annual Review of Anthropology* 46:99–115.
13. Villmoare, B, et al. (2015) Early *Homo* at 2.8 Ma from Ledi-Geraru, Afar, Ethiopia. *Science* 347:1352–1355.
14. DiMaggio, EN, et al. (2015) Late Pliocene fossiliferous sedimentary record and the environmental context of early *Homo* from Afar, Ethiopia. *Science* 347:1355–1359.
15. Robinson, JR; Rowan, J. (2017) Holocene paleoenvironmental change in southeastern Africa (Makwe Rockshelter, Zambia): implications for the spread of pastoralism. *Quaternary Science Reviews* 156:57–68.
16. Bennett, MR, et al. (2009) Early hominin foot morphology based on 1.5-million-year-old footprints from Ileret, Kenya. *Science* 323:1197–1201.
17. Antón, SC, et al. (2014) Evolution of early *Homo*: an integrated biological perspective. *Science* 345:1236828.

18. Herries, AIR, et al. (2020) Contemporaneity of *Australopithecus, Paranthropus,* and early *Homo erectus* in South Africa. *Science* 368:eaaw7293.
19. Teaford, MF; Ungar, PS. (2000) Diet and the evolution of the earliest human ancestors. *Proceedings of the National Academy of Sciences of the United States of America* 97:13506–13511.
20. Maslin, M. (2016) *The Cradle of Humanity: How the Changing Landscape of Africa Made Us So Smart.* Oxford University Press, Oxford.
21. Stringer, C. (2016) The origin and evolution of *Homo sapiens. Philosophical Transactions of the Royal Society B: Biological Sciences* 371:20150237.
22. Hublin, J-J, et al. (2017) New fossils from Jebel Irhoud, Morocco and the pan-African origin of *Homo sapiens. Nature* 546:289–292.
23. Grün, R, et al. (2020) Dating the skull from Broken Hill, Zambia, and its position in human evolution. *Nature* 580:372–375.
24. Rightmire, GP. (1983) The Lake Ndutu cranium and early *Homo sapiens* in Africa. *American Journal of Physical Anthropology* 61:245–254.
25. Dirks, PHGM, et al. (2017) The age of *Homo naledi* and associated sediments in the Rising Star Cave, South Africa. *Elife* 6:e24231.
26. Berthaume, MA, et al. (2018) Dental topography and the diet of *Homo naledi. Journal of Human Evolution* 118:14–26.
27. Owen-Smith, N. (1979) Assessing the foraging efficiency of a large herbivore, the kudu. *South African Journal of Wildlife Research* 9:102–110.
28. Owen-Smith, RN. (2002) *Adaptive Herbivore Ecology: From Resources to Populations in Variable Environments.* Cambridge University Press, Cambridge.
29. Owen-Smith, N; Cooper, SM. (1989) Nutritional ecology of a browsing ruminant, the kudu (*Tragelaphus strepsiceros*), through the seasonal cycle. *Journal of Zoology* 219:29–43.
30. Laden, G; Wrangham, R. (2005) The rise of the hominids as an adaptive shift in fallback foods: plant underground storage organs (USOs) and australopith origins. *Journal of Human Evolution* 49:482–498.
31. Marean, CW. (1997) Hunter–gatherer foraging strategies in tropical grasslands: model building and testing in the East African Middle and Later Stone Age. *Journal of Anthropological Archaeology* 16:189–225.
32. O'Brien, EM; Peters, CR. (1999) Landforms, climate, ecogeographic mosaics, and the potential for hominid diversity in Pliocene Africa. In Bromage, TG; Schrenk, F (eds) *African Biogeography, Climate Change and Human Evolution.* Oxford University Press, Oxford, pp. 115–137.
33. Ungar, PS; Sponheimer, M. (2011) The diets of early hominins. *Science* 334:190–193.
34. Ungar, PS. (2012) Dental evidence for the reconstruction of diet in African early *Homo. Current Anthropology* 53:S318–S329.
35. Sponheimer, M; Lee-Thorp, JA. (2003) Differential resource utilization by extant great apes and australopithecines: towards solving the C_4 conundrum. *Comparative Biochemistry and Physiology Part A: Molecular & Integrative Physiology* 136:27–34.
36. Van der Merwe, NJ, et al. (2008) Isotopic evidence for contrasting diets of early hominins *Homo habilis* and *Australopithecus boisei* of Tanzania. *South African Journal of Science* 104:153–155.

37. Scott, RS, et al. (2005) Dental microwear texture analysis shows within-species diet variability in fossil hominins. *Nature* 436:693–695.
38. Dominy, NJ, et al. (2008) Mechanical properties of plant underground storage organs and implications for dietary models of early hominins. *Evolutionary Biology* 35:159–175.
39. Sponheimer, M; Lee-Thorp, JA. (1999) Oxygen isotopes in enamel carbonate and their ecological significance. *Journal of Archaeological Science* 26:723–728.
40. Sponheimer, M, et al. (2006) Isotopic evidence for dietary variability in the early hominin *Paranthropus robustus*. *Science* 314:980–982.
41. Henry, AG, et al. (2012) The diet of *Australopithecus sediba*. *Nature* 487:90–93.
42. Patterson, DB, et al. (2019) Comparative isotopic evidence from East Turkana supports a dietary shift within the genus *Homo*. *Nature Ecology & Evolution* 3:1048–1056.
43. Gibbon, RJ, et al. (2014) Cosmogenic nuclide burial dating of hominin-bearing Pleistocene cave deposits at Swartkrans, South Africa. *Quaternary Geochronology* 24:10–15.
44. Blumenschine, RJ, et al. (1987) Characteristics of an early hominid scavenging niche [and comments and reply]. *Current Anthropology* 28:383–407.
45. Blumenschine, RJ; Cavallo, JA. (1992) Scavenging and human evolution. *Scientific American* 267:90–97.
46. Domínguez-Rodrigo, M. (2001) A study of carnivore competition in riparian and open habitats of modern savannas and its implications for hominid behavioral modelling. *Journal of Human Evolution* 40:77–98.
47. Houston, DC. (1979) The adaptations of scavengers. In Sinclair, ARE; Norton-Griffiths, M (eds) *Serengeti: Dynamics of an Ecosystem*. University of Chicago Press, Chicago, pp. 263–286.
48. Mills, MGL. (1990) *Kalahari Hyaenas*. Unwin Hyman, London.
49. Schaller, GB; Lowther, GR. (1969) The relevance of carnivore behavior to the study of early hominids. *Southwestern Journal of Anthropology* 25:307–341.
50. Ogutu, JO, et al. (2012) Dynamics of ungulates in relation to climatic and land use changes in an insularized African savanna ecosystem. *Biodiversity and Conservation* 21:1033–1053.
51. Owen-Smith, N; Mills, MGL. (2006) Manifold interactive influences on the population dynamics of a multispecies ungulate assemblage. *Ecological Monographs* 76:73–92.
52. Teaford, MF, et al. (2002) Paleontological evidence for the diets of African Plio-Pleistocene hominins with special reference to early *Homo*. In Ungar, PS; Teaford, MF (eds) *Human Diet: Its Origin and Evolution*. Bergin & Garvey, Westport, pp. 143–166.
53. Thompson, JC, et al. (2019) Origins of the human predatory pattern: the transition to large-animal exploitation by early hominins. *Current Anthropology* 60:1–23.
54. Pobiner, BL. (2020) The zooarchaeology and paleoecology of early hominin scavenging. *Evolutionary Anthropology: Issues, News, and Reviews* 29:68–82.
55. Cooper, SM, et al. (1999) A seasonal feast: long-term analysis of feeding behaviour in the spotted hyaena (*Crocuta crocuta*). *African Journal of Ecology* 37:149–160.
56. Bramble, DM; Lieberman, DE. (2004) Endurance running and the evolution of *Homo*. *Nature* 432:345–352.

57. Roach, NT, et al. (2013) Elastic energy storage in the shoulder and the evolution of high-speed throwing in *Homo*. *Nature* 498:483–486.
58. Bunn, HT; Gurtov, AN. (2014) Prey mortality profiles indicate that Early Pleistocene *Homo* at Olduvai was an ambush predator. *Quaternary International* 322:44–53.
59. Domínguez-Rodrigo, M, et al. (2014) On meat eating and human evolution: a taphonomic analysis of BK4b (Upper Bed II, Olduvai Gorge, Tanzania), and its bearing on hominin megafaunal consumption. *Quaternary International* 322:129–152.
60. Domínguez-Rodrigo, M. (2002) Hunting and scavenging by early humans: the state of the debate. *Journal of World Prehistory* 16:1–54.
61. Domínguez-Rodrigo, M; Pickering, TR. (2003) Early hominid hunting and scavenging: a zooarcheological review. *Evolutionary Anthropology: Issues, News, and Reviews* 12:275–282.
62. Parkinson, JA. (2018) Revisiting the hunting-versus-scavenging debate at FLK Zinj: a GIS spatial analysis of bone surface modifications produced by hominins and carnivores in the FLK 22 assemblage, Olduvai Gorge, Tanzania. *Palaeogeography, Palaeoclimatology, Palaeoecology* 511:29–51.
63. Pante, MC, et al. (2018) The carnivorous feeding behavior of early *Homo* at HWK EE, Bed II, Olduvai Gorge, Tanzania. *Journal of Human Evolution* 120:215–235.
64. Pobiner, BL, et al. (2008) New evidence for hominin carcass processing strategies at 1.5 Ma, Koobi Fora, Kenya. *Journal of Human Evolution* 55:103–130.
65. Oliver, JS, et al. (2019) Bovid mortality patterns from Kanjera South, Homa Peninsula, Kenya and FLK-Zinj, Olduvai Gorge, Tanzania: evidence for habitat mediated variability in Oldowan hominin hunting and scavenging behavior. *Journal of Human Evolution* 131:61–75.
66. Organista, E, et al. (2019) Taphonomic analysis of the level 3b fauna at BK, Olduvai Gorge. *Quaternary International* 526:116–128.
67. Pickering, TR, et al. (2008) Testing the 'shift in the balance of power' hypothesis at Swartkrans, South Africa: hominid cave use and subsistence behavior in the Early Pleistocene. *Journal of Anthropological Archaeology* 27:30–45.
68. Kuman, K, et al. (2018) The Oldowan industry from Swartkrans cave, South Africa, and its relevance for the African Oldowan. *Journal of Human Evolution* 123:52–69.
69. Ferraro, JV, et al. (2013) Earliest archeological evidence of persistent hominin carnivory. *PLoS One* 8:e62174.
70. Patterson, DB, et al. (2017) Ecosystem evolution and hominin paleobiology at East Turkana, northern Kenya between 2.0 and 1.4 Ma. *Palaeogeography, Palaeoclimatology, Palaeoecology* 481:1–13.
71. Van Pletzen-Vos, L, et al. (2019) Revisiting Klasies River: a report on the large mammal remains from the Deacon excavations of Klasies River main site, South Africa. *South African Archaeological Bulletin* 74:127–137.
72. Organista, E, et al. (2016) Did *Homo erectus* kill a *Pelorovis* herd at BK (Olduvai Gorge)? A taphonomic study of BK5. *Archaeological and Anthropological Sciences* 8:601–624.

73. De Heinzelin, J, et al. (1999) Environment and behavior of 2.5-million-year-old Bouri hominids. *Science* 284:625–629.
74. Domínguez-Rodrigo, M, et al. (2005) Cutmarked bones from Pliocene archaeological sites at Gona, Afar, Ethiopia: implications for the function of the world's oldest stone tools. *Journal of Human Evolution* 48:109–121.
75. McPherron, SP, et al. (2010) Evidence for stone-tool-assisted consumption of animal tissues before 3.39 million years ago at Dikika, Ethiopia. *Nature* 466:857–860.
76. Sahle, Y, et al. (2017) Hominid butchers and biting crocodiles in the African Plio–Pleistocene. *Proceedings of the National Academy of Sciences of the United States of America* 114:13164–13169.
77. Dominguez-Rodrigo, M. (2009) Are all Oldowan sites palimpsests? If so, what can they tell us about hominin carnivory? In Hovers, E; Braun, DR (eds) *Interdisciplinary Approaches to the Oldowan*. Springer, Berlin, pp. 129–147.
78. Antón, SC; Snodgrass, J. (2012) Origins and evolution of genus *Homo*: new perspectives. *Current Anthropology* 53:S479–S496.
79. Aiello, LC; Wheeler, P. (1995) The expensive-tissue hypothesis: the brain and the digestive system in human and primate evolution. *Current Anthropology* 36:199–221.
80. Isler, K; van Schaik, CP. (2009) The expensive brain: a framework for explaining evolutionary changes in brain size. *Journal of Human Evolution* 57:392–400.
81. Schwartz, GT. (2012) Growth, development, and life history throughout the evolution of *Homo*. *Current Anthropology* 53:S395–S408.
82. Kaplan, H, et al. (2000) A theory of human life history evolution: diet, intelligence, and longevity. *Evolutionary Anthropology: Issues, News, and Reviews* 9:156–185.
83. Hawkes, K, et al. (1998) Grandmothering, menopause, and the evolution of human life histories. *Proceedings of the National Academy of Sciences of the United States of America* 95:1336–1339.
84. Hawkes, K, et al. (2003) Human life histories: primate tradeoffs, grandmothering socioecology, and the fossil record. In Kappela, P; Pereira, ME (eds) *Primate Life Histories and Socioecology*. University of Chicago Press, Chicago, pp. 204–227.
85. Meindl, RS, et al. (2018) Early hominids may have been weed species. *Proceedings of the National Academy of Sciences of the United States of America* 115:1244–1249.
86. Owen-Smith, N, et al. (2005) Correlates of survival rates for 10 African ungulate populations: density, rainfall and predation. *Journal of Animal Ecology* 74:774–788.
87. Packer, C, et al. (2011) Fear of darkness, the full moon and the nocturnal ecology of African lions. *PLoS One* 6:e22285.
88. Wrangham, RW, et al. (1999) The raw and the stolen: cooking and the ecology of human origins. *Current Anthropology* 40:567–594.
89. Wrangham, R. (2017) Control of fire in the Paleolithic: evaluating the cooking hypothesis. *Current Anthropology* 58:S303–S313.
90. Hlubik, S, et al. (2019) Hominin fire use in the Okote member at Koobi Fora, Kenya: new evidence for the old debate. *Journal of Human Evolution* 133:214–229.

91. Brain, CK; Sillent, A. (1988) Evidence from the Swartkrans cave for the earliest use of fire. *Nature* 336:464–466.
92. Bellomo, RV. (1994) Methods of determining early hominid behavioral activities associated with the controlled use of fire at FxJj 20 Main, Koobi Fora, Kenva. *Journal of Human Evolution* 27:173–195.
93. Gowlett, JAJ. (2016) The discovery of fire by humans: a long and convoluted process. *Philosophical Transactions of the Royal Society B: Biological Sciences* 371:20150164.
94. Berna, F, et al. (2012) Microstratigraphic evidence of in situ fire in the Acheulean strata of Wonderwerk Cave, Northern Cape province, South Africa. *Proceedings of the National Academy of Sciences of the United States of America* 109:E1215–E1220.

Chapter 18: Cultural Evolution: From Tools to Art and Genes

Humans had established their physical form with the emergence of long-legged, large-brained *Homo ergaster/erectus* by 1.7 Ma in the early Pleistocene. Apart from a modest expansion in brain volume (Figure 18.1),[1] little further change in morphology occurred over almost the next million years. Then, after 0.9 Ma, a marked surge in cranial capacity towards modern levels took place, accompanied by changes in the shape of the skull, in Europe as well as in Africa. This was around the time when the period between glacial peaks lengthened to ~100 kyr, producing even greater extremes of cold and aridity. A new species name was applied, shared between the two continents: *H. heidelbergensis*. The European lineage evolved into the stockily built Neanderthal people (*H. neanderthalensis*), with a brain volume matching that of modern humans. In Africa, the skull had assumed almost modern form by ~200 ka. While the brain vault of the Neanderthals extended backwards from a sloping forehead, that of the early modern humans domed upwards above a flattened face with reduced brow ridges. For whatever reasons, modern humans also developed a protruding chin.

While there are few fossils to document changes in human skulls, the products of mental competence in the form of stone artefacts are abundant throughout Africa. Hence the focus of this chapter shifts from paleoanthropology to archaeology – the study of evolving forms of the cultural material left behind by people, rather than their bones and teeth. In some places worked stones are accompanied by fossilised bones of ungulates, indicating how these artefacts had been used as tools to process herbivore remains for consumption of marrow within and meat attached. A new form of tool use emerged around 70 ka when geometric scratches on stones suggest artistic expression. Changes in the shape of some stone flakes indicate a cultural advance with more profound consequences: the hafting of stone points onto spears and later arrows.

People left Africa, in two waves. A minor one preceding 100 ka took humans only as far as the Levant in the Middle East. The second major dispersal, beginning around 60 ka, spread *H. sapiens* through other continents, replacing the Neanderthals and other earlier human inhabitants. Later, some

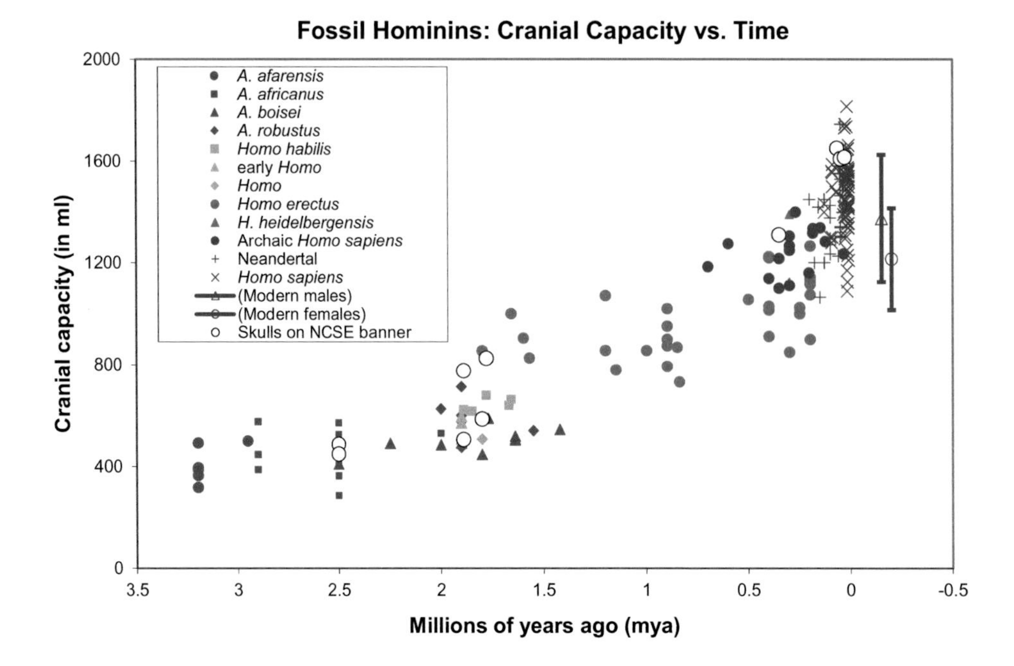

Figure 18.1 Progressive expansion in brain volume of hominins through the Pleistocene (chart by Nick Matzke at www.ncseweb.org from phylo.wikidot.com/fun-with-hominin-cranial-capacity-through-time).

descendants of these people returned to Africa, bringing domesticated ungulates and cultivated plants. Herding and farming lifestyles spread through eastern and southern Africa, displacing stone-age hunter-gatherers. Genetic markers indicate some of the movements of various groups of people within Africa.

In this chapter, I will outline the adaptive shifts in cultural expression in the form of stone tools, rock art, and lifestyles that took place within savanna regions of Africa. These cultural advances enabled humans to persist and expand in numbers through times when several animal forms that had survived prior climatic oscillations faded into extinction.

From Earlier to Later Stone-working Technology

The earliest generally accepted stone artefacts come from the Awash valley in north-eastern Ethiopia, dated to 2.6 Ma,[2] slightly pre-dating other worked stones found nearby.[3] This dating follows shortly after the appearance of the earliest fossil assigned to *Homo*, a mandible found in the same region of Ethiopia from 2.83 Ma.[4] These early stone artefacts take the form of unretouched flakes with sharp edges struck from large cores (Figure 18.2A).[5]

Figure 18.2 Stone tool industries. (A) Oldowan core and flakes; (B) Acheulian tools bifacially shaped; (C) Middle Stone Age finely crafted blades and points (images supplied by Human Origins Program, NMNH, Smithsonian Institution).

This form of stone working was first discovered in association with fossils assigned to *Homo habilis* in the lowest beds at Olduvai Gorge dated to ~2 Ma and accordingly named the Oldowan technology. Stone tools of similar antiquity have been found in the Cradle of Humankind as well as in Wonderwerk Cave in South Africa[6,7] and as far north as Algeria.[8]

A date of 3.3 Ma has been claimed for stones discovered west of Lake Turkana in Kenya, proposed to represent an even earlier 'Lomekwian' culture.[9] These stones, some as heavy as 15 kg, were accompanied by large flakes broken from them. However, this exceptionally early date 700 kyr before the next most recent artefacts has been challenged, questioning whether they were really found 'in situ'.[10,11] Animal bones found nearby do not show signs of percussion, reinforcing doubt about their use as tools by hominins who were still in the ape-man stage. No stone tools have been found in association with

Au. africanus in South Africa dated earlier than 2.2 Ma.[7] Nevertheless, early ape-men probably did use unworked stones to pound tough plant parts and, at some stage, to break open animal bones to extract the marrow.[12] Chimpanzees and capuchin monkeys employ stones to pound nuts.[13] Unworked stones applied for such purposes would not be distinguishable from other rocks lying around. The deployment of stones to soften plant and perhaps animal food resources would have contributed to the trend towards less-robust dentition diverging the lineage evolving into *Homo* from that giving rise to *Paranthropus* around 2.9 Ma.

The inferred use of the Oldowan flakes was to scrape the flesh from the bones of animals scavenged, while cores (cobblestones) were probably used to crack open limb bones so as to extract the marrow. Although robust-jawed ape-men were around at that time, it is doubtful that they were the makers of the stone flakes. A stick or a bone would rather be used to dig up underground plant parts. Bones with wear indicating that they were used for digging have been found in Kromdraai Cave in South Africa where the only hominin fossils present were *Paranthropus*, dated shortly after 2.0 Ma.[14]

Around 1.7 Ma, a distinct technology named the Acheulian appeared almost synchronously in the West Turkana region of Kenya, at Olduvai and in South Africa.[15,16,17] This was associated with the appearance of *H. ergaster* in Africa and *H. erectus* in Eurasia.[18] Acheulian stone artefacts feature cores shaped bifacially (on both sides) to serve as large cutting tools, like handaxes, cleavers or picks. Large flakes knapped from the cores were modified for additional functions (Figure 18.2B). At Olduvai Gorge and elsewhere, the Acheulian technology made its appearance quite abruptly, alongside Oldowan tools, without intermediate forms, suggesting ingress by the tool makers.[19] Oldowan tools tend to be concentrated at sites together with aggregations of large herbivore fossils, while Acheulian tools are generally more widely scattered.[20] The presence of cut marks beneath tooth marks on limb bones of ungulates has been interpreted to mean that carcasses were processed intact and thus had been obtained by hunting rather than scavenged.[21] Hence the appearance of the Acheulian tool assemblages suggests a shift towards hunting as the primary means of obtaining meat in the diet. Both stone cultures coexisted for several thousand years before the Oldowan faded out. Together, the Oldowan and Acheulian industries define the Earlier Stone Age (ESA) of Africa.

Between 600 and 200 ka, flakes became more finely shaped for a greater variety of uses, defining the progressive transition into the Middle Stone Age (MSA; Figure 18.2C).[17] During this period, humans attained almost fully modern appearance. Faunal assemblages were consistently dominated by medium–large ungulates, from wildebeest to buffalo size.[22] The frequency of human cut marks on bones relative to marks inflicted by carnivore teeth increased. In the south-western Cape, animals butchered included prime-aged wildebeest while old and young animals predominated among buffalo, suggesting that the latter had been scavenged from predator kills.

The crucial period between 500 and 320 ka leading towards modern humans is missing from sedimentary layers both at Olorgesaile and Lake Baringo in Kenya.[23] This included a time of extreme aridity around 440 ka followed by a rapid transition into an interglacial peaking around 400 ka (Figure 19.1). A sediment core drilled from a lake basin south of Olorgesaile documents heightened environmental variability after 500 ka with chloridoid grasses present near a shallow saline lake during the glacial maximum.[24] The herbivore fauna had shifted towards smaller species with more browsers and mixed feeders. MSA sites dated earlier than 195 ka are scarce in eastern Africa and mostly situated on the margins of rivers or lakes in rift valley settings.[25] After 130 ka, when milder interglacial conditions ensued, sites with tools become more widely spread on hillsides and in caves and tools became more varied.[26] Artefacts representing MSA technology have been found north of the Sahara fringing the Mediterranean Sea.[27] The earliest stone tools assigned to the Later Stone Age (LSA) of eastern Africa come from Munga Cave, dated before 50 ka.[28] By 25 ka, when the most recent glacial advance neared its maximum, LSA technology had become widely established across numerous sites in eastern Africa.[29]

In South Africa, the transition into the MSA is represented by the 'Fauresmith' industry, including finely worked flakes and points but fewer large cutting tools.[30] Blades with sharp cutting edges were hafted onto wooden shafts to serve as spear points, an innovation recorded as early as ~500 ka at Kathu Pan in north-western South Africa.[31] The Florisbad skull dated roughly to 260 Ma confirms the presence of early modern humans in interior South Africa during this period. However, very few archaeological sites in South Africa, and none in neighbouring Botswana,[32] document the period between 200 and 130 ka when glacial conditions generated extreme aridity. However, this time span is represented at Makapansgat, located in the warmer northern bushveld and at Border Cave on the southern Swaziland border, not far inland from the coast.[33]

Along the southern Cape coast, various cave sites preserve a rich archaeological record. At Pinnacle Point this goes back as early as 160 ka,[34,35] but elsewhere it spans only the period after 130 ka when conditions were warming. The lowering of sea level during the cold extreme prior to 130 ka had exposed the grassy Agulhas plain stretching up to 100 km beyond the current shoreline and supporting concentrations of grazing ungulates. The caves provided shelter from the local winter wetness and a reliable protein source year-round in the form of shellfish plus seals either scavenged or hunted.[35,36,37] Deposits within these caves have yielded bone tools, as well as geometric engravings on soft stones of ochre, dated earlier than 70 ka. Other artefacts include shells strung into necklaces, beads and patterns carved on bones. Finely constructed ('microlith') projectile points made of flint or bone became prominent in deposits dated to 71 ka. By 65 ka, ostrich eggs became used as vessels to carry water. Lower deposits contain the remains of numerous large antelope, while upper layers show a shift towards more medium- or small-sized ungulates, as the coastal plain contracted.[38] The relative representation of browsers also increased.[39] The surrounding vegetation includes numerous plants with underground storage organs and there are signs that such starchy foods were cooked.[40] Following submergence of the coastal plain by rising sea levels after 15 ka during the Holocene interglacial, the meat component of the diet of the cave inhabitants consisted mainly of shellfish along with tortoises.[35]

In South Africa, stone artefacts representing the later MSA have been assigned to successive stages labelled Klasies River (115–80 ka), Still Bay (80–65 ka) and Howieson's Poort (70–65 ka) industries.[41] At Border Cave, MSA technology continued to be in use after 130 ka. In Sibudu Cave located close to the KwaZulu-Natal coast, arrow points made of stone or bone are dated earlier than 60 ka and show signs that poison was applied to their tips.[42,43] Similarly daubed bone points have been identified at Pinnacle Point dated 71 ka.[44] The earliest record for the thin blades initiating the LSA come from Border Cave, dated to 45 ka.[45,46] Digging sticks, bone points and a poison applicator are also preserved there.[47] Warthog and bushpig bones predominate at this and other cave sites, replacing the preceding abundance of bones of larger mammals such as eland and hartebeest.[48] After ~45 ka, stone tool assemblages representing the LSA become widely distributed through inland sites across South Africa.[32,46] During the most extreme phase of the LGM around 20 ka, large mammals increased markedly in their representation at archaeological sites, suggesting that humans concentrated their hunting on them. This deviates from the pattern shown during the previous glacial maximum, when evidence of human habitation was mostly missing from the interior.

In central Africa, the cultural transition from the ESA into the MSA is represented by industries labelled the Sangoan, from a bay in Lake Victoria, and the Lupemban, from a site in eastern Congo DRC.[33] Hand axes fade out and small flake tools become common. Deposits from northern Zambia dated to ~265 ka are affiliated technologically with the transitional period. Numerous sites in Zambia exhibit abundant backed flakes and scrapers used for hunting, exemplifying the Nachikufan industry. The prominence of these tool designs emphasises the continued dependence of modern humans on meat.[49] Cave assemblages from south-western Zimbabwe exhibit similar features of MSA technology, but have not been dated.[33] In Malawi, few sites, if any, show human presence between 135 and 75 ka, when the mega-drought that prevailed through this region of south-central Africa reduced Lake Malawi to puddles.

Stone tools found in the Ogooue River valley in south-west Gabon, where grassland–forest mosaics currently prevail, date back as far as 400 ka and retain technology typical of the ESA.[33] MSA artefacts have been found in the Congo basin and as far west as Senegal, dated later than 130 ka. They also occur in eastern regions of the Sahara, which were well vegetated and contained lakes around that time. Associations between stone tools and the processing of large herbivore remains are widely evident there between 130 and 60 ka.

Modern humans with MSA culture appeared in the Levant region of Israel around 100 ka at a time when the Sahara as well as Arabia were quite green. Settlement sites spread further into the Arabian Peninsula, only to fade out after drier conditions took hold around 75 ka. It is possible that some people moved further into Asia along the coast.[50] However, the main exodus of humans from Africa took place later, around 60 ka. This movement continued along a coastal route to reach Australia by 55 ka or soon thereafter.

Bow Wave Out of Africa

Much emphasis has been placed on the cultural innovations documented in the southern Cape around 70 ka and what they reveal about human cognition. These include body adornments with ochre and shell beads, the beginnings of art and widening of tool kits to include bone as well as stone substrates. This timing, shortly before the wave of movement took people beyond Africa around 60 ka, has founded speculation that the people who colonised Eurasia originated from the Cape coastal region.[36,51] Klein[52] suggests that some revolutionary gene enabled a major cognitive advance, perhaps by facilitating language capabilities. However, I suspect that the key innovation had profound ecological implications: the advent of hunting by bow and arrow (Figure 18.3). Recent findings have pushed the appearance of arrow points made of bone or stone earlier than 60 ka at Klasies River and Pinnacle Point in the Cape plus Sibudu Cave in KwaZulu-Natal.[53] These arrow tips show signs of the application of poison, as employed by San hunters historically, making them lethal without needing to penetrate deeply between the ribs of their prey. Stone shaped for projectiles launched as spears were recognised in MSA assemblages at Kathu Pan somewhat earlier at 500 ka,[31] but many thrusts with a spear from close quarters would be needed to draw a lethal amount of blood

Figure 18.3 Modern hunter-gatherers. (A) San men hunting with bow-and-arrow in Botswana (photo: Ariadne Van Zandbergen); (B) hunters returning with an antelope carcass (photo: GreatStock); (C) San women on a foraging excursion seeking plant resources (photo: Alamy); (D) Hadza hunters searching for animals, holding their large bows (photo: Brian Wood).

from prey perhaps chased to exhaustion. A single prick from a poison-tipped arrow inflicted by a lone hunter could be sufficient to cause death, although the prey may need to be tracked for a few hours, until the poison did its job. The implications for food security are huge – a healthy animal as large as a giraffe could be killed by a single hunter. Following the use of poison-tipped arrows, ungulates of all sizes and ages were almost equally vulnerable to being hunted.

This advanced capability would have facilitated the dispersal of people from Africa's southern coast to culturally infiltrate regions of south-central Africa that had been de-populated during the mega-drought preceding 70 ka.[54,55] Empowered as hunters by the use of bows and arrows, people then dispersed onward through the Horn of Africa, crossing the Red Sea into Arabia and beyond. When they eventually encountered the Neanderthals, or the Denisovans, the short thrusting spears that the latter possessed were no match for arrows launched from a distance by the human ingressors. This enabled *H. sapiens* to replace *H. neanderthalensis* throughout Europe within a few thousand years.[56] But if poison for arrow tips could not be procured in Europe, such projectiles would be less effective for hunting large animals.

Archaeological sites in eastern Africa have not yielded recognisable arrow points dated earlier than 13 ka, but bows don't preserve and bone points are easily overlooked.[57,58] People with a sophisticated armoury of projectile weapons remained present through the southern African interior as the LGM advanced towards its coldest extreme ~20 ka. While food resourced from plants became extremely meagre, there was relatively little shortage of animals to hunt, perhaps concentrated in habitat refugia.

Several large herbivores that had persisted through previous glacial extremes went extinct during or shortly after the LGM in Africa (Chapter 14) as well as in Europe. While it seems difficult to credit predation by humans as playing a role in Africa,[59] some contribution from hunters recently armed with poison-tipped arrows cannot be ruled out. During times of extreme aridity, the dependence of humans on hunting for subsistence would have been amplified. Large herbivores would also have become restricted in their distribution to ecological islands of favourable vegetation. Similar spatial contexts may have fostered the demise of other large mammals worldwide during the late Pleistocene, following colonisation by humans.[60]

The implications of projectile weapons for the success of hunting are profound. They explain why white rhinos became so vulnerable to human hunters, despite their huge size. This mega-grazer disappeared from the broad extent of savanna Africa between the Zambezi and Nile rivers (Chapter 14), while remaining hugely abundant south of the Zambezi River into historic times.[61,62] White rhinos were present as recently as 50 ka at the archaeological site of Mumba in northern Tanzania, but not later.[63] Nevertheless, they are depicted in rock art at Konda in central Tanzania ascribed to Sandawe hunter-gatherers.[64] The rhinos that survived historically in the south are acutely fearful of the slightest whiff of human odour, even though these individuals had never experienced hunting (before the recent poaching wave). Indeed, fear of humans is deeply engraved in the nature of all of Africa's large mammals, even the carnivores.

Nevertheless, it does seem incredible that humans could have been responsible for the extinctions of large herbivores that did take place across Africa, from its southern to northern extremities, during the late Pleistocene, including a species as large and thick-skinned as Reck's elephant. The timing of the extinctions coincides with the advancement of glacial extremes, as discussed in Chapter 14, but it is too glib to dismiss human hunters as having made no contribution to the demise of animals that had long been their prey, at a time when their capability to deploy poison-tipped arrows had emerged and was most needed. Some interaction between accentuated human predation and worsening nutritional stress cannot be excluded as a factor contributing to these large mammal extinctions.

Genomic Evolution

Continuing genetic mutations enable connections among populations to be tracked through time, thereby identifying approximately when populations separated and thus no longer shared new mutations. Genetic information may be obtained from mitochondrial DNA (mtDNA), inherited solely from the mother; Y-chromosome DNA, inherited patrilineally; or from fragments of autosomal DNA subject to single nucleotide polymorphisms (SNPs). Under tropical conditions, opportunities to extract complete genomes from ancient DNA preserved in organic matter are rare. In order to place genetic changes in a time context, assumptions must be made about the rate at which mutations accumulate. Estimates differ, by a factor of two, depending on whether derived from new mutations appearing between recent human generations or, as earlier, calibrated from the fossil record of time since divergence of gene pools.[65] Recent findings based on accurate dating of specimens established by radioactive ^{14}C decay are congruent with the fossil-calibrated estimates.[66] Accordingly, I will adjust reported times of divergence based on recent nucleotide mutations between parents and offspring downward.

The metapopulation structure revealed from mitochondrial gene mutations is largely consistent with the distinct language groups identified among Africa's people. Greatest genetic diversity is retained by the Khoe-San click-speakers, indicating that they are closest to the ancestral population that gave rise to all modern humans.[67] Nevertheless, they are less diverse genetically than the great apes. This suggests that the lineage leading to modern humans underwent one or more population bottlenecks, during which numbers were low enough for genes to be lost.[25,36,68] It seems that, at some stage, humans were reduced to as few as 15,000–40,000 individuals dispersed in small, somewhat separated, subpopulation units.[69]

The groupings of people affiliated by language today became isolated at various times back in the past. The earliest separation was between people of Khoe-San affiliation found in the south and groups living further north in Africa. This became effective around 160 ka or even earlier, shortly after humans had attained their modern cranial form.[67,70] Technologically, it came after the transition into the MSA.[70] However, a genetic connection remained between the Khoe-San and the forest-inhabiting 'pygmy' or Batwa grouping, as well as with the remnant Hadza and Sandawe people inhabiting central Tanzania, until around 100–120 ka. The Hadza and Sandawe are allied with Bantu speakers physically as well as genetically, but retain a language incorporating click consonants along with forms of rock art resembling those of the San. They seem to be a cultural relict of hunter-gatherers who merged genetically with the Bantu-speaking immigrants into eastern Africa after 5 ka. Genetic isolation of the Batwa people from Bantu-language speakers originating in western Africa developed after 75 ka, while Bantu speakers had diverged from their common ancestor shared with Nilotic and Cushitic speakers by 34 ka. Eastern Africa has become particularly mixed genetically because of the ingress of Bantu people from the west plus Cushitic and Nilotic people from the north over the past several thousand years, absorbing the original hunter-gatherers genetically.[67]

It has been suggested, from changes in mtDNA haplotypes, that all modern humans further north originated from a subpopulation of Khoe-San people living near the vast Makgadikgadi–Okavango wetland that existed in Botswana around 200 ka.[71] This region evidently remained a relatively moist oasis while extreme glacial conditions prevailed between 190 and 130 ka. The relatively few number of archaeological sites suggests that people had mostly vacated the interiors of both southern and eastern Africa during this time. Moreover, the area to the north extending from Malawi through southern Tanzania had become depopulated during the 'mega-drought' conditions experienced there from 130 until 75 ka,[72] opening space for recolonisation. Some movement of people from the south into this region is signalled by the appearance there of a mtDNA haplotype typical of the Khoe-San close to 70 ka, shortly preceding the time of great exodus into Eurasia,[55,73] but the genetic profile of the people who continued moving onward beyond Africa does not resemble that of the southerners. Rather, it seems that they had taken up some of the cultural innovations introduced from the south, in body adornment, decorated artefacts, microlith stone and bone tools and, I contend, also hafted arrow points.[58] These technological advancements propelled the continued northward dispersal into Arabia and onward during the unusually moist conditions that prevailed in the Horn of Africa for a while after 70 ka. The intercontinental expansion was facilitated by the newfound capacity to hunt more effectively by launching poisoned arrows from bows. The material evidence for this advancement is lacking, but not surprisingly.

In summary, the genetic findings indicate that groups of people underwent spatial shuffling during the MSA while humans assumed their modern physical form and developed distinct language groups. There is evidence of a population bottleneck between 190 and 130 ka when extremes of glacial aridity took hold through both southern and eastern Africa. While there might have been previous bottlenecks during earlier glacial advances, there was not one during the most recent one, after humans had become armed with highly effective projectile weapons for hunting. The population segment that obtained refuge in caves along the southern Cape coast may have contributed culturally to the great exodus from Africa, but not their specific genomes.

How Do Modern Hunter-Gatherers Subsist Year-round?

Fortunately, the hunter-gatherer lifestyle persisted into modern times in two regions of Africa. Case studies there illustrate how effectively hunting overcomes food shortages potentially arising during the dry season. The San people were formerly widely present throughout southern Africa south of the Zambezi and Kunene rivers, as testified by their rock art on cave walls and other substrates. Following European settlement and, indeed, active extermination, they became restricted to arid savanna regions of Botswana and Namibia, where studies were undertaken. The arrow-tips and other tools that they had formerly made from stone or bone had been replaced by tools made of iron co-opted from the Bantu pastoralists who had settled among them by that time. The Hadza people persist in a small region of Tanzania south of Olduvai not far from the Serengeti plains. Although they likewise incorporate click sounds in their language, they are only distantly related genetically to the San, and quite distinct physically. Wild ungulates are no longer as abundant in the places where they occurred in the past, partly due to the activities of the herders living alongside the hunter-gatherers in this region.

Studies on the San were focused on both the !Kung or Ju/'hoansi groups (the extra symbols represent various click consonants) living in western Botswana at a time when these people still depended largely on hunting and gathering.[74] Male hunters brought back mainly smallish animals like warthogs, duikers, porcupines and springhares. Larger ungulates were hunted opportunistically when they appeared nearby, using bows equipped with poison-tipped arrows (Figure 18.3A). The poison mix did not kill instantly, meaning that a wounded animal had to be followed by its tracks until it came to a standstill and could be killed with spears. While a video vividly shows endurance hunting of an unwounded kudu until it could run no further, this method is no longer used today. Dogs may assist with the hunt, and snares are also deployed. Kill rates of large antelope amounted to merely two animals per hunter per year. Women gathered a mix of fruits, nuts, tubers and bulbs. Meat from various sources provided about 30 percent of the diet, underground plant tissues around 25 percent, and other plant parts formed the remainder. Nuts produced by mongongo trees (*Schinziophyton rautanenii*) were especially sought because they could be stored for consumption during the dry season. Rodents smaller than springhares and insects were not eaten.

Hunting success was best during the late dry season when edible plant parts were scarce, because animals were weaker. Nevertheless, base camps were shifted every few weeks as local plant resources became depleted. Groups occupying temporary camps typically comprised 10–30 individuals moving over annual ranges covering about 150 km^2. This is 5–10 times larger than the home ranges typically traversed by chimpanzee and gorilla groups (Table 16.1). Foraging excursions covered round-trip distances increasing seasonally from 3 to 19 km, vastly further than apes move daily. Time taken up by foraging was similar for women gathering plant foods and men hunting for animals, amounting to less than 20 h per week (8 h per day over 2.5 days per week). Additional time was taken up by chores at camp. People usually slept in the open in front of temporary huts, lighting fires as a deterrent against carnivores.

The Hadza people (Figure 18.3D) live south-east of the Serengeti plains near Lake Eyasi. At the time of the studies their population density was low, with around 1000 people occupying a range of 4000 km^2.[75,76] Only about 350 of them still operated exclusively as hunter-gatherers. Base camps occupied by about 30 people were moved every 1–2 months over an annual home range of about 120 km^2. Foraging activities took up about 4 h per day for women and 6 h per day for men on average, while daily travel distances averaged 5.5 and 8.3 km, respectively. Women sought both large fruits (especially those of baobab trees) and berries of wild raisins and dug for tubers using sharpened sticks. Men hunted using bows with poison-tipped arrows, capable of killing an animal as large as a giraffe. They also scavenged from carcasses of large animals killed by lions when opportunities were presented. After the lions appeared satiated, they were driven off by firing an arrow. Unguarded leopard kills were retrieved from trees. During the dry season, men waited at waterholes for animals coming to drink at night. Hunters returned with carcasses of large ungulates only about once per month, but secured smaller prey like springhares more frequently. Hunting of large mammals would have been more successful in the past when wild animals were more plentiful. Meat provided about 25 percent of the diet, greater for men and less for women. Dry-season conditions were not particularly harsh, because berries remained available and animals were easier to ambush near waterholes. Lions seemed to pose little threat, although people had been killed and eaten by lions at night in the past.

Studies on both of these groups have emphasised their leisurely lifestyles, with much less time spent on securing food than the 40+ hour week devoted to work by modern people in the developed world. However, this contrast is somewhat exaggerated. Maintenance work around camp by the hunter-gatherers is inadequately taken into account, and not all of the 40-hour working week of modern people is taken up by actual work. Today, people also spend time outside of work hours on home maintenance and on excursions to stores to purchase food and other needs. Nevertheless, the hunting and gathering groups that were observed did seem to have more time available for socialising and sleeping than those studying them recall having back in their home places.

Foraging excursions by women were lengthened during the dry season when less food was sourced from plants, but easier access to animals for meat meant less time pressure for men. However, studies were obviously conducted where people had survived, which means places having adequate supplies of fruits or nuts that could be stored plus ready access to animals that could be hunted. They do not represent extreme drought conditions when people might starve. Locating animals of whatever size to hunt during such times would be of crucial importance for survival. Neither the San nor the Hadza groups that were studied lived in places where they could expect to encounter many carcasses of animals that had died besides those killed by carnivores.

Selection pressures imposed by times of starvation are evident from the seemingly superfluous amounts of body fat stored by modern humans, both hunter-gatherers and modern city dwellers.[77] Mean body fat levels average 13 percent for men and 25 percent for woman, compared with merely 3 percent for baboons. Even human babies carry much fat. This is obviously a legacy of past times when women were unable to nurture the survival and growth of their offspring without these body reserves. Fat deposits around the buttocks feature prominently among women of both the Khoe herders and San gatherers, located where they least compromise thermoregulation under hot conditions. Frequent famines must have featured in the ecology of at least this region of Africa.

Rock Art

Cave paintings and rock engravings made by hunter-gatherers are widespread throughout savanna regions of Africa, but styles vary.[78] The earliest rock art, found on loose stones within caves on the southern Cape coast, took the form of geometric scratches, dated to 77 ka.[79] The earliest animal images preserved in Africa come from stones in Apollo Cave in southern Namibia, dated to 27 ka. This is later than the earliest cave paintings in France, dated at 33 ka, but rock paintings made under shallow overhangs in Africa are much less likely to be preserved than those made deep within limestone caves in Europe. Dates for rock art in northern Africa range from 12 to 2.5 ka, after the last glacial advance, when the Sahara region was green and thronged with animals and people. Paintings on rock walls in southern Africa have yielded an earliest date of 6 ka, from sites in Botswana. Cave paintings in the

Drakensberg/Maloti region of South Africa gave earliest dates of 2.5 ka, although human habitation in caves there goes back as far as 80 ka.[80] Rock engravings are more durable and probably have earlier origins than cave paintings, but seem impossible to date.

Geometric shapes painted using fingers predominate in the region broadly surrounding the Central African rainforest.[78] They contain some animal depictions, but these are relatively crude. This art form has been ascribed to ancestors of the Batwa people, represented today by groups living in rainforest regions, but who were more widespread prior to the expansion of Bantu speakers. In the south, this style of rock art occurs across a broad region extending from southern Tanzania, Malawi, northern Mozambique and parts of northern Botswana into Angola. Paintings locally depicting either geometric shapes or antelope are scattered through Zambia. Where human remains have been associated with the geometric art, they show features typical of people inhabiting west-central Africa rather than Khoe-San. The Hadza/Sandawe people exhibit cultural connections with the Khoe-San in art and language, but physically resemble neighbouring Bantu-speakers. The original language of the Batwa peoples is unknown, because everywhere they have adopted the language of local Bantu-speakers. Also enigmatic are the Damara people inhabiting northern Namibia, who are dark-skinned like the neighbouring Himba/Herero people but speak a click language, and lack cultural affinities with any local ethnic group. Genetically, they are allied with western Bantu immigrants represented by the Himba and Herrero.[81] Thus languages, cultures and genes can become reassorted partially independently.

The rock art ascribed to San hunter-gatherers south of the Zambezi and Kunene rivers features particularly human figures finely painted with brushes.[82] Many cave paintings depict ‘anthropomorphs’, combining animal and human features, plus imagery potentially visualised during trance dances. Among the animal depictions, antelope and other large herbivores are most common, with different species prominent in particular regions (Figure 18.4). Eland feature especially in the Drakensberg, replaced by hartebeest or kudu along with giraffe in South Africa’s northern bushveld. Giraffe and kudu along with elephants are especially common in south-western Zimbabwe and springbok in Namibia. Animal depictions found near Konda in central Tanzania, with giraffe once again conspicuous, followed by hartebeest and elephant, are ascribed to Sandawe people, neighbours of the Hadza who no longer live as hunter-gatherers.[64] Black rhinos are especially prominent among rock engravings in southern Africa. The most recent paintings in South Africa depict the arrival of European people, often on horseback, along with their livestock.

Naturalistic paintings and engravings, some of them huge, also occur widely through northern Africa inland from the Mediterranean coast,

Figure 18.4 Large ungulates depicted in rock art in southern Africa. (A) Eland in Drakensberg; (B) kudu in Zimbabwe; (C) giraffe and zebra in Zimbabwe; (D) reedbuck in Drakensberg; (E) hartebeest in Drakensberg; (F) roan antelope petroglyph in northern Cape.

particularly in Libya and Algeria. They are intriguingly similar in style to the art produced by the San people, despite the geographic separation. Animals commonly depicted there include giraffe, elephant, hippo, white rhino, and various unidentified antelope, along with the long-horned buffalo that became extinct 5000 years ago. People are shown hunting with bows and arrows and also standing among evidently domesticated cattle.

Assemblages of rock paintings are notable not only for the animal species shown, but also for those absent. In southern Africa, wildebeest and blesbok are rarely depicted, although among the most common ungulates there until quite recently. Bones of wildebeest are well represented in some fossil assemblages, with signs of butchery.[83] Perhaps wildebeest and blesbok attracted less attention because they are found in open grassland providing little cover for stalking hunters.

The San rock art emphasises the cultural importance of Africa's large herbivores in the lives of their human compatriots, especially their eminence as a food resource, but why do animals feature so much less among the geometric images associated with the Batwa or their affiliates? The Mbuti pygmies hunt forest-inhabiting duikers using nets, but perhaps none of these species is sufficiently prominent to be given ritual significance.

Pastoralism and Cultivation

Humans who spread from Africa through Eurasia had established a hunter-gatherer lifestyle in their continent of origin. Their hunting prowess with arrows as well as thrown spears enabled them to replace the Neanderthal people, who were dependent on mortally wounding large mammals and rivals using short stabbing spears. The fascination that the strange herbivores found in northern climates provided to the African immigrants is evident on cave walls through the limestone regions of Spain and France. Eventually, descendants of these people underwent a radical change in lifestyle in the 'Fertile Crescent' of the Middle East around 10 ka, when conditions had warmed considerably following the LGM. People established settlements, domesticated animals and cultivated crops. Aspects of this lifestyle soon filtered from there back into Africa.

Rock paintings show the presence of aurochs (ancestors of cattle) in southern Egypt around 10,500 years ago, but these animals were evidently hunted and not yet domesticated.[84] Domestic caprines (sheep and/or goats) first appeared in the Mediterranean region of northern Africa around 8000 years ago, as shown by rock engravings and paintings at various sites.[85] They were followed shortly after by cattle, representing the *taurus* form domesticated in Europe rather than the *indicus* (or zebu) originating in India. These domesticated ungulates became spread through the western Sahara region, which remained well vegetated until around 4500 years ago. Intensifying desert conditions then forced people with livestock to move southwards into the Sahel region of western Africa. The domesticated herbivores supplied a year-round source of food and liquid, in the form of milk. Plants were first cultivated in western Africa shortly after herding had spread. The first crop plant,

A

B

Figure 18.5 Herding livestock. (A) Sheep in Kenya; (B) cattle in Uganda.

pearl millet, appeared in western Africa around 3500 years ago, followed by sorghum, wild rice and finger millet.[86]

A separate southward movement of people with livestock took place in the north-east ~5000 years ago. These herders followed the Nile valley through present-day Sudan and settled near Lake Turkana in northern Kenya, where they incorporated fish in their diet.[87,88] There they mixed genetically with the indigenous foraging people. A thousand years passed before herding appeared further south in the vicinity of Lake Victoria, initially only with sheep and goats (Figure 18.5A), and expanded from there into southern Kenya and northern Tanzania shortly after 3000 years ago. The susceptibility of cattle to diseases acquired from wild ungulates may have inhibited movement by these herders southwards into moister savanna regions.[89] Back then the herding people still used stone implements. Starting between 3000 and 2500 years ago, people representing the Bantu language group migrated eastwards from the highlands located near Cross River close to the current Cameroon–Nigeria border into Uganda, following the northern rim of the Congo Basin, and onwards.[90] They brought pottery, grindstones, and later iron-working. How they integrated among the local hunter-gatherers remains unclear. The rock art found in eastern Africa does not depict the appearance of the domesticated ungulates, contrary to the images portrayed in the far north and south of the continent. Nilotic pastoralists, represented today by the Maasai and related groups, are the most recent indigenous immigrants, entering eastern Africa from the north only around 250 years ago.

Domestic livestock, in the form of sheep, appeared in Namibia around 2300 years ago.[91,92] The herding people, affiliated physically and linguistically with the Khoe-San, apparently acquired livestock from people living to the north of Angola. Cultural links between them and the Batwa hunter-gatherers are suggested by the presence of geometric rock art through this region. The Khoe herders, still with LSA implements, expanded their presence southwards, either by moving or cultural infiltration among the San, as far as the Western Cape, acquiring cattle at some stage. Cattle-herders with Khoe affiliations were evidently present in northern Botswana around 1700 years ago.[49] The Khoe settlers in the Cape had acquired the genes for lactose tolerance in adulthood, probably from close contacts with herders further north, enabling them to digest milk.[93]

In the east, a southwards movement of Bantu-speaking herders got underway before 2000 years ago, probably following a route through the grassy highlands so as to avoid the miombo woodland belt where tsetse flies transmitted sleeping sickness (trypanosomiasis) to cattle and humans.[89] This was during a period when cooler and hence more arid conditions were associated with an expansion in C_4 grasses in the diets of wild and domestic ungulates.[94] Bones of caprines accompanied by pottery have been found in Zimbabwe and northern Mozambique dated to around 2250 years ago, closely synchronous with the dispersal in the west.[92] Crops were grown and metal-working took place in Zimbabwe around 1800 years ago, while cattle were present in Malawi shortly after.

By around 1650 years ago, farming people had arrived in South Africa, bringing iron-working[33] and accompanied by sheep and goats, and later by cattle (Figure 18.5B). They settled in river valleys where soils were most easily worked. By 1250 years ago, farmers accompanied by livestock had reached the Fish River in the Eastern Cape. There they came into contact with Khoe herders, who had occupied this transitional zone between summer- and winter-dominant rainfall, providing both grass for cattle and shrublets for sheep plus goats. A further influx by agro-pastoral people along with abundant cattle took place around 1050 years ago. Those settling in the north established stone-walled structures at Mapungubwe alongside the Limpopo River and later built more elaborate stone constructions at Great Zimbabwe and elsewhere. The Mapungubwe settlement disintegrated around 700 years ago when conditions became too dry, while the Zimbabwe structures were abandoned around 500 years ago, supposedly because the local concentration of people exceeded what environmental resources could support.

In parts of Africa too dry to enable cultivation, such as northern Kenya and most of Botswana and Namibia, people subsisted on livestock products in the form of meat, blood and milk, supplemented by plants harvested from the wild. In wetter savannas, crops of sorghum, millet, and later maize formed the main basis of sustenance, augmented by meat and milk from livestock. Although wild ungulates were no longer hunted routinely, ceremonial drives took place and people turned to wildlife as a food buffer during severe droughts after livestock had died. In Botswana, people dug pits to trap animals as large as elephants. No wild animals were domesticated in Africa, probably because the indigenous ungulates were wary of close contact with humans, storing long experience with these meat-seeking primates in their genes. As time continued, more of Africa's grasslands became consumed by domestic ungulates rather than by all of the surviving wild ungulates combined.

European explorers and hunters entered Africa in increasing numbers during the eighteenth and nineteenth centuries. They found abundant wildlife present alongside indigenous people from Maasailand in eastern Africa through much of Zimbabwe, Botswana, Namibia and Zululand to the Eastern Cape.[61,62] Herders avoided infections of their animals with sleeping sickness transmitted by tsetse flies by keeping their cattle away from densely wooded localities. Populations of wild ungulates were decimated only after the introduction of firearms and their deployment for commercial trading as well as sport hunting. Humans no longer lived so accommodatingly along with the wild animals that they had earlier hunted merely for subsistence.

Overview

The adaptions in physical features that led to the establishment of *Homo ergaster* by ~1.7 Ma were accompanied by cultural innovations consolidating its ecological niche as facultative scavenger and later as active hunter. While hyenas developed powerful jaw muscles to access the marrow within bones of carcasses abandoned by, or stolen from, feline carnivores, hominins employed stones as percussion tools to crack them. Stones that were probably used to bash open nuts and tubers by australopithecine ape-men became co-opted for this new task. Flakes that had been waste became deployed to scrape off the scraps of flesh remaining on the outside of the bones. But these hominins did not become obligate carnivores. Their staple food dependency surely remained plant parts, especially those secured from underground by digging. The robust ape-men remained solely dependent on a plant resource base. Early humans diverged by adding the flesh component particularly during the dry season when plant parts became scarce and what remained required much effort to extract. Meanwhile, during this time of the year the carcass remains of animals that had starved or been killed by sabretooth cats and hunting hyenas became more abundant, awaiting picking. To access this food source sufficiently safely, the human scavengers needed to be active during the heat of midday, when other carnivores dozed. Hence this must have been the time in evolutionary development when body hair became reduced to a crown on the head plus a few other patches. This partly scavenging lifestyle apparently persisted from when constructed rocks and flakes made their appearance during the climatic transition into the Pleistocene ice ages, until the shift to more specifically shaped stone artefacts characterising the Acheulian technology accompanied the appearance of *H. ergaster* during the next major climatic transition, for around a million years.

Progressively cooling glacial conditions made plant resources even more scarce during the dry seasons when grazing ungulates became increasingly concentrated around remaining water sources at the time of the year when they were most weakened. The trophic (food-related) niche of the early humans thus expanded to incorporate hunting by pursuit as well as scavenging, building on their capabilities to tolerate heat loads that could prove fatal for animals unable to shed their fur coats. Stone tools became used to butcher the abundant meat secured by own kills in addition to what could be scavenged, requiring more substantial pieces than flakes. The weapons used to kill weakened or exhausted ungulates must have taken the form of sharpened sticks used as thrown or thrust spears, not preserved in fossil accumulations. The group coordination needed for hunting excursions from home bases by males promoted further but small increases in cranial capacity, but little further advancement in tool technology, over a further million years. Of course, females also needed to coordinate their excursions from shared home bases plus the management of infants at different stages.

The onset of widened oscillations between glacial and interglacial cold initiating the Middle Pleistocene ~0.8 Ma prompted a surge in cranial capacity, probably reflecting even greater dependency on hunting and perhaps not solely during dry seasons. Tool kits became even more finely crafted and lighter for greater mobility. Artefacts shaped for body penetration became hafted onto the tips of spears. Then a crucial next step in hunting technology occurred: the launching of sharp-tipped arrows from bows over greater distances than spears could cover, accompanied by the application of poisons to enhance their lethality. But the prey targeted did not die instantly, and the ability of the hunters to follow tracks through the heat of midday until animals could be dispatched with spears was a supporting physiological adaptation. Furthermore, arrows could be launched by hunters lying in ambush to kill even healthy ungulates, so that hunting was no longer mainly a dry-season activity. Other cultural innovations also emerged, including decorative beads and body adornments and the beginnings of art. These expressions took place near the coastline of southern Africa during a period when it seems that most of the subcontinental interior had been abandoned due to the severity of the preceding glacial maximum. When language capabilities emerged can only be conjectured. I envisage a progressive advancement of communication initiated by sounds and gestures facilitating social coordination when australopiths launched into open savannas to seek predator kills, progressively refined into the shaped vibrations that define languages. The cultural advances surely contributed to the population surge that carried a group of people beyond Africa and throughout the rest of the world.

Their increased hunting prowess allowed African inhabitants to survive more successfully during the LGM than people had through previous ones. It may have contributed to the demise of several large grazers clustered locally in places where food and water still remained. The advances inaugurating LSA tool technology were retained into modern times by some groups of people.

Meanwhile, a back-migration took place of people dependent on herding domesticated animals in place of hunting wild ones, supported by the cultivation of crop plants in place of those harvested from the wild. Notably, the domestic animals brought into Africa were mainly grazers and the crop plants grown mainly grasses. The agro-pastoralists spread through Africa, displacing the hunter-gatherers while incorporating their genes. People discovered how to smelt iron ore to obtain metal to make even more robust tools, notably including hoes as well as spear and arrow tips. Having acquired greater subsistence security from their livestock and cultivars, hunting pressure on wild ungulates became alleviated, although these animals remained a fallback resource during severe droughts and disease outbreaks. Wild herbivores living in grassy savannas remained a central feature of human subsistence and culture from when the earliest humans added scavenged carcass remains to their diets until the present time. Africa retains most of its Pleistocene diversity of large mammals because these animals had evolved along with humans while the latter honed their skill as hunters. African animals learnt early on not to become too familiar with upright apes, so none of them became domesticated. Once people living in Africa had gained the food security conferred by herding domestic ungulates, they tolerated, and even revered, the wild ungulates that coexisted in savanna grasslands.

Today, cerebral humans gape at the big mammals that Africa still possesses, which featured so centrally in our ecological trajectory from forest-inhabiting, fruit-eating apes to bipedal hunters. In the next chapter, I will explore more fundamentally how evolutionary transitions among the plethora of candidate hominins eventually led to modern humans in Africa. The final chapter will look ahead to the place of wild herbivores in evolving human culture.

Suggested Further Reading

Barham, L; Mitchell, P. (2008) *The First Africans. African Archaeology from the Earliest Toolmakers to the Most Recent Foragers*. Cambridge University Press, Cambridge.

Plummer, T. (2004) Flaked stones and old bones. Biological and cultural evolution at the dawn of technology. *Yearbook of Physical Anthropology* 47:118–164.

References

1. Rightmire, GP. (2004) Brain size and encephalization in early to mid-Pleistocene *Homo*. *American Journal of Physical Anthropology: The Official Publication of the American Association of Physical Anthropologists* 124:109–123.
2. Braun, DR, et al. (2019) Earliest known Oldowan artifacts at >2.58 Ma from Ledi-Geraru, Ethiopia, highlight early technological diversity. *Proceedings of the National Academy of Sciences of the United States of America* 116:11712–11717.
3. Semaw, S, et al. (1997) 2.5-million-year-old stone tools from Gona, Ethiopia. *Nature* 385:333–336.
4. Villmoare, B, et al. (2015) Early *Homo* at 2.8 Ma from Ledi-Geraru, Afar, Ethiopia. *Science* 347:1352–1355.
5. Kuman, K. (2014) Oldowan industrial complex. In Smith, C (ed.) *Encyclopedia of Global Archaeology*. Springer, New York, pp. 5560–5569.
6. Chazan, M, et al. (2012) The Oldowan horizon in Wonderwerk Cave (South Africa): archaeological, geological, paleontological and paleoclimatic evidence. *Journal of Human Evolution* 63:859–866.
7. Kuman, K, et al. (2018) The Oldowan industry from Swartkrans cave, South Africa, and its relevance for the African Oldowan. *Journal of Human Evolution* 123:52–69.
8. Sahnouni, M, et al. (2018) 1.9-million- and 2.4-million-year-old artifacts and stone tool-cutmarked bones from Ain Boucherit, Algeria. *Science* 362:1297–1301.
9. Harmand, S, et al. (2015) 3.3-million-year-old stone tools from Lomekwi 3, West Turkana, Kenya. *Nature* 521:310–315.
10. Dominguez-Rodrigo, M; Alcalá, L. (2019) Pliocene archaeology at Lomekwi 3? New evidence fuels more skepticism. *Journal of African Archaeology* 17:173–176.
11. Archer, W, et al. (2020) What is 'in situ'? A reply to Harmand et al. (2015). *Journal of Human Evolution* 142:102740.
12. Panger, MA, et al. (2002) Older than the Oldowan? Rethinking the emergence of hominin tool use. *Evolutionary Anthropology: Issues, News, and Reviews* 11:235–245.
13. Boesch, C; Boesch-Achermann, H. (2000) *The Chimpanzees of the Taï Forest: Behavioural Ecology and Evolution*. Oxford University Press, Oxford.
14. Stammers, RC, et al. (2018) The first bone tools from Kromdraai and stone tools from Drimolen, and the place of bone tools in the South African Earlier Stone Age. *Quaternary International* 495:87–101.
15. Plummer, T. (2004) Flaked stones and old bones: biological and cultural evolution at the dawn of technology. *American Journal of Physical Anthropology* 125:118–164.
16. Chazan, M, et al. (2008) Radiometric dating of the Earlier Stone Age sequence in excavation I at Wonderwerk Cave, South Africa: preliminary results. *Journal of Human Evolution* 55:1–11.
17. Kuman, K. (2014) The Acheulean industrial complex. In Smith, C (ed.) *Encyclopedia of Global Archaeology*. Springer, New York, pp. 7–18.

18. De la Torre, I. (2016) The origins of the Acheulean: past and present perspectives on a major transition in human evolution. *Philosophical Transactions of the Royal Society B: Biological Sciences* 371:20150245.
19. Uribelarrea, D, et al. (2019) A geoarchaeological reassessment of the co-occurrence of the oldest Acheulean and Oldowan in a fluvial ecotone from lower middle Bed II (1.7 ma) at Olduvai Gorge (Tanzania). *Quaternary International* 526:39–48.
20. Dominguez-Rodrigo, M; Pickering, TR. (2017) The meat of the matter: an evolutionary perspective on human carnivory. *Azania: Archaeological Research in Africa* 52:4–32.
21. Organista, E, et al. (2019) Taphonomic analysis of the level 3b fauna at BK, Olduvai Gorge. *Quaternary International* 526:116–128.
22. Smith, GM, et al. (2019) Subsistence strategies throughout the African Middle Pleistocene: faunal evidence for behavioral change and continuity across the Earlier to Middle Stone Age transition. *Journal of Human Evolution* 127:1–20.
23. Tryon, CA; McBrearty, S. (2002) Tephrostratigraphy and the Acheulian to Middle Stone Age transition in the Kapthurin formation, Kenya. *Journal of Human Evolution* 42:211–235.
24. Potts, R, et al. (2020) Increased ecological resource variability during a critical transition in hominin evolution. *Science Advances* 6:eabc8975.
25. Basell, LS. (2008) Middle Stone Age (MSA) site distributions in eastern Africa and their relationship to Quaternary environmental change, refugia and the evolution of *Homo sapiens*. *Quaternary Science Reviews* 27:2484–2498.
26. Blinkhorn, J; Grove, M. (2018) The structure of the Middle Stone Age of eastern Africa. *Quaternary Science Reviews* 195:1–20.
27. Scerri, EML. (2017) The North African Middle Stone Age and its place in recent human evolution. *Evolutionary Anthropology: Issues, News, and Reviews* 26:119–135.
28. Lahr, MM;Foley, RA. (2016) Human evolution in late Quaternary eastern Africa. In Jones, SC; Stewart, BA (eds) *Africa from MIS 6-2*. Springer, Dordrecht, pp. 215–231.
29. Kusimba, SB. (1999) Hunter–gatherer land use patterns in Later Stone Age East Africa. *Journal of Anthropological Archaeology* 18:165–200.
30. Kuman, K, et al. (2020) The Fauresmith of South Africa: a new assemblage from Canteen Kopje and significance of the technology in human and cultural evolution. *Journal of Human Evolution* 148:102884.
31. Wilkins, J, et al. (2012) Evidence for early hafted hunting technology. *Science* 338:942–946.
32. Robbins, LH, et al. (2016) The Kalahari during MIS 6-2 (190–12 ka): archaeology, paleoenvironment, and population dynamics. In Jones, SC; Stewart, BA (eds) *Africa from MIS 6-2*. Springer, Dordrecht, pp. 175–193.
33. Barham, L; Mitchell, P. (1992) *The First Africans: African Archaeology from the Earliest Toolmakers to Most Recent Foragers*. Cambridge University Press, Cambridge.
34. Marean, CW, et al. (2007) Early human use of marine resources and pigment in South Africa during the Middle Pleistocene. *Nature* 449:905–908.
35. Wurz, S. (2016) Development of the archaeological record during the middle stone age. In Knight, J; Grab, SW (eds) *Quaternary Environmental Change in Southern Africa: Physical and Human Dimensions*. Cambridge University Press, Cambridge, pp. 371–384.

36. Marean, CW. (2015) An evolutionary anthropological perspective on modern human origins. *Annual Review of Anthropology* 44:533–556.
37. Marean, CW. (2016) The transition to foraging for dense and predictable resources and its impact on the evolution of modern humans. *Philosophical Transactions of the Royal Society B: Biological Sciences* 371:20150239.
38. Reynard, JP; Henshilwood, CS. (2019) Environment versus behaviour: zooarchaeological and taphonomic analyses of fauna from the Still Bay layers at Blombos Cave, South Africa. *Quaternary International* 500:159–171.
39. Reynard, JP; Wurz, S. (2020) The palaeoecology of Klasies River, South Africa: an analysis of the large mammal remains from the 1984–1995 excavations of Cave 1 and 1A. *Quaternary Science Reviews* 237:106301.
40. Larbey, C, et al. (2019) Cooked starchy food in hearths ca. 120 kya and 65 kya (MIS 5e and MIS 4) from Klasies River Cave, South Africa. *Journal of Human Evolution* 131:210–227.
41. d'Errico, F, et al. (2017) Identifying early modern human ecological niche expansions and associated cultural dynamics in the South African Middle Stone Age. *Proceedings of the National Academy of Sciences* 114:7869–7876.
42. Backwell, L, et al. (2008) Middle stone age bone tools from the Howiesons Poort layers, Sibudu Cave, South Africa. *Journal of Archaeological Science* 35:1566–1580.
43. Lombard, M. (2011) Quartz-tipped arrows older than 60 ka: further use-trace evidence from Sibudu, KwaZulu-Natal, South Africa. *Journal of Archaeological Science* 38:1918–1930.
44. Bradfield, J, et al. (2020) Further evidence for bow hunting and its implications more than 60 000 years ago: results of a use-trace analysis of the bone point from Klasies River Main site, South Africa. *Quaternary Science Reviews* 236:106295.
45. Grün, R; Beaumont, P. (2001) Border Cave revisited: a revised ESR chronology. *Journal of Human Evolution* 40:467–482.
46. Bousman, CB; Brink, JS. (2018) The emergence, spread, and termination of the Early Later Stone Age event in South Africa and southern Namibia. *Quaternary International* 495:116–135.
47. d'Errico, F, et al. (2012) Early evidence of San material culture represented by organic artifacts from Border Cave, South Africa. *Proceedings of the National Academy of Sciences of the United States of America* 109:13214–13219.
48. Klein, RG. (1979) Stone Age exploitation of animals in southern Africa: Middle Stone Age people living in southern Africa more than 30,000 years ago exploited local animals less effectively than the Later Stone Age people who succeeded them. *American Scientist* 67:151–160.
49. Mitchell, PJ. (2016) Later Stone Age hunter-gatherers and herders. In Knight, J; Grab, SW (eds) *Quaternary Environmental Change in Southern Africa.* Cambridge University Press, Cambridge, pp. 385–396.
50. Bae, CJ, et al. (2017) On the origin of modern humans: Asian perspectives. *Science* 358: eaai9067.
51. Henn, BM, et al. (2011) Hunter-gatherer genomic diversity suggests a southern African origin for modern humans. *Proceedings of the National Academy of Sciences of the United States of America* 108:5154–5162.

52. Klein, RG. (2019) Population structure and the evolution of *Homo sapiens* in Africa. *Evolutionary Anthropology: Issues, News, and Reviews* 28:179–188.
53. O'Driscoll, CA; Thompson, JC. (2018) The origins and early elaboration of projectile technology. *Evolutionary Anthropology: Issues, News, and Reviews* 27:30–45.
54. Shea, JJ; Sisk, ML. (2010) Complex projectile technology and *Homo sapiens* dispersal into western Eurasia. *PaleoAnthropology* 2010:100–122.
55. Rito, T, et al. (2019) A dispersal of *Homo sapiens* from southern to eastern Africa immediately preceded the out-of-Africa migration. *Scientific Reports* 9:1–10.
56. Garcea, EAA. (2012) Successes and failures of human dispersals from North Africa. *Quaternary International* 270:119–128.
57. Langley, MC, et al. (2016) Poison arrows and bone utensils in late Pleistocene eastern Africa: evidence from Kuumbi Cave, Zanzibar. *Azania: Archaeological Research in Africa* 51:155–177.
58. Shipton, C, et al. (2018) 78,000-year-old record of Middle and Later Stone Age innovation in an East African tropical forest. *Nature Communications* 9:1–8.
59. Faith, JT, et al. (2018) Plio–Pleistocene decline of African megaherbivores: No evidence for ancient hominin impacts. *Science* 362:938–941.
60. Owen-Smith, N. (1999) The interaction of humans, megaherbivores, and habitats in the late Pleistocene extinction event. In MacPhee, RDE (ed.) *Extinctions in Near Time*. Kluwer, New York, pp. 57–69.
61. Delegorgue, A. (1990) *Adulphe Delegorgue's Travels in Southern Africa*. University of Natal Press, Durban.
62. Selous, FC. (1881) *A Hunter's Wanderings in Africa*. R. Bentley & Son, London.
63. Tryon, CA. (2019) The Middle/Later Stone Age transition and cultural dynamics of late Pleistocene East Africa. *Evolutionary Anthropology: Issues, News, and Reviews* 28:267–282.
64. Leakey, M. (1983) *Africa's Vanishing Art. The Rock Paintings of Tanzania*. Doubleday, New York.
65. Scally, A; Durbin, R. (2012) Revising the human mutation rate: implications for understanding human evolution. *Nature Reviews Genetics* 13:745–753.
66. Fu, Q, et al. (2013) A revised timescale for human evolution based on ancient mitochondrial genomes. *Current Biology* 23:553–559.
67. Fan, S, et al. (2019) African evolutionary history inferred from whole genome sequence data of 44 indigenous African populations. *Genome Biology* 20:1–14.
68. Yost, CL, et al. (2018) Subdecadal phytolith and charcoal records from Lake Malawi, East Africa imply minimal effects on human evolution from the ~74 ka Toba supereruption. *Journal of Human Evolution* 116:75–94.
69. Campbell, MC; Tishkoff, SA. (2010) The evolution of human genetic and phenotypic variation in Africa. *Current Biology* 20:R166–R173.
70. Schlebusch, CM, et al. (2017) Southern African ancient genomes estimate modern human divergence to 350,000 to 260,000 years ago. *Science* 358:652–655.
71. Chan, EKF, et al. (2019) Human origins in a southern African palaeo-wetland and first migrations. *Nature* 575:185–189.

72. Blome, MW, et al. (2012) The environmental context for the origins of modern human diversity: a synthesis of regional variability in African climate 150,000–30,000 years ago. *Journal of Human Evolution* 62:563–592.
73. Rito, T, et al. (2013) The first modern human dispersals across Africa. *PLoS One* 8:e80031.
74. Lee, R. (2003) *The Dobe Ju/'hoansi*. 3rd ed. Wadsworth Thomson, Belmont.
75. Marlowe, FW; Berbesque, JC. (2009) Tubers as fallback foods and their impact on Hadza hunter-gatherers. *American Journal of Physical Anthropology: The Official Publication of the American Association of Physical Anthropologists* 140:751–758.
76. Marlowe, F. (2010) *The Hadza: Hunter-Gatherers of Tanzania*. University of California Press, Berkeley.
77. Antón, SC; Josh Snodgrass, J. (2012) Origins and evolution of genus *Homo*: new perspectives. *Current Anthropology* 53:S479–S496.
78. Smith, B. (2006) The rock art of sub-Saharan Africa. In Blundell, G (ed.) *Origins. The Story of the Emergence of Humans and Humanity in Africa*. Double Storey Books, Cape Town, pp. 93–101.
79. Henshilwood, CS, et al. (2018) An abstract drawing from the 73,000-year-old levels at Blombos Cave, South Africa. *Nature* 562:115–118.
80. Stewart, BA, et al. (2016) Follow the Senqu: Maloti–Drakensberg paleoenvironments and implications for early human dispersals into mountain systems. In Jones, SC; Stewart, BA (eds) *Africa from MIS 6-2*. Springer, Dordrecht, pp. 247–271.
81. Oliveira, S, et al. (2018) Matriclans shape populations: insights from the Angolan Namib Desert into the maternal genetic history of southern Africa. *American Journal of Physical Anthropology* 165:518–535.
82. Eastwood, EB; Eastwood, C. (2006) *Capturing the Spoor: An Exploration of Southern African Rock Art*. New Africa Books, Claremont.
83. Faith, JT, et al. (2015) Paleoenvironmental context of the Middle Stone Age record from Karungu, Lake Victoria Basin, Kenya, and its implications for human and faunal dispersals in East Africa. *Journal of Human Evolution* 83:28–45.
84. Brass, M. (2018) Early North African cattle domestication and its ecological setting: a reassessment. *Journal of World Prehistory* 31:81–115.
85. Mitchell, P. (2006) Rediscovering Africa. In Blundell, G (ed.) *Origins. The Story of the Emergence of Humans and Humanity in Africa*. Double Storey Books, CapeTown, pp. 116–165.
86. Marshall, F; Hildebrand, E. (2002) Cattle before crops: the beginnings of food production in Africa. *Journal of World Prehistory* 16:99–143.
87. Chritz, KL, et al. (2019) Climate, ecology, and the spread of herding in eastern Africa. *Quaternary Science Reviews* 204:119–132.
88. Prendergast, ME, et al. (2019) Ancient DNA reveals a multistep spread of the first herders into sub-Saharan Africa. *Science* 365:eaaw6275.
89. Gifford-Gonzalez, D. (2000) Animal disease challenges to the emergence of pastoralism in sub-Saharan Africa. *African Archaeological Review* 17:95–139.

90. Li, S, et al. (2014) Genetic variation reveals large-scale population expansion and migration during the expansion of Bantu-speaking peoples. *Proceedings of the Royal Society B: Biological Sciences* 281:20141448.
91. Sadr, K. (2015) Livestock first reached southern Africa in two separate events. *PLoS One* 10: e0134215.
92. Lander, F; Russell, T. (2018) The archaeological evidence for the appearance of pastoralism and farming in southern Africa. *PLoS One* 13:e0198941.
93. Breton, G, et al. (2014) Lactase persistence alleles reveal partial East African ancestry of southern African Khoe pastoralists. *Current Biology* 24:852–858.
94. Robinson, JR; Rowan, J. (2017) Holocene paleoenvironmental change in southeastern Africa (Makwe Rockshelter, Zambia): implications for the spread of pastoralism. *Quaternary Science Reviews* 156:57–68.

Chapter 19: Reticulate Evolution Through Turbulent Times

Evolution takes the form of changes in lineages through time. These changes are captured by names given to species, traditionally defined morphologically with reference to type specimens.[1] A conceptual structure has been developed for identifying features that have been derived from some ancestral form, perhaps shared with sister species, and those that are novel, distinguishing the species. Supporting this approach is the biological species concept, based on whether individuals can interbreed and produce fertile offspring. If they can, they share a gene pool and their populations follow a common evolutionary trajectory. Populations may become isolated if males and females fail to recognise one another as mates or if the offspring produced are not viable, for whatever reason. Genetic divergence, through the accumulation of mutations over time, however neutral in their effects, can be used to infer that populations have been isolated for sufficiently long to be assumed distinct at the species level. Of course, one cannot be sure that they would not interbreed if they came into contact later. A similar caveat applies spatially. Should populations separated geographically be regarded as distinct species, or merely subspecies, morphologically different in some features but without barriers to mating if connected? Ecologically, species can coexist only if they differ sufficiently in niche occupation as defined by resources exploited, physiological tolerance or mechanisms for evading predation. Some of these issues were addressed in Chapters 10 and 14 with respect to Africa's ungulates.

But looking far back in time, all we have to go by are the morphological features shown by fossilised remains, generally fragmented. Almost invariably, these features change through time. Paleontologists must judge whether the morphological changes are sufficiently meaningful to define new species, or perhaps even new genera, indicating more fundamental changes in lifestyles or 'grades'. Species come and go, but many of these disappearances are not extinctions, merely the transformation of one 'chronospecies' into another over time. There is a need to distinguish genuine extinctions when lineages disappear, leaving no descendants, from persistence in somewhat changed form. Some branches of the evolutionary tree terminate while others ramify. Drawing on these concepts, we seek the primate antecedent that gave rise to both humans and chimpanzees, and earlier also gorillas. The further back in time we go, the more branches coalesce.

Figure 17.1 shows around 20 hominin species eventually culminating in only one, but many of the name changes distinguish chronospecies while others represent geographically separated counterparts. The most remarkable feature is how a connecting lineage has propagated unbroken through time, adapting sufficiently to changing environments to warrant name changes at various stages, to culminate in *Homo sapiens*. How was this brought about, despite huge environmental fluctuations involving shifts in temperature, precipitation, vegetation cover and forms of coexisting animals? Actually, it happened because of these fluctuations, and particularly those distinctive of Africa, as this chapter will explain.

Eight major adaptive transitions can be recognised in the lineage linking some forest-dwelling ape to modern humans:

1. The emergence of the distinct genus *Ardipithecus* with features facilitating bipedal locomotion, around 5.7 Ma.

2. The transformation of *Ardipithecus* into *Australopithecus* with greater bipedal competence, by 3.5 Ma.

3. The separation of *Homo*, with more generalised dentition supported by a stone tool kit, from *Paranthropus* with dental specialisations for chewing tough foods, around 2.8 Ma.

4. The appearance of *H. erectus/ergaster* with fully competent walking and substantially bigger brain, accompanied by shaped stone implements, around 1.7 Ma.

5. The surge in brain volume towards that of modern humans via *H. heidelbergensis*, beginning around 0.8 Ma.

6. The establishment of a finely crafted tool kit, plus other cultural artefacts, by big-brained 'modern' *H. sapiens*, between 600 and 200 ka.

7. The development of projectile points launched from bows and daubed with poison in finely elaborated tool kits around 70 ka, predisposing the expansion of humans beyond Africa.

8. The spread of pastoralism and cultivation through Africa after 5 ka.

It is apparent that these adaptive shifts tended to be coupled with times of major environmental transitions: (1) establishment of seasonally dry savannas between 12 and 5 Ma, (2) expansion of C_4 grasslands during this period, (3) establishment of recurrent glacial advances bringing greater aridity around

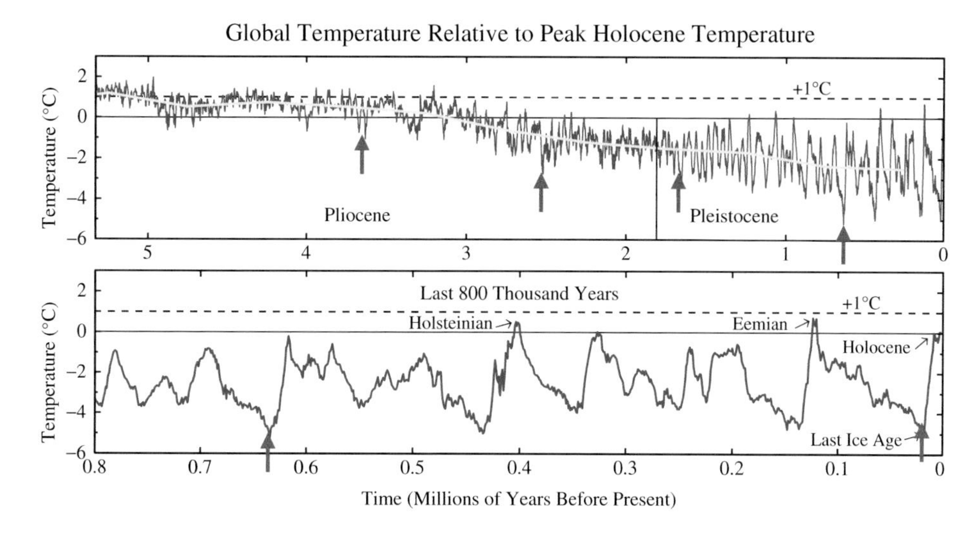

Figure 19.1 Global temperature trend through the Pliocene, Pleistocene and Holocene, noting with red arrows when unprecedented extremes were reached (image from giss.nasa.gov).

2.6 Ma, (4) tectonic disruptions accentuating local aridity ~1.8 Ma, (5) widening climatic oscillations after 0.8 Ma, (6) extremes of aridity during the last-but-one glacial maximum 140 ka, and (7) amelioration of local aridity during the Holocene after 20 ka (Figure 19.1). It is intriguing that there are blanks in the fossil record just around the times when some of these transitions took place. Unconformities in sedimentary layers indicate conditions either too arid to form deposits or so wet as to wash away sediments. Puzzlingly, some of these adaptive advances seem to precede, rather than follow after, the climatic extremes attained.

During these environmental transitions many species, genera and even families went terminally extinct. These included most of the Oligocene/Miocene giants among the herbivores, plus various sabretooth cats and hyenas among the carnivores. Meanwhile, the stem bovid generated the dazzling diversity of grazers plus browsers that we see still extant today. However, not all of these forms survived; several of the largest grazers faded out around the time of the most recent glacial maximum. The product of the primate lineage transforming australopithecine ape-men into modern humans endured, precariously at times, but ultimately became empowered to dominate the whole world.

What happened to the other recognised hominin species? Were they terminal side-branches, like the robust ape-men clearly were? Or chronospecies, connected through time? More fundamentally, were they really functional species, as defined biologically by genetic isolation? Were there indeed times when three or four species of hominins coexisted, side-by-side in the same places at the same time,[2] violating ecological principles of niche separation? Looking beyond Africa, why did super-adaptable *Homo erectus*, the geographic counterpart of *H. ergaster*, eventually fade out in Asia, and why did *Homo sapiens* replace both the Neanderthals and Denisovans in Europe?

Preceding chapters have established an ecologically plausible scenario, consistent with the fragmented fossilised and cultural evidence that we have to build on. I have reported it using the names that have been given to putative species and genera, but perhaps this labelling is obscuring how these species connected in enduring lineages. What is the relationship between species turnover and lineage persistence?

Multi-regional and Reticulate Evolution

It is striking how evolutionary advances took place fairly synchronously in eastern and southern Africa, thousands of kilometres apart. *Au. afarensis* dates back to 3.8 Ma in the Afar region of north-eastern Ethiopia and *Au. promethius*, probably ancestral to *Au. africanus*, to 3.7 Ma in Sterkfontein Cave in South Africa (Chapter 17). The earliest fossils assigned to *Homo ergaster* from the Omo–Turkana Basin and a South African cave site both date to shortly after 2.0 Ma. *H. heidelbergensis* emerges around 0.8 Ma in Germany as well as in Africa. Middle Stone Age technology appears shortly before 200 ka across Africa from south to north. Finely crafted stone flakes typifying the Later Stone Age became established around 45 ka in both southern and eastern African sites. The synchronous spread of these evolutionary advances through Africa was undoubtedly enabled by the continuity of savanna grasslands straddling the equator, without mountain chains or river barriers blocking movements of animals and people.

The notion of synchronous multi-regional evolution was formerly advanced in support of beliefs that modern humans originated in Asia rather than in Africa. Supposedly, local adaptive innovations developed in parallel in different localities in response to similar selective pressures. The concept has recently been resurrected as an 'African Multi-regional' scenario.[3] Nevertheless, it seems unlikely that the isolation of geographically separated populations endured sufficiently long in Africa, given the widely variable conditions in local regions.[4,5] An alternative scenario has been characterised as reticulate evolution.[6,7] It is envisaged that small local populations evolved independently but, rather than remaining isolated, were periodically reconnected to share genes, like a braided stream or reticulate web of linkages.[3,5] Such a population structure could generate a polymorphic mixture of features distinguishing local populations, especially during transitional periods while the novel adaptations spread.

It is quite plausible that *Paranthropus*, grubbing for underground plant parts as a food staple, coexisted alongside early *Homo*, scavenging or hunting animals as well. These hominin forms seem too distinct in appearance and lifestyle to have exchanged genes. But it is not tenable ecologically for *H. habilis* and *H. erectus* and *H. ergaster* and *H. rudolfensis* to have coexisted in time and space as distinct species, as has been inferred.[2] If they were indeed syntopic, as claimed, their distinguishing features are more likely to reflect polytypic variation within a broadly distributed population occupying the *H. ergaster/erectus* niche. Whether *H. erectus* geographically apart in Eurasia should be distinguished specifically from *H. ergaster* in Africa is moot. Representatives of *Australopithecus* and *Paranthropus* in eastern and southern Africa can acceptably be labelled as district species because we lack information about the geographic variation in between. *H. naledi* might have squeezed in alongside *H. sapiens* by being physically smaller and thus more selective among vegetation components exploited.

The morphological features used to distinguish the species represented in Figure 17.1 are based mostly on skulls, jaws and teeth. Taxa in the same grade eat the same sorts of food and share the same mode of locomotion.[8] The form of the dentition can represent adaptive shifts in food resources exploited, notably between early *Homo* and *Paranthropus*, sufficient to enable coexistence. Brain size has multiple adaptive ramifications. However, other features such as the protrusion of the snout and the compaction of the face and brow ridges seem neutral ecologically, although potentially guiding mate recognition. It seems more likely that these cranial features drifted, as tends to happen in small temporarily isolated populations, without preventing the local populations from coalescing genetically at some later stage. As outlined in Chapter 14, subspecies of hartebeest are based on distinctions in horn shape and coat colour, without being judged sufficiently isolated to preclude later genetic merging. Geographically separated populations of baboons interbreed and thereby merge gene pools, whether distinguished as species or subspecies.[9]

Adversity Versus Variability Selection

Potts[10,11] related the advances shown in the lineage leading to *Homo* to the need to adapt to widening environmental fluctuations in temperature and aridity, which he labelled 'variability selection'. He proposed that hominins responded to the highly variable climatic conditions in African savannas by enhancing their adaptive versatility, for example by broadening their diet to cope with both meat and plant matter. However, it is the harshness of the dry extremes that would have formed the selective filter, more than the mesic upswings. Typically, fitness curves flatten towards benign conditions, but fall

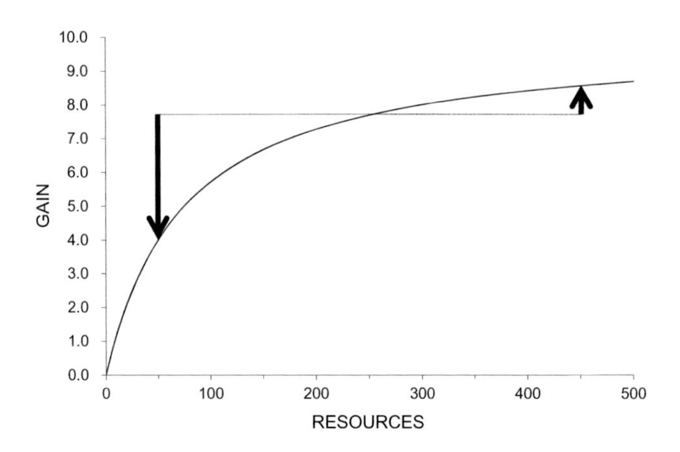

Figure 19.2 Jensen's inequality, showing how variation in rainfall above the mean makes less difference to resource gains, and hence fitness, than equivalent variation below the mean, because of the curvilinearly saturating trend (taken from Owen-Smith (2002) *Adaptive Herbivore Ecology: From Resources to Populations in Variable Environments*).

increasingly steeply with intensifying insufficiency of food supplies, producing 'Jensen's inequality' (Figure 19.2). Hence variation in rainfall or other conditions above the mean has less effect on resource gains, and hence fitness, than variation below the mean. The maximum rate of population growth is constrained by life-history features when conditions are benign (Chapter 12), but rates of population decline in lean times are in free fall. When rates of recovery fail to compensate for the crashes, populations go into decline, ratcheting downwards towards eventual extinction with each successive knock of adversity.[12] I interpret this pattern as *adversity selection.*

There is increasing recognition of how rapidly adaptive shifts in local populations can occur in response to environmental changes, undermining notions of gradually progressive advancements.[13] Adaptive shifts can be especially abrupt during climatic extremes, which impose strong selection for survival-enhancing features of morphology or physiology. The beaks of finches in the Galapagos changed size and shape within a few decades in response to climatic extremes affecting characteristics of the seeds that they sought.[14] Guppies altered their colour patterns within less than a decade in response to the addition or removal of predators.[15] The morphological, physiological and behavioural features of hominins could have shifted within several centuries or a few millennia under strong selection during times of unprecedented adversity.

For hominins, the extremes of aridity reached during glacial advances imposed the selective filter. Recall the associations in time between the major adaptive transitions among hominins noted above and the surges towards unprecedented extremes in low temperatures manifested globally around these times (Figure 19.1). In African savannas it was not temperature that formed the prime selective filter, but rather the extreme shortages of plant resources that developed during the dry season at times when global conditions were coldest. Moreover, regional rainfall was not necessarily coupled solely with global temperature changes. There were additional influences from tectonic upheavals affecting the locations of rain shadows and diverting monsoon winds, while the ITCZ also shifted its seasonal positioning, affecting local rainfall (Chapters 1 and 2).

During severe droughts, large herbivore populations can crash to less than half of their prior abundances.[16,17,18] If food runs out before the end of the dry season, mortality losses as great as 80–90 percent of local populations can ensue.[19,20] This is powerful selection for any attribute conferring greater chances of being among the survivors. Such extremes of adversity probably recurred multiple times around each glacial maximum. Die-offs recorded during the currently prevailing interglacial interlude do not adequately convey the devastation of the extremes of aridity attained in the past, potentially resulting in local or even continental extinctions. We do know that the two most recent glacial advances were both associated with multiple extinctions among large grazers. These included Reck's elephant, gorgops hippo, giant zebra, long-horned buffalo, two blesbok, giant gelada and giant warthog in eastern Africa, and long-horned buffalo, giant wildebeest, a blesbok, a large zebra and two species of springbok in southern Africa. Amazingly, the hominin lineage leading to us made it through every episode of adversity, albeit precariously, although not unchanged.

I conceptualised the dynamic niche as a trajectory through hyperspace to explain how herbivores cope with varying food availability over the seasonal cycle (Figure 19.3).[21] Population persistence depends crucially on how animals find something to eat during savanna dry seasons, while avoiding being eaten by carnivores. Large herbivores achieve this by shifting diets and habitats occupied in functionally distinctive ways, exploiting environmental heterogeneity (Figure 19.3). Survival is multiplicative; if at any stage survival prospects become zero, that is it – local extinction results. For global extinctions, populations must run out of resources, or become subject to non-sustainable levels of predation, everywhere across the species distribution range, a much more restrictive requirement. The wider a species is distributed, the less the chance of terminal extinction. The Africa-wide distribution of the lineage labelled *Homo* dependent on its dietary flexibility surely contributed to its lineage endurance. How it escaped being eliminated by predation, being neither fierce nor fast, remains less apparent.

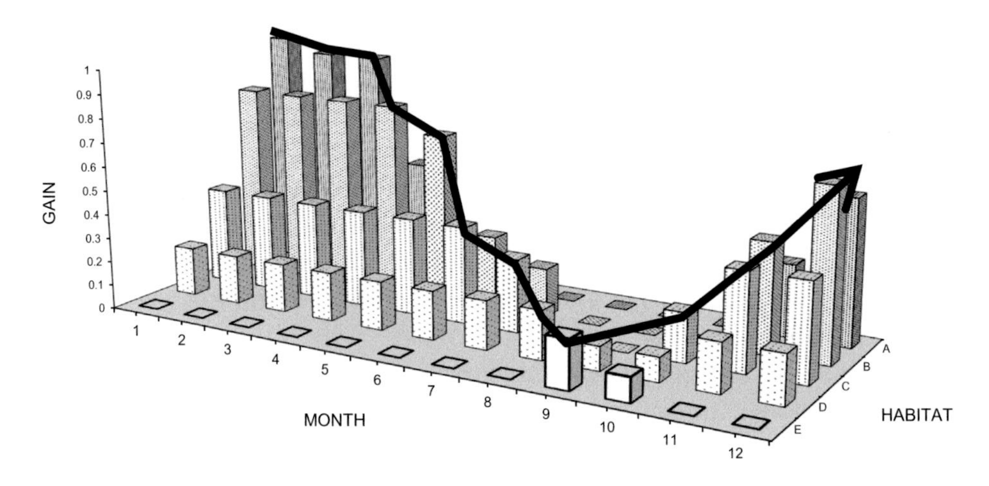

Figure 19.3 Dynamic niche concept, illustrating how herbivores cope with seasonal variation in environmental conditions through exploiting spatial heterogeneity. Five habitat units are distinguished discretely by the seasonally varying resource gains that they potentially convey, represented by the height of the blocks. Herbivores shift their habitat and resource selection to enable survival through the seasonal cycle. Resource gains could be translated into fitness by considering additionally reductions in survival brought about by predation in each habitat type (taken from Owen-Smith (2002) *Adaptive Herbivore Ecology: From Resources to Populations in Variable Environments*).

Why Only in Africa?

The reasons why the crucial evolutionary transformations from forest-dwelling primates to savanna-inhabiting humans could only have occurred in Africa did indeed originate from geo-tectonics. The uplift of Africa and subsequent rifting and volcanics established the physical cradle. It produced unusually low rainfall for the tropics coupled with soils that remained comparatively fertile, at least for nourishing large herbivores. Had Africa been mostly low-lying, it would have become largely a degraded semi-desert like most of Australia. Had the uplift taken place in the west, tropical regions of Africa would have been as moist and infertile as South America, thronged with huge mega-grazers rather than a rich assemblage of medium–large ruminants. Tropical Asia is mostly too low and wet for savanna vegetation to be extensive and lacks an abundant grazing fauna. An intellectual debt must be acknowledged to the geologists who saw that tectonics was crucially responsible for promoting human evolution.[22] I have explored the connections in this book and shown that their perception is justified.

The crucial feature of Africa's climates is the wide prevalence of seasonal dryness. This underlies the spatial predominance of savanna vegetation, with grasses coexisting between and beneath trees. Ruminants radiating during the Miocene adapted especially to digest the fibrous C_4 grasses during the dry seasons. This capability enabled these grazers to attain vastly greater abundances than browsers having their food base restricted to the few leaves remaining within reach on trees during dry seasons. A diet of dry grass made the grazers dependent on access to surface water, concentrating their numbers within reach of perennial water sources during the dry months. This opened opportunities for savanna-dwelling ape-men to incorporate animal flesh into their diet, compensating for the lack of plant food and the effort involved in extracting what remained underground during the dry season. The dietary shift set up an evolutionary train of adaptations for endurance locomotion, including bared skin for efficient sweating and stone tools for extracting marrow and meat.

South America's large grazers were mostly too big to be exploited sustainably for flesh in tropical climates where meat soon rots. Australia and tropical Asia were both deficient in grazers. Nowhere outside of Africa were large herbivores sufficiently abundant to nurture the seasonal dependency of comparatively puny primates on scrounging from carnivore kills or running down their own prey.

The size structure of Africa's large herbivore fauna was also crucially important.[23] Carcasses of small antelope get consumed completely by their mammalian carnivore killers. Those of megaherbivores that have died remain attended by carnivores until the meat turns putrid, while the huge marrow-containing bones are not easily transported to safer sites for processing, let alone their skulls. The unique feature of Africa's large herbivore fauna is the abundance and diversity of medium-large ruminants, weighing 50–500 kg, which was established by 5 Ma. The dominance of the grazers among them was promoted specifically by the prevalence of dry/eutrophic savannas, or 'sweetveld', with relatively palatable grasses, a savanna subdivision not recognised in other continents. Tools developed by hominins to extract and pulverise tough plants became deployed to break open the bones and scrape flesh off the ungulate carcasses abandoned by the big fierce killers, exploiting a time window when large carnivores were mostly inactive.

However, simply adding meat to the diet was not enough. It needed to be obtained reliably, even after herbivore populations had crashed in very dry years. Ingenuity was required to locate where herbivores remained and dispatch them reliably using increasingly effective tools, like poison-tipped arrows. Brains expanded to accommodate this anticipatory planning and contributed to cultural evolution and language development. The greater cerebral competence needed to cope with climatic uncertainty and periodic extremes of aridity exerted strong selection for adaptations that increased survival odds through the crunch periods. Several of the large grazers that had taken form during the Pliocene and Pleistocene fell by the wayside as glacial conditions reached ever greater levels of cold and aridity. Adaptive shifts appeared abruptly, seemingly synchronised with climatic extremes. The ecologically adept hominin survivors dispersed from their local habitats to spread their crucial adaptations through Africa and beyond. After some modern humans acquired domestic livestock from across the Mediterranean Sea and became herders, savanna grasslands enticed them southward amid the wealth of wild herbivores that they no longer needed to hunt, except on ceremonial occasions or during times after livestock herds crashed during droughts.

Space to move, by both people and wildlife, is currently being suppressed as modern humans saturate the Earth, from the equator to the poles. What are the prospects for survival of the faunal diversity that contributed so fundamentally to our origins? This is the topic that I will address in the concluding chapter.

Suggested Further Reading

Potts, R. (2012) Environmental and behavioural evidence pertaining to the evolution of early *Homo*. *Current Anthropology* 53(Suppl. 6):S299–S317.

Scerri, EML., et al. (2018) Did our species evolve in subdivided populations across Africa, and why does it matter? *Trends in Ecology and Evolution* 33:582–592.

Stringer, C. (2016) The origin and evolution of *Homo sapiens*. *Philosophical Transactions of the Royal Society B: Biological Sciences* 371:20150237

References

1. Wood, BK; Boyle, E. (2016) Hominin taxic diversity: fact or fantasy? *American Journal of Physical Anthropology* 159:37–78.
2. Bobe, R; Carvalho, S. (2019) Hominin diversity and high environmental variability in the Okote Member, Koobi Fora Formation, Kenya. *Journal of Human Evolution* 126:91–105.
3. Stringer, C. (2016) The origin and evolution of *Homo sapiens*. *Philosophical Transactions of the Royal Society B: Biological Sciences* 371:20150237.
4. Klein, RG. (2019) Population structure and the evolution of *Homo sapiens* in Africa. *Evolutionary Anthropology: Issues, News, and Reviews* 28:179–188.
5. Scerri, EML, et al. (2018) Did our species evolve in subdivided populations across Africa, and why does it matter? *Trends in Ecology & Evolution* 33:582–594.
6. Arnold, ML. (2009) *Reticulate Evolution and Humans: Origins and Ecology*. Oxford University Press, Oxford.
7. Winder, IC; Winder, NP. (2014) Reticulate evolution and the human past: an anthropological perspective. *Annals of Human Biology* 41:300–311.

8. Wood, B. (2010) Reconstructing human evolution: achievements, challenges, and opportunities. *Proceedings of the National Academy of Sciences of the United States of America* 107:8902–8909.
9. Fischer, J, et al. (2019) The natural history of model organisms: insights into the evolution of social systems and species from baboon studies. *Elife* 8:e50989.
10. Potts, R. (1998) Environmental hypotheses of hominin evolution. *American Journal of Physical Anthropology: The Official Publication of the American Association of Physical Anthropologists* 107:93–136.
11. Potts, R. (2013) Hominin evolution in settings of strong environmental variability. *Quaternary Science Reviews* 73:1–13.
12. Ogutu, JO; Owen-Smith, N. (2003) ENSO, rainfall and temperature influences on extreme population declines among African savanna ungulates. *Ecology Letters* 6:412–419.
13. Reznick, DN, et al. (2019) From low to high gear: there has been a paradigm shift in our understanding of evolution. *Ecology Letters* 22:233–244.
14. Grant, PR, et al. (2017) Evolution caused by extreme events. *Philosophical Transactions of the Royal Society B: Biological Sciences* 372:20160146.
15. Reznick, DN, et al. (1997) Evaluation of the rate of evolution in natural populations of guppies (*Poecilia reticulata*). *Science* 275:1934–1937.
16. Young, TP. (1994) Natural die-offs of large mammals: implications for conservation. *Conservation Biology* 8:410–418.
17. Dublin, HT; Ogutu, JO. (2015) Population regulation of African buffalo in the Mara-Serengeti ecosystem. *Wildlife Research* 42:382–393.
18. Smit, IPJ; Bond, WJ. (2020) Observations on the natural history of a savanna drought. *African Journal of Range & Forage Science* 37:119–136.
19. Spinage, CA; Matlhare, JM. (1992) Is the Kalahari cornucopia fact or fiction? A predictive model. *Journal of Applied Ecology* 29:605–610.
20. Walker, BH, et al. (1987) To cull or not to cull: lessons from a southern African drought. *Journal of Applied Ecology* 24:381–401.
21. Owen-Smith, RN. (2002) *Adaptive Herbivore Ecology: From Resources to Populations in Variable Environments*. Cambridge University Press, Cambridge.
22. Gani, MR; Gani, NDS. (2008) Tectonic hypotheses of human evolution. *Geotimes* 53:34–39.
23. Owen-Smith, N. (2013) Contrasts in the large herbivore faunas of the southern continents in the late Pleistocene and the ecological implications for human origins. *Journal of Biogeography* 40:1215–1224.

Chapter 20: Prospects For a Lonely Planet

Figure 20.1 Assemblage of large grazing herbivores thronging Ngorongoro Crater, Tanzania.

Whither Africa's Large Mammals?

Other animals have been part of our world since long before our lineage became human. We initially feared carnivores, but later boldly collected bones from the remains of their prey. Over time we developed the capacity to kill big herbivores ourselves, ultimately even those as large as elephants. Our ancestors depicted these animals on the walls of the caves and rock shelters that they occupied, both in Africa and in Europe, and wove them into folklore and ritual. Today, many people spend large sums of money to travel to places where they can still view big wild animals living under fairly natural conditions. Some of those who have become super-rich invest their wealth in buying wildlife preserves where they can retire far from city life.

My own involvement with Africa's large mammals has deep psychological roots. I avidly read books about various animals as a schoolboy and longed to meet up with them where they lived, a wish fulfilled when at last my family visited Kruger National Park. Then I aspired to be allowed to step aside from the confines of a vehicle to share their world more intimately. Amazingly, I achieved this wish professionally. My research activities enabled me to investigate what these animals ate and follow where they wandered, on foot or overland by vehicle. I was able to visit wildlife reserves for 'work' reasons through much of Africa, absorbing experiences that money could not buy. I have drifted through moonlit nights trailing courting white rhinos, while gazing incredulously at satellites blinking across the starry sky. I have got to know kudu families individually, registering their births and deaths annually over a decade. I followed closely behind habituated young kudus, employing digital technology to record each bite they took from various plant species between sunrise and sunset. Guided by GPS tracking devices, I traced sites where sable antelope had stopped to feed or to rest in the back woods that they inhabited. I searched for wildebeest and gemsbok across the vast Kalahari from spotter planes so as to tag some to record how they coped with extremes of temperature and aridity. I have shed some of my fear upon meeting lions and leopards, because those encountered either ran away or tried to hide, rather than attempting to eat me. I have dodged round elephants in order to inspect what they had done to trees, large and small. My life has been hugely enriched through sharing the worlds experienced by these wild animals, the *'umwelt'* within which my hominin ancestors evolved. My fascination with these animals surely has deep genetic roots.

But nowadays, large wild mammals are confined mostly to designated national parks and other formally protected areas. The domesticated ungulates that have displaced their wild counterparts are represented by just six species: cattle, sheep, goats, horses, donkeys and pigs. Only donkeys had an African origin. Pastoral lifestyles tolerating wild animals have become displaced by farms and feedlots fattening domestic animals for slaughter. Visitors to Africa's national parks gaze out of car windows, enthralled by the proximity of fearsome lions, leopards, rhinos and elephants. Eco-tourists spend enormous amounts of money to be allowed to spend an hour near a group of gorillas, or chimpanzees (Figure 20.2). If rich enough, they may even be allowed to embark on a 'walking safari' led by a game ranger with a gun. Do the many people who have not had such experiences with wildlife miss them? Quite a few do so, judging by the prevalence of luxurious game lodges hosting them. But the vast majority of humankind remains oblivious to the rich natural heritage that their distant ancestors experienced so intimately.

Do we really need to set aside land to conserve the big game heritage that still persists in Africa, at the expense of human settlements? Early Europeans and Native Americans did away with most of their large mammals shortly after their settlement on these continents. Some of their descendants are now trying to restore what they have lost by 're-wilding', i.e. bringing back some of the big animals or their nearest equivalents. What would this achieve? The environments that these animals formerly inhabited cannot be reinstated, because they were partly created by the megaherbivores that had inhabited them, long gone. African counterparts could be assembled in vast zoos, but to the detriment of incentives for Africans to conserve the animals that they have uniquely retained in their back country. With climates shifting and human settlements pressing up against the boundaries of protected areas, wildlife is becoming squeezed out.[1] Governments tend to value wild animals primarily as a drawcard attracting eco-tourists and hence the local employment that they generate, plus hefty fees paid into state coffers. But this should not be denigrated. The experience is enriching, even if engaged only through a car window or the back of a safari vehicle, because animals remain genuine in what they do, most of the time.

However, there remain vast areas of Africa where people choose not to live because these places do not offer prospects for more than bare sustenance. People are shifting progressively towards city living, opening more lightly inhabited back country where wildlife can be retained. In Europe, deer and wolf numbers are resurging, partly because fewer people want to hunt. Although a similar tendency is developing in Africa, far too many people remain mired in rural poverty with few options to move. This situation is likely to be worsened by global climate change and viral pandemics. I write under the lockdown imposed to restrict the spread of Covid-19.

If Africa's wildlife is to be preserved across sufficiently vast areas in our changing world, the pressing priority is for people living alongside wild animals to benefit economically, improve their livelihoods, educate their children, empower women with controls over their fertility and settle within cities. If material needs are addressed effectively, the spiritual value of wildlife to humankind can be retained into the new era labelled the 'Anthropocene'.

We all had an African ancestry. Only in Africa can we aspire to experience this magical world thronged with animals large and small, continuing to exist in reality rather than merely in recesses of our minds (Figure 20.2).

Figure 20.2 Animal encounters. Looking across the evolutionary chasm separating human from ape: gorilla watching in Uganda – one primate contemplating another.

Reference

1. Veldhuis, MP, et al. (2019) Cross-boundary human impacts compromise the Serengeti–Mara ecosystem. *Science* 363:1424–1428.

Appendix Scientific Names of Extant Animal and Plant Species Mentioned in the Book Chapters (Ecologically Conservative with Regard to Species Recognition)

(a) Animals

Common name	Scientific name	Family or tribe	Notes
Aardvark	*Orycteropus afer*	Orycteropodidae	Feeds on ants and termites
Aardwolf	*Proteles cristata*	Hyaenidae	Feeds on ants and termites
Blesbok	*Damaliscus dorcas*	Alcelaphini	Highveld endemic
Buffalo, African	*Syncerus caffer*	Bovini	Bulk grazer
Cheetah	*Acinonyx jubatus*	Felidae	Cursorial predator
Dikdik	*Madoqua* sp.	Neotragini	Smallest antelope
Dog, African wild	*Lycaon pictus*	Canidae	Cursorial predator
Eland, common	*Tragelaphus oryx*	Tragelaphini	Largest antelope
Elephant, African	*Loxodonta africana*	Elephantidae	Largest herbivore
Gazelle, Grant's	*Nanger granti*	Antilopini	Mainly browser

Common name	Scientific name	Family or tribe	Notes
Gazelle, Thomson's	*Eudorcas thomsonii*	Antilopini	Mainly grazer
Gemsbok	*Oryx gazella*	Hippotragini	Found in arid savannas
Gerenuk	*Litocranius walleri*	Antilopini	Browser
Giraffe	*Giraffa camelopardalis*	Giraffidae	Tallest browser
Hartebeest	*Alcelaphus buselaphus*	Alcelephini	Numerous subspecies
Hartebeest, Hunter's	*Beatragus hunteri*	Alcelaphini	Rare antelope
Hippo	*Hippopotamus amphibius*	Hippopotamidae	Aquatic refuge
Hirola	*Beatragus hunteri*	Alcelaphini	Rare anelope
Hyena, brown	*Hyaena brunnea*	Hyaenidae	Mostly scavenger
Hyena, spotted	*Crocuta crocuta*	Hyaenidae	Hunts and scavenges
Hyena, striped	*Hyaena hyaena*	Hyaenidae	Mostly scavenger
Impala	*Aepyceros melampus*	Aepycerotini	Mixed feeder
Jackal, black-backed	*Canis mesomelas*	Canidae	Small carnivore
Jackal, side-striped	*Canis adjustus*	Canidae	Partly omnivorous
Kob	*Kobus kob*	Reduncini	Wetland grazer
Kob, white-eared	*Kobus kob leucotis*	Reduncini	Wetland grazer

continues

Common name	Scientific name	Family or tribe	Notes
Kudu, greater	*Tragelaphus strepsiceros*	Tragelaphini	Large browser
Kudu, lesser	*Tragelaphus buxtoni*	Tragelaphini	Browser
Lechwe	*Kobus leche*	Reduncini	Wetland grazer
Leopard	*Panthera pardus*	Felidae	Ambush carnivore
Lion	*Panthera leo*	Felidae	Ambush carnivore
Nilgai	*Boselaphus tragocamelus*	Boselaphini	Indian antelope
Nyala	*Tragelaphus angasi*	Tragelaphini	Mixed feeder
Oribi	*Ourebia ourebi*	Neotragini	Smallest grazer
Oryx, South African	*Oryx gazella*	Hippotragini	Arid country grazer
Pangolin	*Smutsia temminckii*	Manidae	Feeds on termites and ants
Porcupine, African	*Hystrix africaeaustralis*	Hystricidae	Large rodent
Puku	*Kobus vardoni*	Reduncini	Wetland grazer
Reedbuck, common	*Redunca arundinum*	Reduncini	Tall grass inhabitant
Reedbuck, mountain	*Redunca fulvorufula*	Reduncini	Inhabits hilly country
Rhino, black	*Diceros bicornis*	Rhinocerotidae	Browser
Rhino, white	*Ceratotherium simum*	Rhinocerotidae	Grazer

Common name	Scientific name	Family or tribe	Notes
Roan antelope	*Hippotragus equinus*	Hippotragini	Tall grass grazer
Sable antelope	*Hippotragus niger*	Hippotragini	Tall grass grazer
Springbok	*Antidorcas marsupialis*	Antilopini	Mixed feeder
Steenbok	*Raphicerus campestris*	Neotragini	Mixed feeder
Suni	*Neotragus moschatus*	Neotragini	Browser
Tiang	*Damaliscus lunatus tiang*	Alcelaphini	Medium height grazer
Topi	*Damaliscus lunatus cokei*	Alcelaphini	Medium height grazer
Tsessebe	*Damaliscus lunatus*	Alcelaphini	Medium height grazer
Warthog, African	*Phacochoerus africanus*	Suidae	Grazer
Waterbuck	*Kobus ellipsiprymnus*	Reduncini	River margin grazer
Wildebeest	*Connochaetes taurinus*	Alcelaphini	Short grass grazer
Wildebeest, black	*Connochaetes gnou*	Alcelaphini	Short grass grazer
Zebra, common	*Equus quagga*	Equidae	Grazer

(b) Plants

Common name	Scientific name	Family, subfamily, or tribe	Notes
Acacia, mountain	*Brachystegia tamarindoides*	Detariodeae	Rocky outcrops
Ana tree	*Faidherbia albida*	Mimosoideae	Floodplains
Baobab tree	*Adansonia digitata*	Bombacaceae	Dry savanna
Bird plum	*Berchemia* spp.	Rhamnaceae	Small berries
Boerbean tree	*Schotia brachypetala*	Detariodeae	Commonly river margins
Bushwillow	*Combretum* spp.	Combretaceae	Mesic savannas
Bushwillow, knobbly	*Combretum mossambicense*	Combretaceae	Understorey shrub
Bushwillow, red	*Combretum apiculatum*	Combretaceae	Mesic savanna
Bushwillow russet	*Combretum hereroense*	Combretaceae	Lower catena
Camphor bush	*Tarchonanthus camphoratus*	Asteraceae	Aromatic foliage
Camwood, sand	*Baphia massaiensis*	Papilionoideae	Understorey shrub
Clusterleaf	*Terminalia* spp.	Combretaceae	Sandy savannas
Clusterleaf, silver	*Terminalia sericea*	Combretaceae	Sandy savannas
Euphorbia, candelabra	*Euphorbia ingens*	Euphorbiaceae	Huge succulent
Fever tree	*Vachellia xanthophloea*	Mimosoideae	Waterlogged soils

Common name	Scientific name	Family, subfamily, or tribe	Notes
Fig, sycamore	*Ficus sycamorus*	Moraceae	Riverine woodland
Gardenia, Transvaal	*Gardenia volkensii*	Rubiaceae	Widespread
Gifblaar	*Dichapetalum cymosum*	Dichapetalaceae	Poisonous geoxyle
Grass, bristle	*Setaria* spp.	Paniceae	Moderately palatable
Grass, buffalo	*Panicum* spp.	Paniceae	Highly palatable
Grass, bushveld signal	*Urochloa mossambicensis*	Paniceae	Highly palatable
Grass, couch	*Cynodon dactylon*	Chlorideae	Highly palatable
Grass, dropseed	*Sporobolus* spp.	Chlorideae	Generally low palatability
Grass, finger	*Digitaria* spp.	Paniceae	Palatable
Grass, Guinea	*Panicum maximum*	Paniceae	Highly palatable
Grass, lemon	*Cymbopogon* spp.	Andropogoneae	Aromatic oil
Grass, love	*Eragrostis* spp.	Chlorideae	Moderate palatability
Grass, pan dropseed	*Sporobolus ioclados*	Chlorideae	Sodium accumulator
Grass, red	*Themeda triandra*	Andropogoneae	Relatively palatable
Grass, rice	*Oryza* spp.	Oryzeae	Wetlands
Grass, rice	*Leersia* spp.	Oryzeae	Wetlands

continues

Common name	Scientific name	Family, subfamily, or tribe	Notes
Grass, stinking	*Bothriochloa* spp.	Andropogoneae	Smelly from oil content
Grass, thatch	*Hyparrhenia* spp.	Andropogoneae	Tall and fibrous
Grass, three-awn	*Aristida* spp.	Aristideae	Arid savannas
Gwarri	*Euclea* spp.	Ebenaceae	Evergreen shrub
Jackalberry tree	*Diospyros mespiliformis*	Ebenaceae	Commonly river margins
Jessebush	*Combretum celastroides*	Combretaceae	Sandveld shrub
Leadwood tree	*Combretum inberbe*	Combretaceae	Heavy wood
Mahobahoba tree	*Uapaca* spp.	Phyllanthaceae	Sugar plum fruit
Mahogany tree	*Trichilia* spp.	Meliaceae	Riverine evergreen
Marula tree	*Sclerocarya birrea*	Anacardiaceae	Plum-like fruits
Mfuti tree	*Brachystegia boehmi*	Detariodeae	Miombo woodlands
Mobola plum tree	*Parinari curatellifolia*	Chrysobalanaceae	Plum-like fruits
Monkey orange tree	*Strychnos* spp.	Logonaceae	Large hard-shelled fruits
Msasa tree	*Brachystegia spiciformis*	Detariodeae	Miombo woodlands
Munondo tree	*Julbernadia globiflora*	Detariodeae	Miombo woodlands

Common name	Scientific name	Family, subfamily, or tribe	Notes
Mutondo tree	*Isoberlinea* spp.	Detariodeae	Miombo woodlands
Myrrh, wild	*Commiphora* spp.	Burseraceae	Aromatic foliage
Olive tree	*Olea* spp.	Oleaceae	Forest tree
Ordeal tree	*Erythrophleum africana*	Detariodeae	Poisonous foliage
Prince of Wales tree	*Brachystegia boehmi*	Detariodeae	Miombo woodlands
Raisin bush	*Grewia* spp.	Tiliaceae	Savanna shrub
Reed	*Phragmites* spp.	Arundinoideae	Riverine sands
Rush	*Typha* spp.	Typhaceae	Wetlands
Sausage tree	*Kigelia africana*	Bignoniaceae	Huge pods
Sedge	*Cyperus* spp.	Cyperaceae	Underground corms
Seringa, wild	*Burkea africana*	Detariodeae	Sandy woodlands
Shepherd's tree	*Boscia albitrunca*	Capparaceae	Dry savanna evergreen
Sickle bush	*Dichrostachys cinerea*	Mimosoideae	Savanna shrub
Snowberry bush	*Flueggia* spp.	Euphorbiaceae	Savanna shrub
Teak, Zambezi or Zimbabwe	*Baikaea plurijuga*	Detariodeae	Kalahari sands
Thorn, ant-gall	*Vachellia drepanolobium*	Mimosoideae	Shub on heavy clay

continues

Common name	Scientific name	Family, subfamily, or tribe	Notes
Thorn, black monkey	*Senegalia burkei*	Mimosoideae	Sandy soils
Thorn, black	*Senegalia mellifera*	Mimosoideae	Multi-stemmed shrub
Thorn, giraffe	*Vachellia erioloba*	Mimosoideae	Dry savannas
Thorn, hook	*Senegalia caffra*	Mimosoideae	Hillsides
Thorn, knob	*Senegalia nigrescens*	Mimosoideae	Clay soils
Thorn, paperbark	*Vachellia sieberiana*	Mimosoideae	Moist savanna
Thorn, splendid	*Vachellia robusta*	Mimosoideae	Commonly riverine
Thorn, sweet	*Vachellia karoo*	Mimosoideae	Widespread southern
Thorn, umbrella	*Vachellia tortilis*	Mimosoideae	Widespread
Thorn, white	*Vachellia polyacantha*	Mimosoideae	River margins
Waterberry tree	*Syzigium* spp.	Myrtaceae	Wet localities
Yellowwood tree	*Podocarpus* spp.	Podocarpaceae	Montane forests
Zebrawood tree	*Brachystegia spiciformis*	Detariodeae	Miombo woodlands

Index

Page numbers in italics refer to figures.

Printed in the United States
by Baker & Taylor Publisher Services